Nonlinear L_p-Norm Estimation

STATISTICS: Textbooks and Monographs

A Series Edited by

Vol. 1: The Generalized Jackknife Statistic, *H. L. Gray and W. R. Schucany*
Vol. 2: Multivariate Analysis, *Anant M. Kshirsagar*
Vol. 3: Statistics and Society, *Walter T. Federer*
Vol. 4: Multivariate Analysis: A Selected and Abstracted Bibliography, 1957-1972, *Kocherlakota Subrahmaniam and Kathleen Subrahmaniam* (out of print)
Vol. 5: Design of Experiments: A Realistic Approach, *Virgil L. Anderson and Robert A. McLean*
Vol. 6: Statistical and Mathematical Aspects of Pollution Problems, *John W. Pratt*
Vol. 7: Introduction to Probability and Statistics (in two parts), Part I: Probability; Part II: Statistics, *Narayan C. Giri*
Vol. 8: Statistical Theory of the Analysis of Experimental Designs, *J. Ogawa*
Vol. 9: Statistical Techniques in Simulation (in two parts), *Jack P. C. Kleijnen*
Vol. 10: Data Quality Control and Editing, *Joseph I. Naus* (out of print)
Vol. 11: Cost of Living Index Numbers: Practice, Precision, and Theory, *Kali S. Banerjee*
Vol. 12: Weighing Designs: For Chemistry, Medicine, Economics, Operations Research, Statistics, *Kali S. Banerjee*
Vol. 13: The Search for Oil: Some Statistical Methods and Techniques, *edited by D. B. Owen*
Vol. 14: Sample Size Choice: Charts for Experiments with Linear Models, *Robert E. Odeh and Martin Fox*
Vol. 15: Statistical Methods for Engineers and Scientists, *Robert M. Bethea, Benjamin S. Duran, and Thomas L. Boullion*
Vol. 16: Statistical Quality Control Methods, *Irving W. Burr*
Vol. 17: On the History of Statistics and Probability, *edited by D. B. Owen*
Vol. 18: Econometrics, *Peter Schmidt*
Vol. 19: Sufficient Statistics: Selected Contributions, *Vasant S. Huzurbazar (edited by Anant M. Kshirsagar)*
Vol. 20: Handbook of Statistical Distributions, *Jagdish K. Patel, C. H. Kapadia, and D. B. Owen*
Vol. 21: Case Studies in Sample Design, *A. C. Rosander*
Vol. 22: Pocket Book of Statistical Tables, *compiled by R. E. Odeh, D. B. Owen, Z. W. Birnbaum, and L. Fisher*
Vol. 23: The Information in Contingency Tables, *D. V. Gokhale and Solomon Kullback*

Vol. 24: Statistical Analysis of Reliability and Life-Testing Models: Theory and Methods, *Lee J. Bain*
Vol. 25: Elementary Statistical Quality Control, *Irving W. Burr*
Vol. 26: An Introduction to Probability and Statistics Using BASIC, *Richard A. Groeneveld*
Vol. 27: Basic Applied Statistics, *B. L. Raktoe and J. J. Hubert*
Vol. 28: A Primer in Probability, *Kathleen Subrahmaniam*
Vol. 29: Random Processes: A First Look, *R. Syski*
Vol. 30: Regression Methods: A Tool for Data Analysis, *Rudolf J. Freund and Paul D. Minton*
Vol. 31: Randomization Tests, *Eugene S. Edgington*
Vol. 32: Tables for Normal Tolerance Limits, Sampling Plans, and Screening, *Robert E. Odeh and D. B. Owen*
Vol. 33: Statistical Computing, *William J. Kennedy, Jr. and James E. Gentle*
Vol. 34: Regression Analysis and Its Application: A Data-Oriented Approach, *Richard F. Gunst and Robert L. Mason*
Vol. 35: Scientific Strategies to Save Your Life, *I. D. J. Bross*
Vol. 36: Statistics in the Pharmaceutical Industry, *edited by C. Ralph Buncher and Jia-Yeong Tsay*
Vol. 37: Sampling from a Finite Population, *J. Hajek*
Vol. 38: Statistical Modeling Techniques, *S. S. Shapiro*
Vol. 39: Statistical Theory and Inference in Research, *T. A. Bancroft and C.-P. Han*
Vol. 40: Handbook of the Normal Distribution, *Jagdish K. Patel and Campbell B. Read*
Vol. 41: Recent Advances in Regression Methods, *Hrishikesh D. Vinod and Aman Ullah*
Vol. 42: Acceptance Sampling in Quality Control, *Edward G. Schilling*
Vol. 43: The Randomized Clinical Trial and Therapeutic Decisions, *edited by Niels Tygstrup, John M. Lachin, and Erik Juhl*
Vol. 44: Regression Analysis of Survival Data in Cancer Chemotherapy, *Walter H. Carter, Jr., Galen L. Wampler, and Donald M. Stablein*
Vol. 45: A Course in Linear Models, *Anant M. Kshirsagar*
Vol. 46: Clinical Trials: Issues and Approaches, *edited by Stanley H. Shapiro and Thomas H. Louis*
Vol. 47: Statistical Analysis of DNA Sequence Data, *edited by B. S. Weir*
Vol. 48: Nonlinear Regression Modeling: A Unified Practical Approach, *David A. Ratkowsky*
Vol. 49: Attribute Sampling Plans, Tables of Tests and Confidence Limits for Proportions, *Robert E. Odeh and D. B. Owen*
Vol. 50: Experimental Design, Statistical Models, and Genetic Statistics, *edited by Klaus Hinkelmann*
Vol. 51: Statistical Methods for Cancer Studies, *edited by Richard G. Cornell*
Vol. 52: Practical Statistical Sampling for Auditors, *Arthur J. Wilburn*
Vol. 53: Statistical Signal Processing, *edited by Edward J. Wegman and James G. Smith*
Vol. 54: Self-Organizing Methods in Modeling: GMDH Type Algorithms, *edited by Stanley J. Farlow*
Vol. 55: Applied Factorial and Fractional Designs, *Robert A. McLean and Virgil L. Anderson*
Vol. 56: Design of Experiments: Ranking and Selection, *edited by Thomas J. Santner and Ajit C. Tamhane*
Vol. 57: Statistical Methods for Engineers and Scientists. Second Edition, Revised and Expanded, *Robert M. Bethea, Benjamin S. Duran, and Thomas L. Boullion*
Vol. 58: Ensemble Modeling: Inference from Small-Scale Properties to Large-Scale Systems, *Alan E. Gelfand and Crayton C. Walker*
Vol. 59: Computer Modeling for Business and Industry, *Bruce L. Bowerman and Richard T. O'Connell*
Vol. 60: Bayesian Analysis of Linear Models, *Lyle D. Broemeling*

Vol. 61: Methodological Issues for Health Care Surveys, *Brenda Cox and Steven Cohen*

Vol. 62: Applied Regression Analysis and Experimental Design, *Richard J. Brook and Gregory C. Arnold*

Vol. 63: Statpal: A Statistical Package for Microcomputers – PC-DOS Version for the IBM PC and Compatibles, *Bruce J. Chalmer and David G. Whitmore*

Vol. 64: Statpal: A Statistical Package for Microcomputers – Apple Version for the II, II+, and IIe, *David G. Whitmore and Bruce J. Chalmer*

Vol. 65: Nonparametric Statistical Inference, Second Edition, Revised and Expanded, *Jean Dickinson Gibbons*

Vol. 66: Design and Analysis of Experiments, *Roger G. Petersen*

Vol. 67: Statistical Methods for Pharmaceutical Research Planning, *Sten W. Bergman and John C. Gittins*

Vol. 68: Goodness-of-Fit Techniques, *edited by Ralph B. D'Agostino and Michael A. Stephens*

Vol. 69: Statistical Methods in Discrimination Litigation, *edited by D.H. Kaye and Mikel Aickin*

Vol. 70: Truncated and Censored Samples from Normal Populations, *Helmut Schneider*

Vol. 71: Robust Inference, *M. L. Tiku, W. Y. Tan, and N. Balakrishnan*

Vol. 72: Statistical Image Processing and Graphics, *edited by Edward J. Wegman and Douglas J. DePriest*

Vol. 73: Assignment Methods in Combinatorial Data Analysis, *Lawrence J. Hubert*

Vol. 74: Econometrics and Structural Change, *Lyle D. Broemeling and Hiroki Tsurumi*

Vol. 75: Multivariate Interpretation of Clinical Laboratory Data, *Adelin Albert and Eugene K. Harris*

Vol. 76: Statistical Tools for Simulation Practitioners, *Jack P. C. Kleijnen*

Vol. 77: Randomization Tests, Second Edition, *Eugene S. Edgington*

Vol. 78: A Folio of Distributions: A Collection of Theoretical Quantile-Quantile Plots, *Edward B. Fowlkes*

Vol. 79: Applied Categorical Data Analysis, *Daniel H. Freeman, Jr.*

Vol. 80: Seemingly Unrelated Regression Equations Models : Estimation and Inference, *Virendra K. Srivastava and David E. A. Giles*

Vol. 81: Response Surfaces: Designs and Analyses, *Andre I. Khuri and John A. Cornell*

Vol. 82: Nonlinear Parameter Estimation: An Integrated System in BASIC, *John C. Nash and Mary Walker-Smith*

Vol. 83: Cancer Modeling, *edited by James R. Thompson and Barry W. Brown*

Vol. 84: Mixture Models: Inference and Applications to Clustering, *Geoffrey J. McLachlan and Kaye E. Basford*

Vol. 85: Randomized Response: Theory and Techniques, *Arijit Chaudhuri and Rahul Mukerjee*

Vol. 86: Biopharmaceutical Statistics for Drug Development, *edited by Karl E. Peace*

Vol. 87: Parts per Million Values for Estimating Quality Levels, *Robert E. Odeh and D. B. Owen*

Vol. 88: Lognormal Distributions: Theory and Applications, *edited by Edwin L. Crow and Kunio Shimizu*

Vol. 89: Properties of Estimators for the Gamma Distribution, *K. O. Bowman and L. R. Shenton*

Vol. 90: Spline Smoothing and Nonparametric Regression, *Randall L. Eubank*

Vol. 91: Linear Least Squares Computations, *R. W. Farebrother*

Vol. 92: Exploring Statistics, *Damaraju Raghavarao*

Vol. 93: Applied Time Series Analysis for Business and Economic Forecasting, *Sufi M. Nazem*
Vol. 94: Bayesian Analysis of Time Series and Dynamic Models, *edited by James C. Spall*
Vol. 95: The Inverse Gaussian Distribution: Theory, Methodology, and Applications, *Raj S. Chhikara and J. Leroy Folks*
Vol. 96: Parameter Estimation in Reliability and Life Span Models, *A. Clifford Cohen and Betty Jones Whitten*
Vol. 97: Pooled Cross-Sectional and Time Series Data Analysis, *Terry E. Dielman*
Vol. 98: Random Processes: A First Look, Second Edition, Revised and Expanded, *R. Syski*
Vol. 99: Generalized Poisson Distributions: Properties and Applications, *P.C. Consul*
Vol. 100: Nonlinear L_p-Norm Estimation, *René Gonin and Arthur H. Money*
Vol. 101: Model Discrimination for Nonlinear Regression Models, *Dale S. Borowiak*
Vol. 102: Applied Regression Analysis in Econometrics, *Howard E. Doran*
Vol. 103: Continued Fractions in Statistical Applications, *K.O. Bowman and L.R. Shenton*
Vol. 104: Statistical Methodology in the Pharmaceutical Sciences, *Donald A. Berry*

ADDITIONAL VOLUMES IN PREPARATION

Nonlinear L_p-Norm Estimation

RENÉ GONIN
Institute for Biostatistics
Medical Research Council
Cape Town, South Africa

ARTHUR H. MONEY
Henley The Management College
Henley-on-Thames
England

CRC Press
Taylor & Francis Group
Boca Raton London New York

CRC Press is an imprint of the
Taylor & Francis Group, an **informa** business

First published 1989 by Marcel Dekker, Inc.

Published 2019 by CRC Press
Taylor & Francis Group
6000 Broken Sound Parkway NW, Suite 300
Boca Raton, FL 33487-2742

First issued in paperback 2019

ISBN 13: 978-0-367-45116-5 (pbk)
ISBN 13: 978-0-8247-8125-5 (hbk)

Visit the Taylor & Francis Web site at
http://www.taylorandfrancis.com

and the CRC Press Web site at
http://www.crcpress.com

Preface

We hope this book will be of use both to the statistician and to the applied mathematician. It is primarily of a computational nature and therefore belongs to the field of statistical computing. The book is self-contained and permits flexibility in the manner in which it can be read. The reader should be able to grasp the underlying computational principles without having had a formal course in statistical computing. Familiarity with linear programming is, however, essential.

The book is organized into two parts, namely, numerical solution procedures and statistical analysis. The original outline for this book arose out of the first author's doctoral thesis. Subsequently the material on nonlinear L_1-norm and L_∞-norm estimation problems was incorporated.

Our main objective is the treatment of nonlinear L_p-norm estimation, which subsumes nonlinear least squares estimation. We realized that an introduction to linear L_p-norm estimation was needed as a preamble to the nonlinear case. Linear L_p-norm estimation is vast enough to warrant a complete text on its own. Limited space permits only a brief review of this problem in Chapter 1. We also felt it important to bring to the statistician's notice useful procedures known to the numerical analyst. We hope that this book will be successful in bridging the gap between the statistician and the numerical analyst.

We have included some FORTRAN programs to assist the reader in solving the nondifferentiable nonlinear L_1- and L_∞-norm estimation problems. For the continuously differentiable nonlinear L_p-norm problem any general unconstrained minimization routine can be used. This is not necessarily the most efficient avenue of solution.

Sections of the manuscript were used by the first author in a course on nonlinear parameter estimation at the University of Michigan in Ann Arbor during 1988.

We would like to express our gratitude to the Medical Research Council in Cape Town for the use of their computer facilities. This entailed not only computational aspects but also the typesetting of the book by means of TEX. The first author would also like to thank the Medical Research Council for financial support related to this project. The second author acknowledges the continued financial support of the University of Cape Town and the Human Sciences Research Council.

We would like to express our sincere appreciation to Jan Pretorius, Director of the Institute for Biostatistics, for creating the environment for pursuing this project. His continued interest and encouragement was an important factor in completing this task.

We would especially like to thank Professor Vince Sposito of Iowa State University for his careful scrutiny of the manuscript and his many valuable comments. Also, many thanks to Professor Cas Troskie of the University of Cape Town for perusing the manuscript.

For the classification of the L_1-norm and L_∞-norm methods we are indebted to Dr. Alistair Watson of Dundee University. Also, our gratitude to Professor Mike Osborne of the Australian National University, who put into perspective for us the numerical methods of nonlinear L_1-norm estimation.

We would like to thank Doctors Joyce Gonin and Sinclair Wynchank for patiently editing the manuscript. Their efforts certainly improved the clarity of the subject matter.

Finally, thanks to our spouses, Joyce and Gillian, for their fortitude in coping with our impatience and idiosyncrasies as p tended to ∞.

René Gonin
Arthur H. Money

Contents

Nonlinear L_p-Norm Estimation

1

L_p-Norm Estimation in Linear Regression

The fundamental problem of fitting equations to a number of observations has for centuries been of interest to scientists. For example, curve fitting models were used by early astronomers to predict the orbits of celestial bodies.

Concerned with the calculation of the distance of a new star from the earth Galileo† makes the following statement through the interlocutor Salviati in his Dialogue of the two chief world systems:

"Then these observers being capable, and having erred for all that, and the errors needing to be corrected in order for us to get the best possible information from their observations, it will be appropriate for us to apply the minimum amendments and smallest corrections that we can — just enough to remove the observations from impossibility and restore them to possibility."

This statement heralded the beginning of the theory of errors in which the model is estimated from a number of inconsistent observations.

1.1. The history of curve fitting problems

One of the first methods for smoothing random errors was based on averages and was known as the Principle of the Arithmetic Mean: suppose we wish to fit a linear regression model $y = \alpha + \beta x$ to the observations (x_i, y_i), $i = 1, \ldots, n$. The estimates of α and β are determined as follows:

Slopes between all possible pairs of points,

$$b_{ij} = \frac{y_j - y_i}{x_j - x_i} \quad \text{with} \quad x_j \neq x_i \quad \text{for} \quad i = 1, \ldots, n-1; \quad j = i+1, \ldots, n$$

---

† Galilei, Galileo (1629). Dialogue concerning the two chief world systems, Ptolemaic and Copernican. (English translation, by Stillman Drake. *Revised 2nd edition UCLA Press 1967).*

are first calculated and the corresponding intercepts a_{ij} in each case are then calculated by substitution. The averages of the intercepts and slopes denoted by a and b respectively are then taken as estimates of α and β.

Cotes (1722) noted in certain models that only the response variable (observations y) is subject to measurement errors and suggested a procedure based on weighted arithmetic means with weights proportional to $|x|$. In the model $y = \beta x + e$, the slope β is estimated by $b = \bar{y}/\bar{x}$, the ratio of the two means, which in turn is equivalent to the *zero sum residuals condition*:

$$\sum_{i=1}^{n}(y_i - bx_i) = 0 \tag{1.1}$$

This is the same as stipulating that the ideal line must pass through the centroid $(\bar{x}, \bar{y})$ of the data.

Euler (1749) and Mayer (1750) independently derived the so-called Method of Averages for fitting a straight line to observed data. The observations are subdivided on some subjective basis into as many subsets as there are coefficients. The grouping is made according to the value of one of the explanatory variables. Those with the largest values of this variable are grouped together and so on. Condition (1.1) is then applied to each observation of the subset.

Boscovich (1757) considered the model $y = \alpha + \beta x + e$ and proposed that two criteria be met when fitting the best straight line†. The estimates a and b of the parameters α and β are determined such that:

Criterion 1: $\sum_{i=1}^{n} |y_i - a - bx_i|$ is a minimum.

Criterion 2: The sum of the positive and negative residuals in the y-variable are equal i.e., $\sum_{i=1}^{n}(y_i - a - bx_i) = 0$.

Criterion 1 is the objective of Least Absolute Value (LAV) estimation (also referred to as L_1-norm estimation), while Criterion 2 ensures that the fitted line passes through the centroid of the data. Boscovich's solution procedure is based on geometric principles. Solution procedures were derived by Simpson (1760) and Laplace (1793,1799).

---

† L_1-norm estimation has in fact been traced back to Galileo (1629), see e.g., Ronchetti (1987).

Laplace (1786) also used the Boscovich principles to test the adequacy of the relationship $y = \alpha + \beta x$ for the data (x_i, y_i). He described an algebraic procedure for determining the coefficients α and β.

Gauss (1806) developed the method of minimizing the squared observation errors in his works on celestial mechanics which subsequently became known as the Method of Least Squares (L_2-norm estimation). Although Gauss had used least squares since 1795, Legendre (1805) was the first to publish the method. He derived the normal equations algebraically without using calculus. He claimed that least squares was superior to other existing methods but gave no proof. In 1809 Gauss derived the normal (Gaussian) law of error which states that the arithmetic mean of the observations of an unknown variable x will be the most probable. Gauss (1820) succinctly writes:

"Determining a magnitude by observation can justly be compared to a game in which there is a danger of loss but no hope of gain ... Evidently the loss in the game cannot be compared directly to the error which has been committed, for then a positive error would represent loss and a negative error a gain. The magnitude of the loss must on the contrary be evaluated by a function of the error of which the value is always positive ... it seems natural to choose the simplest (function), which is, beyond contradiction, the square of the error".

Laplace (1818) examined the distributional properties of the estimator, b, of the parameter β in the simple regression model $y = \beta x + e$ when L_1-norm estimation is used. He assumed that all the errors e had the same symmetric distribution about zero and derived the density function $f(e)$ of the errors e. He also showed that the slope b was asymptotically normally distributed with mean zero and variance $\{4f(0)^2 \sum_{i=1}^{n} x_i^2\}^{-1}$. This is a well known result for the sample median in the location model $y = \beta + e$.

Cauchy (1824) examined the fitting of a straight line $y = \alpha + \beta x$ to data and proposed minimizing the maximum absolute residual. This he achieved by means of an iterative procedure. Chebychev (1854) in his work on the approximation of functions also proposed the estimation of parameters by means of minimizing the maximum absolute difference between the observed function and the estimated function. This minimax procedure later became known as Chebychev approximation or L_∞-norm approximation.

Edgeworth (1883) questioned the universal use of the normal law of errors and also examined the problem of outliers. This problem was considered further by Doolittle (1884). Edgeworth (1887a,b) used the first Boscovich principle and abandoned the zero sum residuals condition which forces the line through the centroid of the data. He, in fact, provided the first workable method for estimating these parameters. He also considered cases where least squares is inappropriate, i.e., where the error distributions are unknown or contaminated (normal) distributions.

The aforementioned authors proposed complex solution procedures for these problems. This possibly accounts for the apparent lack of interest in applying LAV estimation in the period 1884 to 1920. Mild interest was revived again in the period 1920 to 1950 by Edgeworth (1923), Rhodes (1930), Singleton (1940) and Harris (1950).

It was only after the appearance of the papers by Charnes et al. (1955) and Wagner (1959,1962) that interest in L_1 and L_∞ estimation was again stimulated. These authors described the linear programming (LP) formulation to the L_1- and L_∞-norm problems respectively. Progress, however, was hampered during this period since computers were not advanced enough to handle very large LP problems.

In general, the parameters of the equations may be estimated by minimizing the sum of the pth powers of the absolute deviations of the estimated values from the observed values of the response variable. This procedure is referred to as L_p-norm estimation. In the rest of this chapter we give a review of important developments in linear L_p-norm estimation.

1.2. The linear L_p-norm estimation problem

Consider the general linear model

$$\boldsymbol{y} = \boldsymbol{X\beta} + \boldsymbol{e} \tag{1.2}$$

where

$\boldsymbol{y}$ is an n-vector of observable random variables

$\boldsymbol{X}$ is an $n \times k$ matrix of known regressor variables

$\boldsymbol{\beta}$ is a k-vector of unknown parameters, and

$\boldsymbol{e}$ is an n-vector of unobservable errors.

1.2.1. Formulation

The linear L_p-norm estimation problem is then defined as:

Find the parameter vector $\boldsymbol{b} = (b_1, \ldots, b_k)'$ which minimizes

$$\sum_{i=1}^{n} |y_i - \boldsymbol{x_i}\boldsymbol{b}|^p = \sum_{i=1}^{n} |\hat{e}_i|^p \tag{1.3}$$

with y_i the response and $\boldsymbol{x_i} = (x_{i1}, \ldots, x_{ik})$ the explanatory variables respectively, and $x_{i1} = 1$ for all i. Hence b_1 is the estimate of the intercept. $\boldsymbol{b}$ is the L_p-norm estimate of $\boldsymbol{\beta}$ and $\hat{\boldsymbol{e}} = (\hat{e}_1, \ldots, \hat{e}_n)'$ is the n-vector of residuals.

Charnes et al. (1955) showed that problem (1.3) (with $p = 1$) can be formulated as a linear programming problem by considering the unrestricted error term to be the difference between two nonnegative variables. Let $\hat{e}_i = u_i - v_i$, where $u_i, v_i \geq 0$, represent the positive and negative deviations respectively. The general L_p-norm problem then becomes:

$$\text{Minimize} \quad \sum_{i=1}^{n} (u_i^p + v_i^p) \tag{1.4}$$

$$\text{subject to} \quad \left.\begin{aligned} \boldsymbol{x_i}\boldsymbol{b} + u_i - v_i &= y_i \\ u_i, v_i &\geq 0 \end{aligned}\right\} \quad i = 1, \ldots, n$$

$$\boldsymbol{b} \text{ unconstrained}$$

Only in the cases $p = 1$ and $p \to \infty$ can linear programming procedures be used. When $p = 2$ the normal equations of least squares are solved. For other values of p unconstrained minimization procedures are used.

1.2.2. Algorithms

Barrodale and Roberts (1970) showed that the linear L_p-norm estimation problem can be formulated as a nonlinear programming (NLP) problem in which the objective function is concave for $0 < p < 1$ and convex for $1 \leq p < \infty$ with the constraints being linear. They suggested for $p > 1$ the use of the convex simplex or Newton's method. For the case $0 < p < 1$ a modification of the simplex method for linear programming was proposed.

Ekblom (1973) rewrites problem (1.3) as the perturbation problem:

Find the parameter vector $\boldsymbol{b}$ which minimizes

$$\sum_{i=1}^{n}[(y_i - \boldsymbol{x_i b})^2 + c^2]^{p/2} \qquad \text{(where } c \text{ is finite)} \tag{1.5}$$

The author used the modified (damped) Newton method to solve problem (1.5). The advantage is that the Hessian (matrix of 2nd-order derivatives) of the perturbed problem remains positive definite as long as $c \neq 0$. A decrease is therefore assured at every iteration. Ekblom then showed that the limiting solution as $c \to 0$ is the solution to the original problem (1.3). More recently Ekblom (1987) suggested that problem (1.5) with $p = 1$ be used for L_1-norm estimation.

Fischer (1981) considered problem (1.3) with $1 < p < 2$. Setting $r_i = \boldsymbol{x_i b} - y_i$ for $i = 1, \ldots, n$, he transformed (1.3) to the following "linearly constrained" problem:

Find the parameters $\boldsymbol{b}$ which minimize

$$\sum_{i=1}^{n}|r_i|^p \qquad \text{subject to} \quad \boldsymbol{x_i b} - r_i = y_i, \quad i = 1, \ldots, n$$

This problem, known as the primal [see Zangwill (1969:Chapter 2)] can be formulated as:

$$\min_{\boldsymbol{b}} \max_{\boldsymbol{\lambda}} L(\boldsymbol{b}, \boldsymbol{\lambda}) = \sum_{i=1}^{n}[|r_i|^p + \lambda_i(\boldsymbol{x_i b} - r_i - y_i)]$$

where $\boldsymbol{\lambda}$ is the vector of undetermined Lagrange multipliers. The dual problem is then:

$$\max_{\boldsymbol{\lambda}} \min_{\boldsymbol{b}} L(\boldsymbol{b}, \boldsymbol{\lambda})$$

Fischer also indicated how a constrained version of Newton's method can be used to solve the dual problem. The solution of the primal problem corresponds to the iteratively reweighted least squares algorithm of Merle and Späth (1973). Fischer showed that his algorithm converges globally at a superlinear rate. He claims that his proof is not as involved as the one by Wolfe (1979) who in any event only succeeded in proving that the method is locally convergent at a linear rate. Fischer also provided some numerical results which he compared to those found by Merle and Späth. He used the number of iterations as the criterion for efficient convergence instead of the number of function evaluations performed. This can be misleading since an iteration is generally dependent on the structure of the algorithm and efficiency of the programme code. The numerical results indicate that Fischer's algorithm is more efficient than the algorithm by Merle and Späth (1973).

1.2.3. L_2-estimation

The least squares problem can be formulated as a quadratic programming problem as in (1.3) but this is unnecessarily cumbersome. However, the solution to the normal equations is easily obtained by using Gaussian elimination [see Dahlquist and Björck (1974)]. In this case the estimates of the β_j's are linear combinations of the elements of $\boldsymbol{y}$. Specifically we have

$$\boldsymbol{b} = (\boldsymbol{X}'\boldsymbol{X})^{-1}\boldsymbol{X}'\boldsymbol{y} \tag{1.6}$$

Note that when $p \neq 2$ the estimate $\boldsymbol{b}$ cannot be expressed as an explicit function of $\boldsymbol{X}$ and $\boldsymbol{y}$.

The Gauss-Markoff theorem indicates that the least squares estimator is the best linear unbiased estimator (BLUE), see Appendix 1A. It is important to note that this theorem refers only to the class of linear estimators. It does not consider nonlinear alternatives (e.g., values of $p \neq 2$). The statistical tractibility of least squares estimators has made least squares the most popular method of estimation.

Typically the least squares estimator is a kind of sample mean estimator [Kiountouzis (1973)]. This can be seen by considering the case $k = 1$ (location model). Hence (1.1) becomes $y = \beta_1 + e$, with least squares estimate $b_1 = \bar{y}$. This mean-estimator can be undesirable when the error distribution has long tails [Blattberg and Sargent (1971)]. An obvious alternative would be the median-type estimator. This in fact coincides with the case $p = 1$ which will now be discussed.

1.2.4. L_1-estimation

The L_1-norm estimation problem can be formulated as

$$
\begin{aligned}
&\min \sum_{i=1}^{n}(u_i + v_i) \\
&\text{subject to} \quad \left.\begin{aligned} \textstyle\sum_{j=1}^{k} b_j x_{ij} + u_i - v_i = y_i \\ u_i,\ v_i \geq 0 \end{aligned}\right\} \quad i = 1, 2, \ldots, n \\
&\qquad b_j \quad \text{unconstrained} \quad j = 1, \ldots, k
\end{aligned}
\tag{1.7}
$$

It therefore follows that the L_1-norm problem can be formulated as an LP problem in $2n + k$ variables and n constraints.

As with any linear programming problem a dual linear programming problem exists. Denote the dual variables by f_i. The dual of (1.7) is formulated as follows [see Wagner (1959)]:

$$
\begin{aligned}
&\max \sum_{i=1}^{n} f_i y_i \\
&\text{subject to} \quad -1 \leq f_i \leq 1 \qquad i = 1, \ldots, n \\
&\qquad \sum_{i=1}^{n} f_i = 0 \\
&\qquad \sum_{i=1}^{n} f_i x_{ij} = 0 \qquad j = 2, \ldots, k
\end{aligned}
\tag{1.8}
$$

Wagner (1959) also showed that by setting $w_i = f_i + 1$ the bounded variable dual problem may be formulated in terms of the nonnegative (upper bounded) variables w_i.

$$
\begin{aligned}
&\max \sum_{i=1}^{n} w_i y_i \\
&\text{subject to} \quad 0 \leq w_i \leq 2 \qquad i = 1, \ldots, n \\
&\qquad \sum_{i=1}^{n} w_i = n \\
&\qquad \sum_{i=1}^{n} w_i x_{ij} = \sum_{i=1}^{n} x_{ij} \qquad j = 2, \ldots, k
\end{aligned}
$$

Barrodale and Roberts (1973) managed to reduce the total number of primal simplex iterations by means of a process bypassing several neighbouring simplex vertices in a single iteration. Subroutine L1 may be found in Barrodale and Roberts (1974). Armstrong et al. (1979) modified the Barrodale and Roberts algorithm by using the revised simplex algorithm of linear programming which reduces storage requirements and uses an LU decomposition to maintain the current basis.

A number of interesting geometrical properties of the solution to the L_1-norm problem have been derived. We shall list a few of these. The interested reader is referred to Appa and Smith (1973) and Gentle et al. (1977) for more detail.

1) If the matrix $\boldsymbol{X}$ is of full column rank, i.e., $r(\boldsymbol{X}) = k$ then there exists a hyperplane passing through k of the points. If $r(\boldsymbol{X}) = r < k$ there exists an optimal hyperplane which will pass through r of the observations.
2) Let n^+ be the number of observations with positive deviations, n^- the corresponding number with negative deviations and n^* the maximum number of observations that lie on any hyperplane. Then

$$|n^+ - n^-| \leq n^*$$

3) For n odd every L_1 hyperplane passes through at least one point.
4) For n odd there is at least one observation which lies on every L_1 hyperplane.

One point that remains to be considered is that of optimality. Optimality conditions for the nonlinear L_1-norm problem are derived in Chapter 2 (see §2.2). The linear case is treated as a corollary of the nonlinear case.

Corollary 1.1 Let $A = \{i | y_i - \boldsymbol{x_i b} = 0\}$, $I = \{1, 2, \ldots, n\} - A$ be the sets of indices corresponding to the active and inactive functions respectively. Then a necessary and sufficient condition for parameters $\boldsymbol{b}$ to be a global L_1-norm solution is the existence of multipliers $-1 \leq \alpha_i \leq 1$ for all i such that

$$\sum_{i \in A} \alpha_i \boldsymbol{x_i} + \sum_{i \in I} sign(y_i - \boldsymbol{x_i b})\boldsymbol{x}_i = 0$$

At a recent conference aspects of statistical data analysis based on the L_1-norm were presented. The proceedings of the conference [see Dodge (1987)] is a must to anyone interested in L_1-norm estimation.

1.2.5. L_∞-estimation

The L_∞-estimation problem is often encountered. An example of where this approach is useful is in stabilizing the propagation of rounding errors in numerical computing. In this case the data follow a uniform distribution and the maximum absolute error should be minimized. Another example is the truncated distribution resulting from standardising examination results where low scores are truncated [Sklar and Armstrong (1983)]. Other examples are the approximation of library functions on digital computers where the objective is to make the largest error as small as possible (the analysis of worst case circumstances). The L_∞-norm criterion is equivalent to choosing the coefficients $\boldsymbol{b} = (b_1, \ldots, b_k)'$ as follows

$$\min_{\boldsymbol{b}} \max_{1 \le i \le n} |y_i - \boldsymbol{x}_i \boldsymbol{b}| \tag{1.9}$$

A comprehensive account of the theory of Chebychev approximation is given by Watson (1980).

Denoting the maximum absolute deviation by d_∞, the LP formulation of (1.9) is:

$$\min d_\infty$$

$$\text{subject to} \quad y_i - d_\infty \le \sum_{j=1}^{k} x_{ij} b_j \le y_i + d_\infty, \qquad i = 1, \ldots, n \tag{1.10}$$

$$d_\infty \ge 0 \quad \text{and} \quad b_j \text{ unconstrained} \qquad j = 1, \ldots, k$$

Let w_i and z_i be the dual variables corresponding to the constraints of (1.10). The dual problem can then be formulated as follows [Wagner (1959)]:

$$\max \sum_{i=1}^{n} y_i (w_i - z_i)$$

$$\begin{aligned} \text{subject to} \quad & \sum_{i=1}^{n} (w_i + z_i) = 1 \\ & \sum_{i=1}^{n} (w_i - z_i) = 0 \\ & \sum_{i=1}^{n} (w_i - z_i) x_{ij} = 0 \quad j = 2, \ldots, k \\ & w_i, z_i \ge 0 \qquad i = 1, \ldots, n \end{aligned} \tag{1.11}$$

A number of geometrical properties of the L_∞-norm solution have been derived by Appa and Smith (1973).

1) For problem (1.10) there exists one optimal hyperplane which is vertically equidistant from at least $k+1$ observations.

2) Any $k+1$ observations determining the optimal hyperplane must lie in the convex hull of n observations.

Barrodale and Phillips (1975) proposed a dual method for solving the primal formulation of the L_∞-norm estimation problem. A FORTRAN code CHEB is also given by these authors. Armstrong and Kung (1980) also proposed a dual method for this problem. By maintaining a reduced basis the algorithm bypasses certain simplex vertices.

Optimality conditions for the L_∞-norm problem also follow from the nonlinear case (§3.2 in Chapter 3) and are given in the following corollary:

Corollary 1.2 Let $A = \{i | y_i - \boldsymbol{x}_i \boldsymbol{b} = d_\infty\}$, $I = \{1, 2, \ldots, n\} - A$ be the sets of indices corresponding to the active and inactive functions respectively. Then a necessary and sufficient condition for parameters $\boldsymbol{b}$ to be a global L_∞-norm solution is the existence of nonnegative multipliers α_i such that

$$\sum_{i=1}^{n} \alpha_i = \sum_{i \in A} \alpha_i = 1$$

$$\alpha_i = 0 \qquad i \in I$$

$$\sum_{i \in A} \alpha_i sign(y_i - \boldsymbol{x}_i \boldsymbol{b}) \boldsymbol{x}_i = 0$$

1.3. The choice of p

Forsythe (1972), in estimating the parameters α and β in the simple regression model $y = \alpha + \beta x$, investigated the use of L_p-norm estimation with $1 < p \le 2$. He argued that since the mean is sensitive to deviations from normality the L_p-norm estimator will be more robust than least squares in estimating the mean. This will be the case when outliers are present. He suggested the compromise use of $p = 1.5$ when contaminated (normal) or skewly distributed error distributions are encountered. The Davidon-Fletcher-Powell (DFP) method was used as a minimization technique.

In a simulation study Ekblom (1974) compared the L_p-norm estimators with the Huber M-estimators. He also considered the case $p \le 1$. He concluded that the Huber estimator is superior to the L_p-norm estimator when the errors follow a contaminated normal distribution. For other error distributions (Laplace, Cauchy) he suggested that $p = 1.25$ be used. The proposal that $p \leq 1$ should be used for skewly distributed (χ^2) errors is interesting and shows that the remark by Rice (1964) that problems where $p < 1$ are not of interest, is unjustified.

Harter (1977) suggested an adaptive scheme which relies on the kurtosis of the regression error distribution. He suggested L_1-norm estimation if the kurtosis $\beta_2 > 3.8$, least squares if $2.2 \leq \beta_2 \leq 3.8$ and L_∞-norm estimation if $\beta_2 < 2.2$. This scheme has been extended by Barr (1980), Money et al. (1982), Sposito et al. (1983) and Gonin and Money (1985a,b) and will be discussed in Chapter 5.

Nyquist (1980,1983) considered the statistical properties of linear L_p-norm estimators. He derived the asymptotic distribution of linear L_p-norm estimators and showed it to be normal for sufficiently small values of p. It is not stated, however, how small p should be. A procedure for selecting the optimal value of p based on the asymptotic variance is proposed which validates the empirical studies by Barr (1980) and Money et al. (1982). Money et al. derived an empirical relationship between the optimal value of p and the kurtosis of the error distribution.

Sposito et al. (1983) derived a different empirical relationship which also relates the optimal value of p to the kurtosis of the error distribution. These authors also showed that the formula of Money et al. (1982) yields a reasonable value of p for error distributions with a finite range and suggested the use of their own formula for large sample sizes ($n \geq 200$) when it is known that $1 \leq p \leq 2$. The following modification of Harter's rule was suggested: Use $p = 1.5$ (Forsythe) if $3 < \beta_2 < 6$, least squares if $2.2 \leq \beta_2 \leq 3$ and L_∞-norm estimation if $\beta_2 < 2.2$. The abovementioned formulae will be the object of study in Chapter 5.

1.4. Statistical Properties of Linear L_p-norm Estimators

We introduce this section by stating the following theorem due to Nyquist (1980,1983). See also Gonin and Money (1985a).

Theorem 1.1 Consider the linear model $\boldsymbol{y} = \boldsymbol{X\beta} + \boldsymbol{e}$. Let $\boldsymbol{b}$ be the estimate of $\boldsymbol{\beta}$ chosen so that :

$$S_p(\boldsymbol{b}) = \sum_{i=1}^{n} |y_i - \boldsymbol{x}_i \boldsymbol{b}|^p$$

is a minimum where $1 \leq p < \infty$. Assume that:

A1: The errors e_i are independently, identically distributed with common distribution $\mathcal{F}$.

A2: The L_1 (and L_∞)-norm estimators are unique (in general L_p-norm estimators are uniquely defined for $1 < p < \infty$).

A3: The matrix $Q = \lim_{n\to\infty} X'X/n$ is positive definite with rank$(Q) = k$.

A4a: $\mathcal{F}$ is continuous with $\mathcal{F}'(0) > 0$ when $p = 1$.

A4b: When $1 < p < \infty$ the following expectations exist:

$$E\{|e_i|^{p-1}\},\ E\{|e_i|^{p-2}\},\ E\{|e_i|^{2p-2}\};\ \text{and also } E\{|e_i|^{p-1} sign(e_i)\} = 0$$

Under the above four assumptions $\sqrt{n}(\boldsymbol{b} - \boldsymbol{\beta})$ is asymptotically normally distributed with mean $\boldsymbol{\beta}$ and variance $\omega_p^2 \boldsymbol{Q}^{-1}$ where

$$\omega_p^2 = \begin{cases} \frac{1}{[2\mathcal{F}'(0)]^2} & \text{if } p = 1; \\ \frac{E[|e_i|^{2p-2}]}{[(p-1)E(|e_i|^{p-2})]^2} & \text{if } 1 < p < \infty. \end{cases}$$

Bassett and Koenker (1978) considered the case $p = 1$ and the following corollary results. [See also Dielman and Pfaffenberger (1982b)].

Corollary 1.3 For the linear model let $\boldsymbol{b}$ be the estimate of $\boldsymbol{\beta}$ such that

$$S_1(\boldsymbol{b}) = \sum_{i=1}^{n} |y_i - \boldsymbol{x}_i \boldsymbol{b}|$$

is a minimum. Under the assumptions A1, A2, A3 and A4a of Theorem 1.1 it follows for a random sample of size n that $\sqrt{n}(\boldsymbol{b} - \boldsymbol{\beta})$ is asymptotically normally distributed with mean $\boldsymbol{O}$ and variance $\lambda^2(\boldsymbol{X}'\boldsymbol{X})^{-1}$ where λ^2/n is the variance of the sample median of residuals.

1.5. Confidence intervals for β

1.5.1. Case $1 < p < \infty$

In view of the asymptotic results of Nyquist (1980,1983) we can construct the following confidence intervals for the components of $\boldsymbol{\beta}$. Specifically the $100(1-\alpha)\%$ confidence interval for β_j is given by

$$b_j \pm z_{\alpha/2}\sqrt{\omega_p^2(\boldsymbol{X}'\boldsymbol{X})^{jj}}$$

Where $(\boldsymbol{X}'\boldsymbol{X})^{jj}$ denotes the jth diagonal element of $(\boldsymbol{X}'\boldsymbol{X})^{-1}$ and $z_{\alpha/2}$ denotes the appropriate percentile of the standard normal distribution.

A major drawback is that ω_p^2 is unknown. However, we have performed a simulation study in which the sample moments of the residual distribution were used to estimate ω_p^2. The estimate $\hat{\omega}_p^2$ was calculated as follows:

$$\hat{\omega}_p^2 = \frac{m_{2p-2}}{[(p-1)m_{p-2}]^2}$$

with $m_r = \frac{1}{n}\sum_{i=1}^{n}|\hat{e}_i|^r$ where $\hat{e}_i$ is the ith residual from the L_p-norm fit.

The simulation was performed using a two parameter model of the form

$$y_i = 10 + \beta_1 x_{i1} + \beta_2 x_{i2}$$

In this model the parameters β_1 and β_2 were fixed at 8 and -6 respectively. Values of (x_{i1}, x_{i2}) for $i = 1, 2, \ldots, n$ ($n = 30, 50, 100, 200, 400$) were selected from a uniform distribution over the interval 0 to 20 and held fixed in all 500 samples. In all the runs a uniform error distribution with mean zero and variance 25 was used.

The exact ω_p^2 for the uniform error distribution is given by $3\sigma^2/(2p-1)$ which is equal to $75/(2p-1)$ for the chosen error distribution. In the simulation the parameters β_1 and β_2 were estimated using $L_{1.5}$ and L_3. The residuals and the appropriate sample moments of $\hat{\omega}_p^2$ were calculated. For $p < 2$ we have a negative exponent and although this is not a moment in the usual statistical sense, the agreement is still good. This approximation of ω_p^2 was found to be adequate for other error distributions as well as for varying values of p. Table 1.1 resulted:

Table 1.1:

p	True ω_p^2	Sample size(n)	Mean $\hat{\omega}_p^2$	Standard Deviation	Coefficient of variation
1.5	37.5	30	33.81	15.7	46.5
		50	35.21	13.4	38.1
		100	36.19	10.0	27.6
		200	36.74	8.1	22.2
		400	37.01	6.2	16.8
3.0	15.0	30	14.31	2.0	14.2
		50	14.65	1.5	10.6
		100	14.85	1.0	7.0
		200	14.92	0.7	4.7
		400	14.95	0.5	3.2

1.5.2. Case $p = 1$

Using the Bassett and Koenker (1978) result, the $100(1-\alpha)\%$ confidence interval for a single parameter β_j is given by

$$b_j \pm z_{\alpha/2}\lambda\sqrt{(\boldsymbol{X}'\boldsymbol{X})^{jj}}$$

However, λ is unknown, Dielman and Pfaffenberger (1982a) therefore suggest the Cox and Hinkley (1974:70) procedure for estimating λ. Alternative approaches for estimating the variance of the median of residuals (λ^2/n) are given in Maritz and Jarrett (1978) and Efron (1979). Alternative procedures for estimating λ have been proposed by McKean and Schrader (1987), Schrader and McKean (1987), Sheather (1987) and Welsh (1987a).

Cramér (1946:369) showed that asymptotically

$$\lambda = \frac{1}{2f(m)} \tag{1.12}$$

where $f(m)$ is the ordinate of the error distribution e at the median, m. Denoting the L_1 estimator as $\boldsymbol{b}$ then $\hat{e}_i = y_i - \boldsymbol{x}_i\boldsymbol{b}$ and $f(m)^{-1}$ can be estimated by

$$\hat{f}(m)^{-1} = \frac{\hat{e}_{(i)} - \hat{e}_{(j)}}{(i-j)/n} \tag{1.13}$$

where $\hat{e}_{(i)}$ denotes the ordered residuals $\hat{e}_i$. Moreover, Cox and Hinkley (1974) stress that i and j should be symmetric about the index of the median sample residual and that the difference between i and j should be kept small.

We shall proceed as follows: Only the nonzero residuals are considered. Thus instead of n use $\tilde{n} = n - k$ in (1.13). The following formulae for i and j can be used:

$$i = \left[\frac{\tilde{n}+1}{2}\right] + v \quad \text{and} \quad j = \left[\frac{\tilde{n}+1}{2}\right] - v \qquad \text{for} \quad \tilde{n} \text{ odd}$$

$$i = \left[\frac{\tilde{n}}{2}\right] + 1 + v \quad \text{and} \quad j = \left[\frac{\tilde{n}}{2}\right] - v \qquad \text{for} \quad \tilde{n} \text{ even}$$

where $[\cdot]$ denotes the truncated integer value and where v is a positive but small integer. Cox and Hinkley also show that $[2\hat{f}(m)]^{-1}$ is a consistent estimator of λ.

1.6. Example

To illustrate the application of L_p-norm estimation we use the econometric example given by Narula and Wellington (1977:189). However, only a subset of the variables is considered. Here the price of a house (PR) is assumed to be linearly dependent on property tax (T) and living space (S). The moment matrix $\boldsymbol{X}'\boldsymbol{X}$ and its inverse $(\boldsymbol{X}'\boldsymbol{X})^{-1}$ are set out below.

$$\boldsymbol{X}'\boldsymbol{X} = \begin{pmatrix} 28.000 & 201.212 & 42.327 \\ 201.212 & 1664.510 & 339.007 \\ 42.327 & 339.007 & 72.088 \end{pmatrix}$$

$$(\boldsymbol{X}'\boldsymbol{X})^{-1} = \begin{pmatrix} 0.324578 & -0.00998516 & -0.143621 \\ -0.00998516 & 0.0145387 & -0.0625078 \\ -0.143621 & -0.0625078 & 0.392154 \end{pmatrix}$$

TABLE 1.2 $\widehat{PR}$ vs $\widehat{PR_{dev}}$ for various values of p

$p = 2$		$p = 1$		$p = 2.5$		$p \to \infty$	
$\widehat{PR}$	$\widehat{PR_{dev}}$	$\widehat{PR}$	$\widehat{PR_{dev}}$	$\widehat{PR}$	$\widehat{PR_{dev}}$	$\widehat{PR}$	$\widehat{PR_{dev}}$
25.78	0.12	25.90	0.00	25.72	0.18	25.95	-0.05
32.75	-3.25	30.14	-0.64	32.89	-3.39	33.91	-4.41
27.24	0.66	25.99	1.91	27.27	0.63	27.90	-0.00
28.04	-2.14	26.48	-0.58	28.09	-2.19	28.81	-2.91
27.77	2.13	27.34	2.56	27.76	2.14	28.14	1.76
23.16	6.74	22.31	7.59	23.14	6.76	23.66	6.24
31.40	-0.50	31.13	-0.23	31.39	-0.49	31.72	-0.82
34.18	-5.28	32.14	-3.24	34.29	-5.39	35.14	-6.24
83.68	1.22	80.62	4.28	84.10	0.80	85.14	-0.24
75.72	7.18	74.08	8.82	76.03	6.87	76.66	6.24
31.03	4.87	30.78	5.12	31.02	4.88	31.34	4.56
34.13	-2.63	31.50	0.00	34.27	-2.77	35.30	-3.80
28.75	2.25	30.36	0.64	28.63	2.37	28.40	2.60
29.95	0.95	30.42	0.48	29.90	1.00	30.01	0.89
26.39	3.61	26.52	3.48	26.34	3.66	26.56	3.44
34.18	-5.28	32.14	-3.24	34.29	-5.39	35.14	-6.24
42.77	-5.87	42.46	-5.56	42.83	-5.93	43.14	-6.24
36.65	5.25	35.79	6.11	36.71	5.19	37.21	4.69
37.79	2.71	38.64	1.86	37.76	2.74	37.73	2.77
42.49	1.41	43.90	0.00	42.45	1.45	42.25	1.65
31.83	5.67	31.57	5.93	31.83	5.67	32.15	5.35
41.41	-3.51	40.25	-2.35	41.51	-3.61	42.08	-4.18
46.19	-1.69	45.54	-1.04	46.28	-1.78	46.69	-2.19
37.39	0.51	34.99	2.91	37.54	0.36	38.48	-0.58
44.56	-5.66	43.72	-4.82	44.66	-5.76	45.12	-6.22
40.37	-3.47	40.85	-3.95	40.37	-3.47	40.45	-3.55
47.17	-1.37	46.82	-1.02	47.26	-1.46	47.57	-1.77
45.65	-4.65	51.72	-10.72	45.37	-4.37	43.76	-2.76

L_∞:

$$\widehat{PR} = 0.33093 + 2.07485T + 15.443S$$

The maximum deviation is $\hat{d}_\infty = 6.238$. The predicted prices and residuals are given in Table 1.2 under the heading $p \to \infty$. We do not give the standard errors of the estimates since the underlying error distribution is unknown.

$L_{2.5}$:

$$\begin{array}{ccccccc} \widehat{PR} & = & 0.26504 & + & 2.38019T & + & 13.7817S \\ & & (1.969) & & (0.417) & & (2.164) \end{array}$$

$\sum_{i=1}^{28} |\hat{e}_i|^{2.5} = 925.643$, mean deviation $= 4.2403$ and $\omega_p^2 = 11.934$, the standard errors of the parameters are given in parentheses. The predicted prices and residuals are given in Table 1.2 under the heading $p = 2.5$.

Least squares:

$$\begin{array}{ccccccc} \widehat{PR} & = & 0.48918 & + & 2.42385T & + & 13.39559S \\ & & (2.309) & & (0.489) & & (2.538) \end{array}$$

$R^2 = 0.924$, $s_e^2 = 16.424$. The predicted prices and residuals are given in Table 1.2 under the heading $p = 2$.

LAV:

$$\begin{array}{ccccccc} \widehat{PR} & = & 1.33081 & + & 3.4249T & + & 7.7424S \\ & & (3.111) & & (0.658) & & (3.419) \end{array}$$

$\sum_{i=1}^{28} |\hat{e}_i| = 89.0981$, mean deviation $= 3.56$. The predicted prices and residuals are given in Table 1.2 under the heading $p = 1$. The estimation of the standard errors for the estimates of the parameters determined by LAV will now be illustrated. First we need to determine λ.

A number of approaches have been proposed. Two of these will be considered. The first is based on the Cox-Hinkley estimator and the other on the α-percent nonparametric confidence interval formula of McKean and Schrader (1987).

Cox-Hinkley

We want to choose v as small as possible. However, for $v = 1$ $\hat{\lambda} = 5.4598$, for $v = 2$ $\hat{\lambda} = 7.605$ for $v = 3$ $\hat{\lambda} = 5.3188$ and for $v = 4$ $\hat{\lambda} = 5.5894$.

McKean and Schrader

McKean and Schrader (1987) consider an estimator which is based on the α-percent nonparametric confidence interval for λ. The estimate of λ is given by

$$\hat{\lambda}_{1-\alpha} = \frac{\sqrt{\tilde{n}}(\tilde{e}_{(\tilde{n}-l+1)} - \tilde{e}_{(l)})}{2z_{\alpha/2}}$$

where asymptotically

$$l = \frac{\tilde{n}+1}{2} - z_{\alpha/2}(\frac{\tilde{n}}{4})^{\frac{1}{2}}$$

[see Hollander and Wolfe (1973:49)]. l is usually rounded to the nearest integer.

Hence an estimate for $\hat{\lambda}$ is:

$$\begin{aligned}\hat{\lambda} &= \frac{5(2.911 + 1.044)}{2 \times 1.96} \\ &= 5.0446\end{aligned}$$

So we can expect to find that the variance of the median residual, $\hat{\lambda}^2/n \approx 1$.

95 % confidence interval ($v = 1,\ \hat{\lambda} = 5.4598$)

Parameter	Lower limit	Upper limit
β_1	-4.77	7.43
β_2	2.13	4.72
β_3	1.04	14.44

The standard errors are given by $\hat{\lambda}\sqrt{(\boldsymbol{X}'\boldsymbol{X})^{jj}}$.

To check for optimality we see that the set of active functions $A = \{1, 12, 20\}$. Note that $k = 3$ in this example. We need to determine $-1 \leq \alpha_i \leq 1$ such that

$$\sum_{i \in A} \alpha_i \boldsymbol{x}_i + \sum_{i \in I} sign(\hat{e}_i)\boldsymbol{x}_i = 0$$

where $\hat{e}_i$ is the ith residual and $\boldsymbol{x}_i$ is the ith row of X.

$$\alpha_1 \begin{pmatrix} 1 \\ 4.92 \\ 1.0 \end{pmatrix} + \alpha_2 \begin{pmatrix} 1 \\ 5.30 \\ 1.55 \end{pmatrix} + \alpha_3 \begin{pmatrix} 1 \\ 9.04 \\ 1.5 \end{pmatrix} - \begin{pmatrix} 1 \\ 5.02 \\ 1.5 \end{pmatrix} + \begin{pmatrix} 1 \\ 4.54 \\ 1.18 \end{pmatrix} \ldots - \begin{pmatrix} 1 \\ 12 \\ 1.2 \end{pmatrix} = \begin{pmatrix} 0 \\ 0 \\ 0 \end{pmatrix}$$

Thus we obtain:

$$\alpha_1 + \alpha_2 + \alpha_3 = 1$$

$$4.92\alpha_1 + 5.3\alpha_2 + 9.04\alpha_3 = 4.14$$

$$1.0\alpha_1 + 1.55\alpha_2 + 1.5\alpha_3 = 1.36$$

The solution of which is

$$\alpha_1 = 0.37 \qquad \alpha_2 = 0.90 \qquad \alpha_3 = -0.27$$

Hence the necessary conditions for optimality are satisfied.

In the L_∞-norm case $A = \{6, 8, 10, 16, 17\}$, thus we solve

$$\alpha_1 \begin{pmatrix} 1 \\ 3.891 \\ 0.998 \end{pmatrix} - \alpha_2 \begin{pmatrix} 1 \\ 5.604 \\ 1.501 \end{pmatrix} + \alpha_3 \begin{pmatrix} 1 \\ 14.460 \\ 3 \end{pmatrix} - \alpha_4 \begin{pmatrix} 1 \\ 5.604 \\ 1.501 \end{pmatrix} - \alpha_5 \begin{pmatrix} 1 \\ 8.246 \\ 1.664 \end{pmatrix} = \begin{pmatrix} 0 \\ 0 \\ 0 \end{pmatrix}$$

Bearing in mind the sign of the residuals $\hat{e}_i$, $i \in A$ we get another equation

$$\alpha_1 + \alpha_2 + \alpha_3 + \alpha_4 + \alpha_5 = 1$$

We then have 4 equations in 5 unknowns, letting $\alpha_4 = 0$ and solving for the others we have

$$\alpha_1 = 0.35 \qquad \alpha_2 = 0.24 \qquad \alpha_3 = 0.15 \qquad \alpha_4 = 0 \qquad \alpha_5 = 0.26$$

We can summarise the solutions for various values of p as follows:

	p			
Coefficients	1.0	2.0	2.5	∞
Intercept	1.331	0.489	0.265	0.331
Tax	3.425	2.424	2.380	2.075
Living space	7.742	13.396	13.782	15.443

From the results in the table above the choice of p is clearly critical.

CONCLUSION

Recent research has concentrated on deriving efficient algorithms for solving linear L_p-norm estimation problems. The distributional properties of these estimators have been examined as well as the consequences of problems such as the presence of heteroscedasticity in the errors. Those problems in which all the variables are subject to error are also currently under investigation.

The nonlinear L_p-norm problem has received little attention except for algorithms which solve nonlinear L_1-norm and L_∞-norm estimation problems. The purpose of this book is to bring together both the numerical and statistical aspects of the nonlinear L_p-norm estimation problem.

Appendix 1A: Gauss-Markoff Theorem

Theorem 1 Consider the general linear model

$$\boldsymbol{y} = \boldsymbol{X\beta} + \boldsymbol{e}$$

as described in §1.2. Under the assumptions that $\mathcal{E}(\boldsymbol{e}) = 0$ and $\mathcal{E}(\boldsymbol{ee}') = \sigma^2\boldsymbol{I}$, the least squares estimator $\boldsymbol{b} = (\boldsymbol{X}'\boldsymbol{X})^{-1}\boldsymbol{X}'\boldsymbol{y}$ is the best linear unbiased estimator (BLUE) of $\boldsymbol{\beta}$ with variance–covariance matrix $\sum_{\boldsymbol{bb}'} = \sigma^2(\boldsymbol{X}'\boldsymbol{X})^{-1}$.

Proof Differentiate the sum of the squared errors, $SSE = (\boldsymbol{y} - \boldsymbol{X\beta})'(\boldsymbol{y} - \boldsymbol{X\beta})$ with respect to $\boldsymbol{\beta}$ and set the derivative equal to zero, hence:

$$-2\boldsymbol{X}'\boldsymbol{y} + 2\boldsymbol{X}'\boldsymbol{X\beta} = 0$$

Denoting the solution by $\boldsymbol{b}$ we obtain, $\boldsymbol{b} = (\boldsymbol{X}'\boldsymbol{X})^{-1}\boldsymbol{X}'\boldsymbol{y}$. The variance–covariance matrix is therefore

$$\sum_{\boldsymbol{bb}'} = \mathcal{E}(\boldsymbol{b} - \boldsymbol{\beta})(\boldsymbol{b} - \boldsymbol{\beta})'$$

Since $(\boldsymbol{b} - \boldsymbol{\beta}) = (\boldsymbol{X}'\boldsymbol{X})^{-1}\boldsymbol{X}'\boldsymbol{e}$, $\boldsymbol{X}$ is nonstochastic and $\mathcal{E}(\boldsymbol{ee}') = \sigma^2\boldsymbol{I}$ it follows that

$$\sum_{\boldsymbol{bb}'} = \sigma^2(\boldsymbol{X}'\boldsymbol{X})^{-1}$$

- *Unbiasedness of* $\boldsymbol{b}$ — Since $\boldsymbol{b} = \boldsymbol{\beta} + (\boldsymbol{X}'\boldsymbol{X})^{-1}\boldsymbol{X}'\boldsymbol{e}$ it follows that $\mathcal{E}(\boldsymbol{b}) = \boldsymbol{\beta}$.
- *Minimum variance* $\sum_{\boldsymbol{bb}'}$ — Let $\tilde{\boldsymbol{b}} = [(\boldsymbol{X}'\boldsymbol{X})^{-1}\boldsymbol{X}' + \boldsymbol{D}']\boldsymbol{y}$ be any other linear estimator with $\boldsymbol{D}$ a $n \times k$ nonstochastic matrix. It will be an unbiased estimator if and only if $\boldsymbol{D}'\boldsymbol{X} = 0$. Then

$$\begin{aligned}\sum_{\tilde{\boldsymbol{b}}\tilde{\boldsymbol{b}}'} = \mathcal{E}(\tilde{\boldsymbol{b}} - \boldsymbol{\beta})(\tilde{\boldsymbol{b}} - \boldsymbol{\beta})' &= \mathcal{E}\left\{[(\boldsymbol{X}'\boldsymbol{X})^{-1}\boldsymbol{X}' + \boldsymbol{D}']\boldsymbol{ee}'[(\boldsymbol{X}'\boldsymbol{X})^{-1}\boldsymbol{X}' + \boldsymbol{D}']'\right\} \\ &= \sigma^2\left\{(\boldsymbol{X}'\boldsymbol{X})^{-1} + \boldsymbol{D}'\boldsymbol{D}\right\} \\ &\geq \sum_{\boldsymbol{bb}'}\end{aligned}$$

The above follows since $\boldsymbol{D}'\boldsymbol{D}$ is positive-semi definite. ∎

Additional notes

§1.1

1 A complete summary of curve fitting spanning a period of approximately four centuries is given by Harter (1974a,1974b,1975a,1975b,1975c,1975d,1976). See also Nyquist (1980) and Farebrother (1987).

2 Gauss' statement is given in Nyquist (1980:11).

3 Kelley (1958) and Stiefel (1960) independently formulated the L_∞-norm problem as an LP problem and proposed solution procedures.

4 Karst (1958) derived an iterative numerical procedure for L_1-estimation for the simple linear regression model.

5 Rapid development of computers in the late 1960's and the work of Barrodale and Roberts (1970,1973,1974) and Barrodale and Phillips (1975) stimulated substantial research on L_1- and L_∞-norm estimation.

6 Papers by Fletcher et al. (1971), Forsythe (1972), Kahng (1972), Ekblom (1973,1974) and Merle and Späth (1973) stimulated research on the more general L_p-norm estimation problem. Ekblom (1974) also considered the case $p < 1$.

§1.2.1

1 It should be noted that this mathematical programming formulation is flexible as it allows linear constraints to be included. For example, certain coefficients may be specified to be nonnegative or weights can be assigned to certain deviations.

2 Descloux (1963) and Fletcher et al. (1974a) showed that the L_∞-norm estimator is the limiting L_p-norm estimator as $p \to \infty$.

3 Fletcher et al. (1974b) demonstrated that the parameters of the continuous L_p-norm approximation problem for $p \geq 2$ are both continuous and differentiable functions of p. This result, however, does not hold for the discrete L_p-norm problem (1.3).

§1.2.2

1 Fletcher et al. (1971) derived a method which takes the structure of problem (1.3) into account. Their method is analogous to Newton's method for solving algebraic equations. They proved that provided certain integrals exist the method converges for all values of $p \geq 2$ and that the convergence rate is quadratic when $p \geq 3$. Prior to this paper the DFP quasi-Newton method was

the method of choice. This is a standard method for unconstrained optimization problems. The method of Fletcher (1970) is, however, more efficient than the DFP method (see Himmelblau (1972) and Himmelblau and Lindsay (1980) for numerical comparisons). However, it is surprising that the homogeneous algorithms due to Jacobson and Oksman (1972) and Jacobson and Pels (1974) have not received more prominence in the literature due to their computational efficiency compared with well known variable metric methods such as the DFP method. Kahng (1972) and Rey (1975) have also presented algorithms based on Newton's method.

2 Merle and Späth (1973) remarked that the errors in the response variables are often nonnormally distributed with unequal variances. They suggested two algorithms; the first for problems where $1 \leq p \leq 2$ and the second for problems where $p > 2$. In the first algorithm, also known as iteratively reweighted least squares, they set zero residuals equal to a small positive constant. The second algorithm is also based on Newton's method as mentioned above. They found that the first algorithm converges on numerical examples (although the convergence is not proved) whilst the second Newton-type algorithm is, of course, known to converge.

3 Schlossmacher (1973) derived a computational procedure for solving linear L_p-norm problems which uses iteratively reweighted least squares. The method temporarily deletes observations which give rise to zero residuals and then reinstates them in subsequent iterations if their residuals become larger. The method will not always yield an optimal solution nor has a convergence proof been derived [Kennedy and Gentle (1980:532)].

4 Schlossmacher's method was extended by Sposito et al. (1977) to the case $1 \leq p \leq 2$ for the simple linear model $y = \alpha + \beta x$. Kennedy and Gentle (1978) showed that a conventional quasi-Newton method should be be used when $1 < p < 2$ and that a modified Newton method works well on problems when $p > 2$. Barr (1980) extended the method of Sposito et al. (1977) to the multiple regression case. He concluded that the method is useful for solving linear L_p-norm regression problems for $p \in [1\,,\,2.6]$. However, he demonstrated that for $p > 2.6$ the DFP method is superior and that when $p \geq 3$ the method will not necessarily converge.

5 Wolfe (1979) examined the first algorithm of Merle and Späth (1973) and went on to prove convergence of their algorithm when $1 < p < 2$. He showed that the rate of convergence is geometric with an asymptotic convergence constant of $2 - p$. A similar result holds for $p = 1$ if the best approximation is unique. This paper supports the empirical findings of Merle and Späth (1973).

Baboolal and Watson (1981) consider the Wolfe (1979) and Merle and Späth (1973) algorithms for the L_1 case to be equivalent to the Barrodale and Roberts (1973,1974) algorithm. The Barrodale and Roberts algorithm is in general superior to the other two.

6 Watson (1982,1983) and Norback and Morris (1980) considered problem (1.3) when both the response and explanatory variables are subject to error. When only the response variable is subject to error the norm of the vertical distances is minimized. Similarly, the norm of the horizontal distances is minimized when the explanatory variables alone are subject to error. It is therefore natural to minimize the norm of the perpendicular distances from the observations to the fitted line when all the variables are subject to error. Two methods based on recent work by Späth (1982) which are locally convergent for $1 < p < 3$ and two algorithms that are globally convergent for all $p > 1$ are presented. The third method is a descent method and the fourth, a Newton-based method. Numerical difficulties were experienced when p was close to 1. Osborne and Watson (1985) developed a finite descent algorithm to overcome this problem. See also Nyquist (1987).

§1.2.3

1 It has been shown by Andrews (1974) that when error distributions are nonnormal, least squares estimation is not ideal. It is demonstrated how dramatically the least squares estimates can change should only a small proportion of extreme observations be present in the data. In such cases more robust procedures are appropriate in the sense that they are statistically more efficient.

2 Sometimes the combination (convex) of least squares and L_1 estimation is of interest, see e.g., Gentle et al. (1977), Arthanari and Dodge (1981), Best and Chakravarti (1987) and Dodge and Jurečkova (1987).

§1.2.4

1 Extensive research in linear L_1-norm estimation has been reported in the statistical and mathematical journals. Comprehensive surveys have been undertaken by Gentle (1977), Narula and Wellington (1982), Dielman and Pfaffenberger (1982a), Dielman (1984) and Narula (1987). An overview of the historical development of L_1 and L_∞-norm estimation in the period 1793–1930 is provided by Farebrother (1987).

2 Bartels et al. (1978) summarize the LAV criterion taking into account its piecewise differentiability. Any L_1-norm estimation problem can be formulated as a linear programming problem, however, they show that any bounded feasible linear programming problem can be formulated as an L_1-norm problem. Various alternative algorithms as well as modifications have been proposed to

improve convergence. Sadovski (1974) followed an alternative approach which is based on the original method of Edgeworth (1888). Bloomfield and Steiger (1980) extend the method by Rhodes (1930) and Karst (1958). Abdelmalek (1980a,1980b) also used the dual formulation and developed an algorithm similar to that of Barrodale and Roberts (1973,1974). Wesolowsky (1981) reported a descent method for L_1-norm estimation. Peters and Willms (1983) consider procedures for up-and down- dating the solution of the linear L_1 regression problem when a column of X is inserted, deleted or if $\boldsymbol{y}$ is changed. The condensed simplex tableau of the Barrodale and Roberts (1973,1974) algorithm is up or down-dated. A FORTRAN listing is included as well as the necessary modification to subroutine L1 of Barrodale and Roberts (1974).

Späth (1986) considers a clustering approach to solving linear L_1 regression problems. Given n observations and that $n \gg k$. A number l (unknown) nonempty clusters or subsets of observations is created so that

$$\bigcup_{i=1}^{l} C_i = \{1, \ldots, n\}$$

Then $\hat{\boldsymbol{\beta}}$ is determined so that the average sum

$$\sum_{j=1}^{l} \min_{\boldsymbol{b}_i} \sum_{i \in C_j} |y_i - \boldsymbol{x}_i \boldsymbol{b}| \qquad \text{is a minimum}$$

Späth feels that this should be done rather than attempting to solve the whole problem, but does not motivate his conjecture or mention whether this problems is equivalent to the original problem. A FORTRAN subroutine CWLL1R in conjunction with subroutine L1NORM by Armstrong et al. (1979) is used.

Osborne (1987) considers a reduced gradient method for L_1-estimation and the rank regression problem. McConnell (1987) shows how the Kabe (1986) method of vanishing Jacobians for solving quadratic programming problems can be used to solve L_1-norm problems. Ekblom (1987) reviews the literature in which the linear L_1-norm problem is solved as a limiting case of the linear L_p-norm problem. The algorithm for computing L_p-norm estimates if applied to the linear L_1-norm problem will encounter difficulties. Ekblom suggests that the embedding approach of El-Attar et al. (1979) for the nonlinear L_1-norm problem can be used for the linear case. Schellhorn (1987) considers homotopy methods for solving the L_1-norm problem.

3 Gentle et al. (1987) report their computational experience with available computer programmes for linear L_1-norm problems and conclude from their sim-

ulation studies that the code by Armstrong et al. (1979) appears to be the best. A comparison of methods for L_1-estimation was reported by Bloomfield and Steiger (1983). They compared their algorithm with the Barrodale and Roberts (1973,1974) and Bartels et al. (1978) methods.

4 A SAS procedure PROC LAV is also available [see Gentle and Lee (1983)]. A FORTRAN subroutine RLAV is available in the IMSL library.

5 The primal and dual formulations (1.7) and (1.8) are often written in matrix notation as follows:

Primal

$$\min \mathbf{1}_n{}'\boldsymbol{u} + \mathbf{1}_n{}'\boldsymbol{v}$$

$$\ni \boldsymbol{X}\boldsymbol{b} + \boldsymbol{I}_n\boldsymbol{u} - \boldsymbol{I}_n\boldsymbol{v} = \boldsymbol{y}$$

$$\boldsymbol{u}, \boldsymbol{v} \geq 0$$

$$\boldsymbol{b} \quad \text{unconstrained}$$

Dual

$$\max \boldsymbol{f}'\boldsymbol{y}$$

$$\ni \boldsymbol{X}'\boldsymbol{f} = \mathbf{o}$$

$$-1 \leq f_i \leq 1 \qquad i = 1, 2 \ldots, n$$

where $\boldsymbol{X}$ is the $n \times k$ design matrix

$$X = \begin{pmatrix} 1 & x_{12} & \ldots & x_{1k} \\ \vdots & \vdots & \vdots & \vdots \\ 1 & x_{n2} & \ldots & x_{nk} \end{pmatrix}$$

6 A further extension of this scheme of skipping extreme point solutions was developed by Sklar and Armstrong (1983). The interested reader is referred to the book by Arthanari and Dodge (1981). Sklar and Armstrong (1982) used the linear least squares solution as an initial basis in the linear programming formulation of the simple L_1- and L_∞-norm estimation problems. They showed that a significant saving in computation can be achieved on a number of well known algorithms. [see also McCormick and Sposito (1976) and Hand and Sposito (1980)].

7 Stepwise selection of variables has been incorporated into L_1-norm and L_∞-norm algorithms. See for example Roodman (1974), Gentle and Hanson (1977),

and Wilson (1979). Armstrong and Kung (1982) and Armstrong and Beck (1985) consider the selection of the best subset of parameters in L_1-norm estimation. This is achieved by branch and bound methodology. Armstrong et al. (1984) provide a FORTRAN program for best subset selection of parameters and it is available from the ACM algorithms distribution service, [see also Narula and Wellington (1985)].

8 For L_1-norm estimation, subject to restrictions on the parameters, conventional linear programming can be used. Barrodale and Roberts (1977,1978) generalized their 1973 algorithm to handle linear constraints. Other approaches can be found in Armstrong and Hultz (1977), Bartels et al. (1978), Bartels and Conn (1980) and Dinkel and Pfaffenberger (1981). A FORTRAN code, Algorithm 563, (by Bartels and Conn), is available from the ACM algorithm service.

§1.2.5

1 For L_∞-norm estimation, subject to constraints on the parameters, Joe and Bartels (1983) derive an exact penalty function method. Abdelmalek (1977) also addressed this problem.

2 Armstrong and Beck (1983) consider the selection of the best subset of parameters in L_∞-norm estimation.

3 A FORTRAN subroutine RLMV is available in the IMSL library.

§1.3

Militký and Čáp (1987) apply the Bayesian approach to adaptive nonlinear L_p-norm regression. This approach in some ways resembles the adaptive scheme of Gonin and Money (1985b) for selecting p (see Chapter 5).

§1.4

1 Using the asymmetrical estimator concept introduced by Sielken and Hartley (1973), Harvey (1978) showed that the linear L_p-norm estimator will always be unbiased for $1 < p < \infty$ given the assumptions that the regression errors are symmetrically distributed, the first moment exists (for Cauchy distributed errors the estimator will obviously be biased) and that $\boldsymbol{X}$ is of full column rank. When $p = 1$ the estimator may not be unique and may be biased. Sielken and Hartley (1973) showed how an unbiased estimator may be obtained by means of linear programming. Sposito (1982) extended this result to the general case where $p \geq 1$. This he achieved by formulating the linear L_p-norm problem as a convex programming problem which can be solved by the convex-simplex method of Zangwill (1969). A computational scheme similar to Sielken and Hartley's is then applied. Hence unbiased L_p-norm estimates for all $p \geq 1$ can be obtained.

2 Farebrother (1985) provides a procedure which yields unbiased L_1- and L_∞-norm estimates of $\boldsymbol{\beta}$ in the linear model $\boldsymbol{y} = \boldsymbol{X}\boldsymbol{\beta} + \boldsymbol{e}$. It is demonstrated that the L_1-problem is the dual of a variant of the L_∞-problem and vice versa. This result is then used to obtain unbiased estimates of $\boldsymbol{\beta}$ for the L_1 and L_∞ estimation problems. These estimates are obtained by solving any one of five pairs of linear programming problems. It is claimed by the author that the suggested procedures are only marginally less efficient in terms of computing costs than their corresponding biased procedures. It is a pity that no numerical examples are given to illustrate the procedure as well as to support the claim. As far as the L_∞ problem is concerned the advantage of this new method over the method of Sielken and Hartley (1973) or Sposito (1982) is not clear.

3 Nyquist (1980) also examined the multicollinear, stochastic regressor and linearly dependent residual cases.

4 Pfaffenberger and Dielman (1983) compared OLS, LAV and Ridge regression procedures to estimate the coefficients of the multiple regression model in the presence of multicollinearity.

5 Asymptotic properties of the L_1-norm estimator were derived by Bassett and Koenker (1978) and Dielman and Pfaffenberger (1983). They also showed that the relative efficiency of the L_1-norm estimator to the least squares estimator is the same as the relative efficiency of the sample median to the sample mean. (Subsequently Koenker and Bassett (1982a) proposed large sample tests of hypotheses for linear LAV regression parameters.)

6 Koenker and Bassett (1982b) investigate the distributional properties of 3 test statistics for testing hypotheses concerning the regression parameters when LAV estimation is used to estimate these parameters. For the simple exclusion hypothesis, the Wald, the Likelihood Ratio, and the Lagrange Multiplier tests were shown to have the same χ^2-distribution asymptotically. They point out that the Lagrange Multiplier test has an advantage over the other two tests in that unlike them its test statistic does not require knowledge or estimation of the scale parameter λ. The Lagrange multiplier test statistic is independent of λ. Koenker (1987) cautions that the test statistic has a discrete distribution and that the χ^2-approximation can be problematical.

This was extended by Koenker (1987) to the general linear hypothesis and the test statistics were again shown to have the same χ^2-distribution asymptotically.

7 Dielman and Pfaffenberger (1984) investigate the simple exclusion hypothesis. By means of a simulation study, the performance of these three tests for samples of size 20, 30, 40, 50, and 100 was measured. It is suggested that for samples of

size 40 or more the Lagrange Multiplier test be used in preference to the Wald test. This seems to be borne out by the remark of Koenker (1987) in 6. The Lagrange multiplier test, however, can also be used for samples of size down to 30. For smaller sample sizes it is suggested that the Wald test is appropriate. In general the Likelihood Ratio test performed poorly.

Koenker (1987) investigates the more general linear hypothesis. By means of a Monte Carlo study the small sample performance of these three tests are examined. The model used is a 2×3 ANOVA with 3 sample sizes 30,60,120 corresponding to 5,10,20 replications per cell. One thousand simulation runs were performed. It is concluded that the performance of these tests was disappointing with the Lagrange multiplier test somewhat superior to the other two.

§1.5.2

1 Dielman and Pfaffenberger (1981) and Sposito and Tveite (1986) suggest that the Cox and Hinkley (1974) procedure for estimating λ should be used.

2 Welsh (1987a) considers kernel estimates of the sparsity function:

$$\frac{\partial \mathcal{F}^{-1}(y)}{\partial y} = \frac{1}{f(\mathcal{F}^{-1}(\frac{1}{2}))}, \qquad 0 < y < 1$$

and also investigates properties of these estimates.

Estimators of the sparsity function can either be based on the empirical conditional quantile function proposed by Bassett and Koenker (1982) or the inverse empirical distribution function of the residuals obtained from the fit. The sparsity function can then be estimated using the histogram approach of Welsh (1987b) or by smoothing and differentiation of the chosen estimator of $\mathcal{F}^{-1}(y)$ as suggested by Koenker and Bassett (1982a).

However, the density estimation of $f(0)$ appears to be unstable for small to moderate sample sizes [see Sheather (1987:214) and Schrader and McKean (1987:310)].

3 Maritz and Jarrett (1978) and Efron (1979) independently derived the bootstrap variance of the sample median of the residuals (λ^2/n) from which λ can be estimated. Ignoring the k zero residuals, only the remaining $\tilde{n} = n - k$ ordered residuals, $\tilde{e}_{(i)}$, are considered in determining the bootstrap estimate $\hat{\lambda}^2_{BS}$ of λ [McKean and Schrader (1987)]. For $\tilde{n} = 2m - 1$ odd:

$$\hat{\lambda}^2_{BS} = \sum_{i=1}^{\tilde{n}} (\tilde{e}_{(i)} - \tilde{e})^2 [B(m-1; \tilde{n}, \frac{i-1}{\tilde{n}}) - B(m-1; \tilde{n}, \frac{i}{\tilde{n}})]$$

where $\tilde{e}$ is the median of the remaining residuals and $B(k; \tilde{n}, p)$ is the binomial cumulative distribution function. Maritz and Jarrett (1978) also give an expression for $\tilde{n}$ even.

McKean and Schrader (1987) state that the bootstrap estimator and the estimator based on the α-percent nonparametric confidence interval perform reasonably well in practice.

4 Stangenhaus (1987) in a Monte Carlo simulation study uses the bootstrap procedure to assess the standard deviation and confidence intervals of the estimates b_j obtained from L_1-norm regression when the sample size is small.

Bibliography Chapter 1

Abdelmalek, N. N. (1977). The discrete linear restricted Chebyshev approximation. *BIT* **17**, pp 249–261.

Abdelmalek, N. N. (1980a). L_1 solution of overdetermined systems of linear equations. *ACM Trans. Math. Software* **6**, pp 220–227.

Abdelmalek, N. N. (1980b). A FORTRAN subroutine for the L_1 solution of overdetermined systems of linear equations. *ACM Trans. Math. Software* **6**, pp 228–230.

Andrews, D.F. (1974). A robust method for multiple linear regression. *Technometrics* **16**, pp 523–531.

Appa, G. and Smith, C. (1973). On L_1 and Chebychev estimation. *Math. Programming* **5**, pp 73–87.

Armstrong, R.D. and Beck, P.O. (1983). The best parameter subset using the Chebychev curve fitting criterion. *Math. Programming* **27**, pp 64–74.

Armstrong, R.D. and Beck, P.O. (1985). Penalty calculations and branching rules in a LAV best subset procedure. *Naval Res. Log. Quart.* **32**, pp 407–415.

Armstrong, R.D., Beck, P.O. and Kung, M.T. (1984). Algorithm 615 — The best subset of parameters in LAV regression. *ACM Trans. Math. Software* **10**, pp 202–206.

Armstrong, R.D., Frome, E.L. and Kung, D.S. (1979). A revised simplex algorithm for the absolute deviation curve fitting problem. *Commun. Statist.-Simula. Computa.* **8**, pp 175–190.

Armstrong, R.D. and Hultz, J.W. (1977). An algorithm for a restricted discrete approximation problem in the ℓ_1 norm. *SIAM J. Numer. Anal.* **14**, pp 555–565.

Armstrong, R.D. and Kung, D.S. (1980). A dual method for discrete Chebychev curve fitting. *Math. Programming* **19**, pp 186–199.

Armstrong, R.D. and Kung, M.T. (1982). An algorithm to select the best subset for a least absolute value regression problem. In: S.H. Zanakis and J.S. Rustagi, eds. *Optimization in Statistics, TIMS Studies in Management Science* **19**, *North Holland, Amsterdam.*

Arthanari, T.S. and Dodge, Y. (1981). Mathematical programming in Statistics. *John Wiley, New York.*

Baboolal, S. and Watson, G.A. (1981). Computational experience with an algorithm for discrete L_1 approximation. *Computing* **27**, pp 245–252.

Barr, G.D.I. (1980). A contribution to adaptive robust estimation. *PhD Thesis. University of Cape Town, Cape Town.*

Barrodale, I. and Phillips, C. (1975). Solution of an overdetermined system of linear equations in the Chebychev norm. *ACM Trans. Math. Softw.* **1**, pp 264–270.

Barrodale, I. and Roberts, F.D.K. (1970). Applications of Mathematical Programming to L_p approximation. In: J.B. Rosen, O.L. Mangasarian and K. Ritter, eds. *Nonlinear programming, Academic Press, New York*, pp 447–463.

Barrodale, I. and Roberts, F.D.K. (1973). An improved algorithm for discrete ℓ_1 linear approximation. *SIAM J. Numer. Anal.* **10**, pp 839–848.

Barrodale, I. and Roberts, F.D.K. (1974). Algorithm 478: Solution of an overdetermined system of equations in the ℓ_1-norm. *Commun. Assoc. Comput. Mach.* **17**, pp 319–320

Barrodale, I. and Roberts, F.D.K. (1977). Algorithms for restricted least absolute value estimation. *Commun. Statist.- Simula. Computa.* **6**, pp 353–363.

Barrodale, I. and Roberts, F.D.K. (1978). An efficient algorithm for discrete l_1 linear approximation with linear constraints. *SIAM J. Numer. Anal.*, **15**, pp 603–611.

Bartels, R.H. and Conn, A.R. (1980). Linearly constrained discrete ℓ_1 problems. *ACM Trans. Math. Softw.* **6**, pp 594–608.

Bartels, R.H., Conn, A.R. and Sinclair, J.W. (1978). Minimization techniques for piecewise differentiable functions: The L_1 solution to an overdetermined linear system. *SIAM J. Numer. Anal.* **15**, pp 224–241.

Bassett, G. and Koenker, R. (1978). Asymptotic theory of least absolute error regression. *J. Amer. Statist. Assoc.* **73**, pp 618–622.

Bassett, G. and Koenker, R. (1982). An empirical quantile function for linear models with iid errors. *J. Amer. Statist. Assoc.* **77**, pp 407–415.

Best, M.J. and Chakravarti, N. (1987). A storage efficient algorithm for calculating a convex combination of least square and LAV regression. *Unpublished manuscript, Dept. of Combinatorics and Optimization, University of Waterloo.*

Blattberg, R. and Sargent,T. (1971). Regression with non-Gaussian stable disturbances. *Econometrica* **39**, pp 501–510.

Bloomfield, P. and Steiger, W.L. (1980). Least absolute deviations curve-fitting. *SIAM J. Stat. Comput.* **1**, pp 290–301.

Bloomfield, P. and Steiger, W.L. (1983). Least absolute deviations: Theory, applications, and algorithms. *Birkhäuser, Boston, Massachusetts.*

Boscovich, R.J. (1757). De litteraria expeditione per pontificiam ditionem et synopsis amplioris operis, ac habentur plura ejus ex exemplaria etiam sensorum impressa, *Bononiensi Scientiarum et Artum Instituto atque Academia Commentarii* **4**, pp 353–396.

Cauchy, A-L. (1824). Sur le système des valeurs qu'il faut attribuer a deux éléments determines par un grand nombre d'observations, pour que la plus grande de toutes les erreurs, abstraction faite du signe, devienne un minimum. *Bulletin de la Societé Philomatique (Paris)*, pp 92–99.

Charnes, A., Cooper, W.W. and Ferguson, R. (1955). Optimal estimation of executive compensation by linear programming. *Management Sci.* **1**, pp 138–151.

Chebychev, P.L. (1854). Théorie des mécanismes connus sous le nom de parallélogrammes. *Reprinted in: Ouvres de P.L. Chebychev ed. A. Markoff and N. Sonin Vol I, (1899) pp 111–143. Imprimerie de l'Académie Impériale des Sciences St. Petersbourg.*

Cotes, R. (1722). Aestimatio errorum in mixta mathesi, per variationes partium trianguli plani et sphaerici. *Opera Miscellanea, (Appended to Cotes, Harmonia Mensur, Cantabrigiae)*, pp 1–22.

Cox, D.R. and Hinkley, D.V. (1974). Theoretical statistics. *Chapman and Hall, London.*

Cramér, H. (1946). Mathematical methods of statistics. *Princeton University Press, Princeton.*

Dahlquist, G. and Björck, A. (1974). Numerical methods. *Translated by N. Anderson, Prentice Hall, New Jersey.*

Descloux, J. (1963). Approximations in L^p and Chebyshev approximations. *J. Soc. Ind. Appl. Math* **11**, pp 1017–1026.

Dielman, T.E. (1984). Least absolute value estimation in regression models: An annotated biliography. *Commun. Statist.– Theor. Meth.* **13**, pp 513–541.

Dielman, T.E. and Pfaffenberger, R.C. (1981). A simulation study of inference procedures for least absolute value regression. *Amer. Statist. Assoc., Business and Economic Statistics Proc.*, pp 537–542.

Dielman, T.E. and Pfaffenberger, R.C. (1982a). LAV (Least Absolute Value) estimation in linear regression: A review. In: S. Zanakis and J. Rustagi, eds. *TIMS Studies in Management Science, North Holland, Amsterdam*, pp 31–52.

Dielman, T.E. and Pfaffenberger, R.C. (1982b). An examination of inference for least absolute value regression. *Amer. Statist. Assoc., Business and Economic Statistics Proc.*, pp 582–585.

Dielman, T.E. and Pfaffenberger, R.C. (1983). LAV estimation with correlated independent variables. *Amer. Statist. Assoc., Business and Economic Statistics Proc.*, pp 709–713.

Dielman, T.E. and Pfaffenberger, R.C. (1984). Tests of linear hypotheses and L_1 estimation: A Monte Carlo comparison. *Amer. Statist. Assoc., Business and Economic Statistics Proc.*, pp 644–647.

Dinkel, J.J. and Pfaffenberger, R. (1981). Constrained ℓ_1 estimation via geometric programming. *European J. Oper. Res.* **7**, pp 299–305.

Dodge, Y. (1987). An introduction to statistical data analysis L_1-norm based. In: Y. Dodge, ed. *Statistical data analysis - Based on the L_1-norm and related methods, North Holland, Amsterdam*, pp 1–21.

Dodge, Y. and Jurečkova, J. (1987). Adaptive combination of least squares and least absolute deviations estimators. In: Y. Dodge, ed. *Statistical data analysis – Based on the L_1-norm and related methods, North Holland, Amsterdam*, pp 275–284.

Doolittle, H.M. (1884). The rejection of doubtful observations (abstract). *Bulletin of the Philosophical Society of Washington (Mathematical Section)* **6**, pp 153–156.

Edgeworth, F.Y. (1883). The law of error. *Phil. Magazine* **16**, pp 360–375.

Edgeworth, F.Y. (1887a). On discordant observations. *Phil. Mag.* **23**, pp 364–375.

Edgeworth, F.Y. (1887b). On observations relating to several quantities. *Hermathena* **6**, pp 279–285.

Edgeworth, F.Y. (1888). On a new method of reducing observations relating to several quantities. *Phil. Mag.* **25**, pp 184–191.

Edgeworth, F.Y. (1923). On the use of medians for reducing of observations. *Phil. Mag.* **66**, pp 1074–1088.

Efron, B. (1979). Bootstrap methods: Another look at the Jackknife. *Ann. Statist.* **7**, pp 1–26.

Ekblom, H. (1973). Calculation of linear best L_p-approximations. *BIT* **13**, pp 292–300.

Ekblom, H. (1974). L_p-methods for robust regression. *BIT* **14**, pp 22–32.

Ekblom, H. (1987). The L_1-estimate as limiting case of an L_p- or Huber-estimate. In: Y. Dodge, ed. *Statistical data analysis – Based on the L_1-norm and related methods, North Holland, Amsterdam*, pp 109–116.

El-Attar, R.A. Vidyasagar, M. and Dutta, S.R.K. (1979). An algorithm for L_1-norm minimization with application to nonlinear L_1-approximation. *SIAM J. Numer. Anal.* **16**, pp 70–86.

Euler, L. (1749). Pièce qui a Remporté le Prix de l'Academie Royale des Sciences en 1748, sur les *Inegalités du Movement de Saturne et de Jupiter, Paris.*

Farebrother, R.W. (1985). Unbiased L_1 and L_∞ estimation: *Commun. Statist.–Theor. Meth.* **14**, pp 1941–1962.

Farebrother, R.W. (1987). The historical development of the L_1 and L_∞ estimation procedures. In: Y. Dodge, ed. *Statistical data analysis – Based on the L_1-norm and related methods, North Holland, Amsterdam*, pp 37–63.

Fischer, J. (1981). An algorithm for discrete linear L_p approximation. *Numer. Math.* **38**, pp 129–139.

Fletcher, R. (1970). A new approach to variable metric algorithms. *Comput. J.* **13**, pp 317–322.

Fletcher, R., Grant, J.A. and Hebden, M.D. (1971). The calculation of linear best L_p approximations. *Comput. J.* **14**, pp 276–279.

Fletcher, R., Grant, J.A. and Hebden, M.D. (1974a). Linear minimax approximation as the limit of best L_p approximation. *SIAM J. Numer. Anal.* **11**, pp 123–136.

Fletcher, R., Grant, J.A. and Hebden, M.D. (1974b). The continuity and differentiability of the parameters of best linear L_p approximations. *J. Approx. Theory* **10**, pp 69–73.

Forsythe, A.B. (1972). Robust estimation of straight line regression coefficients by minimizing pth power deviations. *Technometrics* **14**, pp 159–166.

Gauss, C.F. (1806). II Comet von Jahr 1805. *Monatliche Correspondenz zur Beförderung der Erd-und Himmelskunde* **14**, pp 181–186.

Gauss, C.F. (1820). Theoria combinationis observationum erroribus minimus obnoxiae, pars prior, printed in *Werke, (Gottingen 1880) IV*, pp 6–7.

Gentle, J.E. (1977). Least absolute values estimation: An introduction. *Commun. Statist.-Simula. Computa.* **6**, pp 313–328.

Gentle, J.E. and Hanson, T.A. (1977). Variable selection under L_1. *Proc. Statist. Comp. Section A.S.A.*, pp 228–230.

Gentle, J.E., Kennedy, W.J. and Sposito, V.A. (1977). On least absolute values estimation. *Commun. Statist.-Theor. Meth.* **6**, pp 839–845.

Gentle, J.E. and Lee, W. (1983). The LAV procedure. *SUGI Supplemental Library Users Guide, SAS Institute, North Carolina*, pp 177–180.

Gentle, J.E., Narula, S.C. and Sposito, V.A. (1987). Algorithms for unconstrained L_1 linear regression. In: Y. Dodge, ed. *Statistical data analysis – Based on the L_1-norm and related methods, North Holland, Amsterdam*, pp 83–94.

Gonin, R. and Money, A.H. (1985a). Nonlinear L_p-norm estimation: Part I – On the choice of the exponent, p, where the errors are additive. *Commun. Statist.-Theor. Meth.* **14**, pp 827–840.

Gonin, R. and Money, A.H. (1985b). Nonlinear L_p-norm estimation: Part II – The aymptotic distribution of the exponent, p, as a function of the sample kurtosis. *Commun. Statist.- Theor. Meth.* **14**, pp 841–849.

Hand, M. and Sposito, V.A. (1980). Using the least squares estimator in Chebychev estimation. *Commun. Statist.-Simula. Computa.* **9**, pp 43–49.

Harris, T. (1950). Regression using minimum absolute deviations. *Amer. Statistician* **4**, pp 14–15.

Harter, H.L. (1974a). The method of least squares and some alternatives I. *Int. Stat. Rev.* **42**, pp 147–174.

Harter, H.L. (1974b). The method of least squares and some alternatives II. *Int. Stat. Rev.* **42**, pp 235–264.

Harter, H.L. (1975a). The method of least squares and some alternatives III. *Int. Stat. Rev.* **43**, pp 1–44.

Harter, H.L. (1975b). The method of least squares and some alternatives IV. *Int. Stat. Rev.* **43**, pp 125–190.

Harter, H.L. (1975c). The method of least squares and some alternatives V. *Int. Stat. Rev.* **43**, pp 269–272.

Harter, H.L. (1975d). The method of least squares and some alternatives — *Addendum to part IV. Int. Stat. Rev.* **43**, pp 273–278.

Harter, H.L. (1976). The method of least squares and some alternatives VI. *Int. Stat. Rev.* **44**, pp 113–159.

Harter, H.L. (1977). The nonuniqueness of absolute values regression. *Commun. Statist.-Simula. Computa.* **6**, pp 829–838.

Harvey, A.C. (1978). On the unbiasedness of robust regression estimators. *Commun. Statist.-Theor. Meth.* **7**, pp 779–783.

Himmelblau, D.M. (1972). Applied nonlinear programming. *McGraw–Hill, New York.*

Himmelblau, D.M. and Lindsay, J.V. (1980). An evaluation of substitute methods for derivatives in unconstrained optimization. *Operations Res.* **28**, pp 668–686.

Hollander, M. and Wolfe, D.A. (1973). Nonparametric statistical methods. *John Wiley, New York.*

Jacobson, D.H. and Oksman, W. (1972). An algorithm that minimizes homogeneous functions of N variables in $N+2$ iterations and rapidly minimizes general functions. *J. Math. Anal. Appl.* **38**, pp 535–552.

Jacobson, D.H. and Pels, L.M. (1974). A modified homogeneous algorithm for function minimisation. *J. Math. Anal. Appl.* **46**, pp 533–541.

Joe, B. and Bartels, R. (1983). An exact penalty method for constrained, discrete, linear ℓ_∞ data fitting. *SIAM J. Sci. Stat. Comput.* **4**, pp 69–84.

Kabe, D.G. (1986). Vanishing Jacobians and mathematical programming. *Industrial Mathematics,* **36**, pp 51–66.

Kahng, S.W. (1972). Best L_p approximation. *Math. Comput.* **26**, pp 505–508.

Karst, O.J. (1958). Linear curve fitting using least deviations. *J. Amer. Statist. Assoc.* **53**, pp 118–132.

Kelley, J.E. (1958). An application of linear programming to curve fitting. *SIAM J. Appl. Math.* **6**, pp 15–22.

Kennedy, W.J. and Gentle, J.E. (1980). Statistical Computing. *Marcel Dekker, New York.*

Kennedy, W.J. and Gentle, J.E. (1978). Comparisons of algorithms for minimum L_p norm linear regression. In: D. Hogben, ed. *Proceedings of Computer Science and Statistics: 10th Annual Symposium on the Interface*, pp 373–378.

Kiountouzis, E.A. (1973). Linear programming techniques in regression analysis. *Appl. Stat.* **22**, pp 69–73.

Koenker, R. (1987). A comparison of asymptotic testing methods for L_1 regression. In: Y. Dodge, ed. *Statistical data analysis – Based on the L_1-norm and related methods, North Holland, Amsterdam*, pp 287–295.

Koenker, R. and Bassett, G. (1982a). Robust tests for heteroscedasticity based on regression quantiles. *Econometrica* **50**, pp 43–61.

Koenker, R. and Bassett, G. (1982b). Tests of linear hypotheses and ℓ_1 estimation. *Econometrica* **50**, pp 1577-1583.

Laplace, P. S. (1786). Exposition du système du monde, Paris.

Laplace, P. S. (1793). Sur quelques points du système du monde, *Mémoires de l'Académie Royale des Sciences de Paris. Reprinted (1895) in Ouvres Complètes de Laplace* **7**, *Gauthier-Villars, Paris*, pp 477–558.

Laplace, P. S. (1799) Traité de Mécanique Céleste, tome II, J.B.M. Duprat, Paris. *English translation: Bowditsch, Hillard, Grey, Little and Wilkins, Boston 1832, Reprinted by Chelsea Publishing Company, New York, 1966.*

Laplace, P. S. (1818). Deuxième supplément à la théorie analytique des probabilités, Paris, Courcier. Reprinted (1887) in *Ouvres Complètes de Laplace* **7**, Gauthier-Villars, Paris, pp 531–580

Legendre, A.M. (1805). Nouvelles méthodes pour la détermination des orbites des comètes Courcier Paris. *Appendice sur la méthode des moindres quarrés*, pp 72–80.

Maritz, J.S. and Jarrett, R.G. (1978). A note on estimating the variance of the sample median. *J. Amer. Statist. Assoc.* **73**, pp 194–196.

Mayer, J. T. (1750). Abhandlung über die Umwälzung des Mondes um seine Axe. *Kosmographische Nachrichten und Sammlungen für 1748,* **1**, pp 52–183.

McConnell, C.R. (1987). On computing a best discrete L_1 approximation using the method of vanishing Jacobians. *Computational Statistics & Data Analysis*, **5**, pp 277–288.

McCormick, G.F. and Sposito, V.A. (1976). Using the L_2 estimator in L_1 estimation. *SIAM J. Numer. Anal.* **13**, pp 337–343.

McKean, J.W. and Schrader, R.M. (1987). Least absolute errors analysis of variance. In: Y. Dodge, ed. *Statistical data analysis – Based on the L_1-norm and related methods, North Holland, Amsterdam*, pp 297–305.

Merle, G. and Späth, H. (1973). Computational experiences with discrete L_p-approximation. *Computing* **12**, pp 315–321.

Militký, J. and Čáp, J. (1987). Application of the Bayes approach to adaptive L_p nonlinear regression. *Computational Statistics & Data Analysis*, **5**, pp 381–389.

Money, A.H., Affleck-Graves, J.F., Hart, M.L. and Barr, G.D.I. (1982). The linear regression model: L_p norm estimation and the choice of p. *Commun. Statist.– Simula. Computa.* **11**, pp 89–109.

Narula, S.C. (1987). The minimum of absolute errors regression. *J. Quality Tech.* **19**, pp 37–45.

Narula, S.C. and Wellington, J.F. (1977). Prediction, linear regression and minimum sum of relative errors. *Technometrics* **19**, pp 185–190.

Narula, S.C. and Wellington, J.F. (1982). The minimum sum of absolute errors regression: A state of the art survey. *Int. Stat. Rev.* **50**, pp 317–326.

Narula, S.C. and Wellington, J.F. (1985). A branch and bound procedure for selection of variables in minimax regression. *SIAM J. Sci. Stat. Comput.* **6**, pp 573–581.

Norback, J.P. and Morris, J.G. (1980). Fitting hyperplanes by minimizing orthogonal deviations. *Math. Programming* **19**, pp 102–105.

Nyquist, H. (1980). Recent studies on L_p-norm estimation. *PhD Thesis. University of Umeå, Sweden.*

Nyquist, H. (1983). The optimal L_p norm estimator in linear regression models. *Commun. Statist.– Theor. Meth.* **12**, pp 2511–2524.

Nyquist, H. (1987). Least orthogonal absolute deviations. *Unpublished manuscript, Dept. of Statistics, University of Umeå, Sweden.* To appear in *Computational Statistics & Data Analysis.*

Osborne, M.R. and Watson, G.A. (1985). An analysis of the total approximation problem in separable norms, and an algorithm for the total ℓ_1 problem. *SIAM J. Sci. Stat. Comput.* **6**, pp 410–424.

Osborne, M.R. (1987). The reduced gradient algorithm. In: Y. Dodge, ed. *Statistical data analysis – Based on the L_1-norm and related methods, North Holland, Amsterdam*, pp 95–107.

Peters, U. and Willms, C. (1983). Up- and down-dating procedures for linear L_1 regression. *OR Spektrum*, **5**, pp 229–239.

Pfaffenberger, R.C. and Dielman, T.E. (1983). Multicollinearity in the multiple regression model: A comparison among least absolute value, ordinary least squares and ridge regression estimators. *Amer. Inst. for Decision Sciences Proc.*, pp 1–4.

Rey, W. (1975). On least pth power methods in multiple regressions and location estimations. *BIT* **15**, pp 174–185.

Rhodes, E.C. (1930). Reducing observations by the method of minimum deviations. *Phil. Mag.* **9**, pp 974–992.

Rice, J.R. (1964). The approximation of functions, vol 1: Linear Theory. *Addison–Wesley, Reading, Massachusetts.*

Ronchetti, E. (1987). Bounded influence inference in regression: a review. In: Y. Dodge, ed. *Statistical data analysis – Based on the L_1-norm and related methods, North Holland, Amsterdam*, pp 65–80.

Roodman, G. (1974). A procedure for optimal stepwise MSAE regression analysis. *Operations Res.* **22**, pp 393–399.

Sadovski, A.N. (1974). L_1-norm fit of a straight line. *Appl. Stat.* **23**, pp 244–248.

Schellhorn J.-P. (1987). Fitting data through homotopy methods. In: Y. Dodge, ed. *Statistical data analysis – Based on the L_1-norm and related methods, North Holland, Amsterdam*, pp 131–137.

Schlossmacher, E.J. (1973). An iterative technique for absolute deviation curve fitting. *J. Amer. Statist. Assoc.* **68**, pp 857–859.

Schrader, R.M. and McKean, J.W. (1987). Small sample properties of least absolute errors analysis of variance. In: Y. Dodge, ed. *Statistical data analysis – Based on the L_1-norm and related methods, North Holland, Amsterdam*, pp 307–321.

Sheather, S.J. (1987). Assessing the accuracy of the sample median: Estimated standard errors versus interpolated confidence intervals In: Y. Dodge, ed. *Statistical data analysis – Based on the L_1-norm and related methods, North Holland, Amsterdam*, pp 203–215.

Sielken, R.L. and Hartley, H.O. (1973). Two linear programming algorithms for unbiased estimation of linear models. *J. Amer. Statist. Assoc.* **68**, pp 639–641.

Simpson, T. (1760). In: Stigler, S.M. (1984). Studies in the history of probability and Statistics XL: Boscovich, Simpson and a 1760 manuscript note on fitting a linear relation. *Biometrika* **71**, pp 615–620.

Singleton, R.R. (1940). A method for minimizing the sum of absolute values of deviations. *Ann. Math. Stats.* **11**, pp 301–310.

Sklar, M.G. and Armstrong, R.D. (1982). Least absolute value and Chebychev estimation utilizing least squares results. *Math. Programming* **24**, pp 346–352.

Sklar, M.G. and Armstrong, R.D. (1983). A linear programming algorithm for the simple model for discrete Chebychev curve fitting. *Comput. & Ops Res.* **10**, pp 237–248.

Späth, H. (1982). On discrete linear orthogonal L_p-approximation. *Z. Angew. Math. Mech.* **62**, pp 354–355.

Späth, H. (1986). Algorithm: Clusterwise linear least absolute deviations regression. *Computing* **37**, pp 371–377.

Sposito, V.A. (1982). On unbiased L_p regression estimators. *J. Amer. Statist. Assoc.* **77**, pp 652–654.

Sposito, V.A., Kennedy, W.J. and Gentle, J.E. (1977). L_p norm fit of a straight line. *Appl. Statist.* **26**, pp 114–118.

Sposito, V.A., Hand, M.L. and Skarpness, B. (1983). On the efficiency of using the sample kurtosis in selecting optimal L_p estimators. *Commun. Statist.-Simula. Computa.* **12**, pp 265–272.

Sposito, V.A. and Tveite, M.D. (1986). On the estimation of the variance of the median used in L_1 linear inference procedures. *Commun. Statist.-Theor. Meth.* **15**, pp 1367–1375.

Stangenhaus, G. (1987). Bootstrap and inference procedures for L_1 regression. In: Y. Dodge, ed. *Statistical data analysis - Based on the L_1-norm and related methods, North Holland, Amsterdam*, pp 323–332.

Stiefel, E. (1960). A note on Jordan elimination, linear programming, and Tchebycheff approximation. *Numer. Math.* **2**, pp 1–17.

Wagner, H.M. (1959). Linear programming techniques for regression analysis. *J. Amer. Statist. Ass.* **54**, pp 206–212.

Wagner, H.M. (1962). Non-linear regression with minimal assumptions. *J. Amer. Statist. Ass.* **57**, pp 572–578.

Watson, G.A. (1980). Approximation theory and numerical methods. *John Wiley, New York.*

Watson, G.A. (1982). Numerical methods for linear orthogonal L_p Approximation. *IMA J. Numer. Anal.* **2**, pp 275–287.

Watson, G.A. (1983). The total approximation problem. In: L.L. Schumaker, ed. *Approximation Theory IV, Academic Press*, pp 723–728.

Welsh, A.H. (1987a). Kernel estimates of the sparsity function. In: Y. Dodge, ed. *Statistical data analysis – Based on the L_1-norm and related methods, North Holland, Amsterdam*, pp 369–377.

Welsh, A.H. (1987b). One step L-estimators for the linear model. *Ann. Statist.* **15**, pp 626–641.

Wesolowsky, G.O. (1981). A new descent algorithm for the least absolute value regression problem. *Commun. Statist.-Simula. Computa.* **10**, pp 479–491.

Wilson, H.G. (1979). Upgrading transport costing methodology. *Transportation J.* **18**, pp 49–55.

Wolfe, J.M. (1979). On the convergence of an algorithm for discrete L_p approximation. *Numer. Math.* **32**, pp 439–459.

Zangwill, W.I. (1969). An algorithm for the Chebyshev problem – with an application to concave programming. *Management Sci.* **14**, pp 58–78.

2

The Nonlinear L_1-Norm Estimation Problem

In this Chapter we focus on the special case of the general *nonlinear* L_p-norm estimation problem where $p = 1$. The nonlinear L_1- (and L_∞-) norm problems have received more attention than the more general nonlinear L_p-norm estimation problem ($1 < p < \infty$). In the main, attention has focused on computational aspects.

From a computational point of view the nonlinear L_1- and L_∞-norm estimation problems are the most difficult of the L_p-norm problems to solve. The reason for this is that the objective functions for these two problems have discontinuous derivatives. Their solution can either be obtained directly by means of numerical minimization (nondifferentiable) or by transformation of the original problem to a nonlinear programming (NLP) problem. It is, however, not recommended that the resulting NLP problem be solved by means of conventional NLP procedures since this is an inefficient avenue of solution.

We now proceed to define the nonlinear L_1-norm estimation problem, to give the conditions under which a solution will be optimal and to discuss algorithms which can be used to solve this problem.

2.1. The nonlinear L_1-norm estimation problem

Given the following data were collected on n occasions:

$$(y_i, x_{i1}, \ldots, x_{im}) \qquad i = 1, \ldots, n$$

where y_i is the response (or dependent) variable and $\boldsymbol{x}_i = (x_{i1}, \ldots, x_{im})$ the explanatory variables. The problem is then to estimate the k parameters $\boldsymbol{\theta} = (\theta_1, \ldots, \theta_k)$ from the nonlinear model

$$y_i = f_i(\boldsymbol{x}_i, \boldsymbol{\theta}) + e_i, \qquad i = 1, \ldots, n \quad (n \geq k) \tag{2.1}$$

with f_i the response function, $\boldsymbol{\theta}$ the unknown parameters and e_i the unobservable error variates. In vector notation we can write:

$$\boldsymbol{y} = \boldsymbol{f}(\boldsymbol{x}, \boldsymbol{\theta}) + \boldsymbol{e}$$

The L_1-norm estimation problem for (2.1) is defined as:

Find parameters $\boldsymbol{\theta}$ which minimize

$$S_1(\boldsymbol{\theta}) = \sum_{i=1}^{n} |y_i - f_i(\boldsymbol{x_i}, \boldsymbol{\theta})| \tag{2.2}$$

It will be assumed that $S_1(\boldsymbol{\theta}) > 0$ for all $\boldsymbol{\theta} \in \Re^k$. *In the event that a $\tilde{\boldsymbol{\theta}}$ exists so that $y_i = f_i(\boldsymbol{x_i}, \tilde{\boldsymbol{\theta}})$, $i = 1, \ldots, n$, then it is immaterial which L_p-norm estimation we use since $S_1(\tilde{\boldsymbol{\theta}}) = S_2(\tilde{\boldsymbol{\theta}}) = S_p(\tilde{\boldsymbol{\theta}}) = S_\infty(\tilde{\boldsymbol{\theta}})$.* See (3.2) and (4.2).

The L_1-norm estimation problem can be reformulated as the NLP problem:

$$\text{Minimize} \sum_{i=1}^{n} u_i \tag{2.3}$$

$$\text{subject to} \quad \left.\begin{aligned} y_i - f_i(\boldsymbol{x_i}, \boldsymbol{\theta}) - u_i &\le 0 \\ -y_i + f_i(\boldsymbol{x_i}, \boldsymbol{\theta}) - u_i &\le 0 \\ u_i &\ge 0 \end{aligned}\right\} \quad i = 1, \ldots, n$$

The above formulation includes the constraints $u_i \ge 0$ for all i to ensure that the region defined by the constraints is connected. As a consequence the derivation of the optimality conditions does not implicitly take into account the fact that $u_i \ge 0$ for all i. The parameters $\boldsymbol{\theta}$ will of course be unconstrained. Observe that the constraints of (2.3) can be stated as $|y_i - f_i(\boldsymbol{x_i}, \boldsymbol{\theta})| = |e_i| \le u_i$ for all i, and the equivalence of (2.3) to (2.2) follows. The NLP formulation (2.3) has $2n$ nonlinear constraints in $n + k$ variables. To solve this NLP problem is cumbersome in view of the number of nonlinear constraints and the additional number of constrained variables involved.

An equivalent formulation for (2.3) is obtained by adding nonnegative v_i and w_i to each constraint:

$$-u_i + v_i + y_i - f_i(\boldsymbol{x_i}, \boldsymbol{\theta}) = 0$$

$$-u_i + w_i - y_i + f_i(\boldsymbol{x_i}, \boldsymbol{\theta}) = 0$$

By first adding and then subtracting these equations it follows that $u_i = \frac{1}{2}(v_i + w_i)$ and $y_i - f_i(\boldsymbol{x_i}, \boldsymbol{\theta}) + \frac{1}{2}(v_i - w_i) = 0$. Let $d_i^+ = \frac{1}{2}v_i$ and $d_i^- = \frac{1}{2}w_i$ then problem (2.3) can be written as

$$\min \sum_{i=1}^{n} \{d_i^+ + d_i^-\} \tag{2.4}$$

$$\text{subject to} \quad \left.\begin{array}{r} y_i - f_i(\boldsymbol{x_i}, \boldsymbol{\theta}) + d_i^+ - d_i^- = 0 \\ d_i^+, d_i^- \geq 0 \end{array}\right\} \quad i = 1, \ldots, n$$

We have therefore succeeded in reducing the $2n$ inequality constraints to n equality constraints whilst increasing the number of variables from $n + k$ to $2n + k$ variables. In a mathematical programming framework this formulation is attractive because of the reduced number of constraints. In deriving optimality conditions, however, the formulation (2.3) will be used. We now discuss optimality conditions for the nonlinear L_1-norm estimation problem. The well-known Karush (1939), Kuhn and Tucker (1951) necessary conditions (feasibility, complementary slackness and stationarity) of NLP will be utilized in the derivation of the optimality conditions. Optimality conditions for the linear L_1-norm problem (see Chapter 1) will be stated as a corollary.

2.2. Optimality conditions for L_1-norm estimation problems

The following concepts will be needed:

- Active function set

$$A = \{i \mid y_i - f_i(\boldsymbol{x_i}, \boldsymbol{\theta}) = 0\}$$

- Inactive function set

$$I = \{i \mid y_i - f_i(\boldsymbol{x_i}, \boldsymbol{\theta}) \neq 0\} = \{1, \ldots, n\} - A$$

- Active constraint sets

$$K_1 = \{i \mid y_i - f_i(\boldsymbol{x_i}, \boldsymbol{\theta}) = u_i\}$$

$$K_2 = \{i \mid y_i - f_i(\boldsymbol{x_i}, \boldsymbol{\theta}) = -u_i\}$$

$$K = K_1 \cup K_2$$

- Feasible region

$$F = \{(\boldsymbol{u}, \boldsymbol{\theta}) \in \Re^n \times \Re^k \mid y_i - f_i(\boldsymbol{x_i}, \boldsymbol{\theta}) - u_i \leq 0, -y_i + f_i(\boldsymbol{x_i}, \boldsymbol{\theta}) - u_i \leq 0\}$$

- The Lagrangian function for problem (2.3)

$$L(\boldsymbol{u}, \boldsymbol{\theta}, \boldsymbol{\lambda}) = \sum_{i=1}^{n} \{(1 - \lambda_i - \lambda_{i+n})u_i + (\lambda_i - \lambda_{i+n})[y_i - f_i(\boldsymbol{x_i}, \boldsymbol{\theta})]\}$$

We shall assume that at least one $u_i > 0$. The following condition or *constraint qualification* will also be used:

Definition 2.1 Let e_i be an ith unit vector. A point $(\boldsymbol{u}^*, \boldsymbol{\theta}^*)$ such that the active constraint gradients evaluated at $(\boldsymbol{u}^*, \boldsymbol{\theta}^*)$

$$\begin{pmatrix} -\boldsymbol{e_i} \\ -\nabla f_i(\boldsymbol{x_i}, \boldsymbol{\theta}^*) \end{pmatrix} \quad i \in K_1 \quad \text{and} \quad \begin{pmatrix} -\boldsymbol{e_i} \\ \nabla f_i(\boldsymbol{x_i}, \boldsymbol{\theta}^*) \end{pmatrix} \quad i \in K_2$$

are linearly independent is termed a *regular* point of the constraint set.

The Karush-Kuhn-Tucker (KKT) theorem will now be stated.

Theorem 2.1 Suppose the functions $f_i(\boldsymbol{x_i}, \boldsymbol{\theta})$ are continuously differentiable with respect to $\boldsymbol{\theta}$ and that $(\boldsymbol{u}^*, \boldsymbol{\theta}^*)$ is a regular local minimum point of problem (2.3). Then the following conditions are necessary for a local minimum point:

(a) (Feasibility). $(\boldsymbol{u}^*, \boldsymbol{\theta}^*) \in F$

(b) (Complementary slackness). There exist multipliers $\lambda_i^*, \lambda_{i+n}^* \geq 0$, $i = 1, \ldots, n$ such that

$$\left.\begin{aligned} \lambda_i^*(y_i - f_i(\boldsymbol{x_i}, \boldsymbol{\theta}^*) - u_i^*) = 0 \\ \lambda_{i+n}^*(y_i - f_i(\boldsymbol{x_i}, \boldsymbol{\theta}^*) + u_i^*) = 0 \end{aligned}\right\} \quad i = 1, \ldots, n$$

(c) (Stationarity).

$$\nabla L(\boldsymbol{u}^*, \boldsymbol{\theta}^*, \boldsymbol{\lambda}^*) = \begin{pmatrix} \nabla_{\boldsymbol{u}} L \\ \nabla_{\boldsymbol{\theta}} L \end{pmatrix} = \boldsymbol{O}$$

For notational convenience denote $h_i(\boldsymbol{\theta}) = y_i - f_i(\boldsymbol{x_i}, \boldsymbol{\theta})$, $\boldsymbol{h}(\boldsymbol{\theta}) = [h_1(\boldsymbol{\theta}), \ldots, h_n(\boldsymbol{\theta})]'$, $\boldsymbol{y} = (y_1, \ldots, y_n)'$ and $\boldsymbol{f}(\boldsymbol{x}, \boldsymbol{\theta}) = [f_1(\boldsymbol{x_1}, \boldsymbol{\theta}), \ldots, f_n(\boldsymbol{x_n}, \boldsymbol{\theta})]'$. Then the active function set is $A = \{i \mid h_i(\boldsymbol{\theta}^*) = 0\}$ and inactive function set is $I = \{i \mid h_i(\boldsymbol{\theta}^*) \neq 0\}$. The KKT conditions can be simplified and will be stated as the following **necessity theorem**.

Theorem 2.2 Suppose $h_i(\boldsymbol{\theta})$ is continuously differentiable with respect to $\boldsymbol{\theta}$ and that $(\boldsymbol{u}^*, \boldsymbol{\theta}^*)$ is a regular point of (2.3) and a local minimum point of (2.2). Then a necessary condition for $\boldsymbol{\theta}^*$ to be a local minimum point is the existence of multipliers $-1 \le \alpha_i \le 1$ for all $i \in A$ such that

$$\sum_{i \in I} sign[h_i(\boldsymbol{\theta}^*)] \nabla h_i(\boldsymbol{\theta}^*) + \sum_{i \in A} \alpha_i \nabla h_i(\boldsymbol{\theta}^*) = 0$$

Proof Let $I_1 = \{i | h_i(\boldsymbol{\theta}^*) > 0\}$ and $I_2 = \{i | h_i(\boldsymbol{\theta}^*) < 0\}$ be the inactive function sets. KKT conditions (b) and (c) state that

$$\left.\begin{aligned} \lambda_i^* \{h_i(\boldsymbol{\theta}^*) - u_i^*\} &= 0 \\ \lambda_{i+n}^* \{h_i(\boldsymbol{\theta}^*) + u_i^*\} &= 0 \\ 1 - \lambda_i^* - \lambda_{i+n}^* &= 0 \\ \textstyle\sum_{i=1}^n (\lambda_i^* - \lambda_{i+n}^*) \nabla h_i(\boldsymbol{\theta}^*) &= 0 \end{aligned}\right\} i = 1, \ldots, n$$

If $i \in I_1$ then $u_i^* \ge h_i(\boldsymbol{\theta}^*) > 0$ since $(u_i^*, \boldsymbol{\theta}^*) \in F$. From complementary slackness $\lambda_i^* \ge 0$ and $\lambda_{i+n}^* = 0$. From (c), $\lambda_i^* = 1$ is the only possibility and hence $u_i^* = h_i(\boldsymbol{\theta}^*)$. Similarly if $i \in I_2$ then $\lambda_{i+n}^* = 1, \lambda_i^* = 0$ and $u_i^* = -h_i(\boldsymbol{\theta}^*)$. If $i \in A$ then it follows that $u_i^* = 0$ and $\lambda_i^* + \lambda_{i+n}^* = 1$.

We can thus write

$$\begin{aligned} \sum_{i=1}^{n} (\lambda_i^* - \lambda_{i+n}^*) \nabla h_i(\boldsymbol{\theta}^*) &= \sum_{i \in I_1} (\mathbf{1}) \nabla h_i(\boldsymbol{\theta}^*) + \sum_{i \in I_2} (-\mathbf{1}) \nabla h_i(\boldsymbol{\theta}^*) + \sum_{i \in A} (\lambda_i^* - \lambda_{i+n}^*) \nabla h_i(\boldsymbol{\theta}^*) \\ &= \sum_{i \in I} sign[h_i(\boldsymbol{\theta}^*)] \nabla h_i(\boldsymbol{\theta}^*) + \sum_{i \in A} (\lambda_i^* - \lambda_{i+n}^*) \nabla h_i(\boldsymbol{\theta}^*) \end{aligned}$$

where $\mathbf{1}$ and $-\mathbf{1}$ are unit vectors. Let $\alpha_i = \lambda_i^* - \lambda_{i+n}^*$. Since $\lambda_i^*, \lambda_{i+n}^* \ge 0$ and $\lambda_i^* + \lambda_{i+n}^* = 1$ it follows that $-1 \le \alpha_i \le 1$ and the proof is complete. ∎

Remarks

(1) If the functions $h_i(\boldsymbol{\theta})$ are convex or linear then the optimality condition of Theorem 2.2 will be necessary as well as sufficient.

(2) At the optimal solution the sets $K_1 = K_2 = \{1, \ldots, n\}$. Hence all the constraints are active at optimality. This fact is used in the Murray and Overton (1981) algorithm (see §2.3.3.1).

In linear LAV estimation the model $\boldsymbol{y} = \boldsymbol{X\theta} + \boldsymbol{e}$ is fitted where $\boldsymbol{X}$ is an $n \times k$ matrix and $\boldsymbol{y}, \boldsymbol{\theta}$ and $\boldsymbol{e}$ are n-, k- and n-vectors respectively. Denote the ith row of $\boldsymbol{X}$ by $\boldsymbol{x}_i = (x_{i1}, \ldots, x_{ik})$ then $h_i(\boldsymbol{\theta}) = y_i - \boldsymbol{x}_i\boldsymbol{\theta}$. Define $A = \{\, i \mid y_i - \boldsymbol{x}_i\boldsymbol{\theta}^* = 0\}$ and $I = \{\, i \mid y_i - \boldsymbol{x}_i\boldsymbol{\theta}^* \neq 0\}$.

Corollary 2.1 In linear LAV estimation a necessary and sufficient condition for $\boldsymbol{\theta}^*$ to be a global LAV solution point is the existence of multipliers $-1 \leq \alpha_i \leq 1$ for all i such that

$$\sum_{i \in A} \alpha_i \boldsymbol{x}_i + \sum_{i \in I} sign(y_i - \boldsymbol{x}_i\boldsymbol{\theta}^*)\boldsymbol{x}_i = 0$$

The following **Necessity Theorem** is stated in terms of directional derivatives $\boldsymbol{d}$ (see Appendix 2B).

Theorem 2.3 Suppose the functions $h_i(\boldsymbol{\theta})$ are continuously differentiable with respect to $\boldsymbol{\theta}$. Then a necessary condition for $\boldsymbol{\theta}^*$ to be a local minimum point of problem (2.2) is that

$$\sum_{i \in I} \boldsymbol{d}' sign[h_i(\boldsymbol{\theta}^*)] \nabla h_i(\boldsymbol{\theta}^*) + \sum_{i \in A} |\boldsymbol{d}' \nabla h_i(\boldsymbol{\theta}^*)| \geq 0 \qquad \text{for all} \quad \boldsymbol{d} \in \Re^k$$

The following theorem constitutes **sufficiency** conditions for optimality.

Theorem 2.4 Suppose the functions $h_i(\boldsymbol{\theta})$ are twice continuously differentiable with respect to $\boldsymbol{\theta}$. Then $\boldsymbol{\theta}^*$ is an isolated (strong) minimum point of $S_1(\boldsymbol{\theta})$ if there exist multipliers $-1 \leq \alpha_i \leq 1$ for all $i \in A$ such that

$$\sum_{i \in I} sign[h_i(\boldsymbol{\theta}^*)] \nabla h_i(\boldsymbol{\theta}^*) + \sum_{i \in A} \alpha_i \nabla h_i(\boldsymbol{\theta}^*) = 0$$

and for every $i \in A$ and $\boldsymbol{d} \neq 0$ satisfying:

$$\begin{aligned} \boldsymbol{d}' \nabla h_i(\boldsymbol{\theta}^*) &= 0 \quad \text{if} \quad |\alpha_i| \neq 1 \\ \boldsymbol{d}' \nabla h_i(\boldsymbol{\theta}^*) &\geq 0 \quad \text{if} \quad \alpha_i = 1 \\ \boldsymbol{d}' \nabla h_i(\boldsymbol{\theta}^*) &\leq 0 \quad \text{if} \quad \alpha_i = -1 \end{aligned}$$

it follows that

$$\boldsymbol{d}' \left[\sum_{i \in I} sign[h_i(\boldsymbol{\theta}^*)] \nabla^2 h_i(\boldsymbol{\theta}^*) + \sum_{i \in A} \alpha_i \nabla^2 h_i(\boldsymbol{\theta}^*) \right] \boldsymbol{d} > 0 \qquad (2.5)$$

Remarks

(1) Observe that the above theorem states that the Hessian of the Lagrangian function $L(\boldsymbol{u}^*, \boldsymbol{\theta}^*, \boldsymbol{\lambda}^*)$ in (2.5) will be positive definite on a subspace tangent to the active functions — also known as the projected Hessian on the null space of the active functions.

(2) If the number of active functions in (2.2) is greater than or equal to k, then we can no longer apply the sufficiency theorem. For if $\boldsymbol{\theta}^*$ is a regular point and $-1 < \alpha_i < 1$ for all $i \in A$, then the $\nabla h_i(\boldsymbol{\theta}^*)$ are linearly independent. The minimum will in any case be unique and will be at the intersection of the active constraints — the Hessian will be positive definite. See also the remark by Charalambous (1979:131).

2.2.1. Examples

Example 2.1

In this example due to El-Attar et al. (1979) we consider the L_1-solution of:

$$\min \sum_{i=1}^{3} |h_i(\boldsymbol{\theta})|$$

where

$$h_1(\boldsymbol{\theta}) = (\theta_1)^2 + \theta_2 - 10$$

$$h_2(\boldsymbol{\theta}) = \theta_1 + (\theta_2)^2 - 7$$

$$h_3(\boldsymbol{\theta}) = (\theta_1)^2 - (\theta_2)^3 - 1$$

The optimal solution is

$$\boldsymbol{\theta}^* = (2.8425033, 1.9201751)$$

$$S_1(\boldsymbol{\theta}^*) = 0.4704242$$

$$h_1(\boldsymbol{\theta}^*) = 0.0$$

$$h_2(\boldsymbol{\theta}^*) = -0.4704242$$

$$h_3(\boldsymbol{\theta}^*) = 0.0$$

The contourlines for the objective function are given in Fig. 2.1.

Thus $A = \{1, 3\}$. At the optimal solution

$$\nabla h_1(\boldsymbol{\theta}^*) = \begin{pmatrix} 2\theta_1^* \\ 1 \end{pmatrix} = \begin{pmatrix} 5.6850066 \\ 1 \end{pmatrix}$$

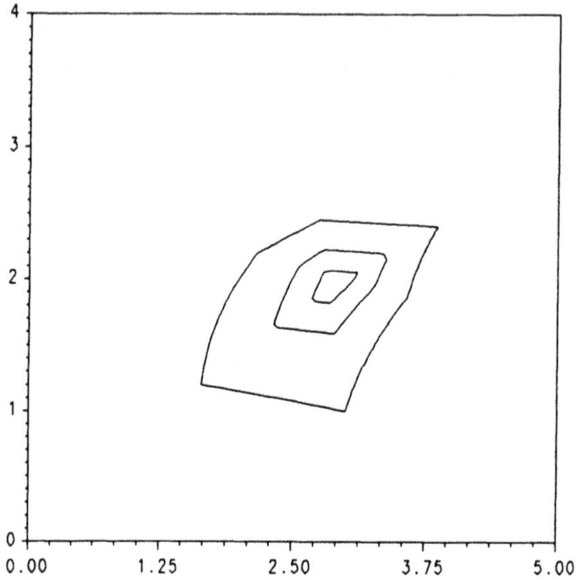

Figure 2.1 Contours of $S_1(\boldsymbol{\theta}) = 1,\ 5,\ 10$.

$$\nabla h_2(\boldsymbol{\theta}^*) = \begin{pmatrix} 1 \\ 2\theta_2^* \end{pmatrix} = \begin{pmatrix} 1 \\ 3.8403512 \end{pmatrix}$$

$$\nabla h_3(\boldsymbol{\theta}^*) = \begin{pmatrix} 2\theta_1^* \\ -3(\theta_2^*)^2 \end{pmatrix} = \begin{pmatrix} 5.6850066 \\ -11.061217 \end{pmatrix}$$

Clearly the first two gradient vectors are linearly independent. We need to solve for the multipliers $\alpha_i \in A$ in Theorem 2.2, i.e.,

$$\alpha_1 \begin{pmatrix} 5.6850066 \\ -1 \end{pmatrix} + \alpha_3 \begin{pmatrix} 5.6850066 \\ -11.061217 \end{pmatrix} - (1) \begin{pmatrix} 1 \\ 3.8403512 \end{pmatrix} = \begin{pmatrix} 0 \\ 0 \end{pmatrix}$$

The unique solution is $\alpha_1 = 0.4797$, $\alpha_3 = -0.3038$ and the necessary optimality condition is satisfied. We observe that for the sufficiency conditions two constraints are active thus

$$\boldsymbol{d}'\nabla h_1(\boldsymbol{\theta}^*) = \boldsymbol{d}'\nabla h_3(\boldsymbol{\theta}^*) = 0 \quad \Rightarrow \boldsymbol{d} = 0$$

is the only solution. The Hessian from (2.5),

$$H = \begin{pmatrix} 0.351802 & 0 \\ 0 & 1.500338 \end{pmatrix}$$

is clearly positive definite.

Example 2.2

This L_1-norm problem is due to Madsen (1975).

$$\min \sum_{i=1}^{3} |h_i(\boldsymbol{\theta})|$$

where

$$h_1(\boldsymbol{\theta}) = (\theta_1)^2 + (\theta_2)^2 + \theta_1\theta_2$$

$$h_2(\boldsymbol{\theta}) = \sin\theta_1$$

$$h_3(\boldsymbol{\theta}) = \cos\theta_2$$

The contourlines for the objective function are given in Fig. 2.2.

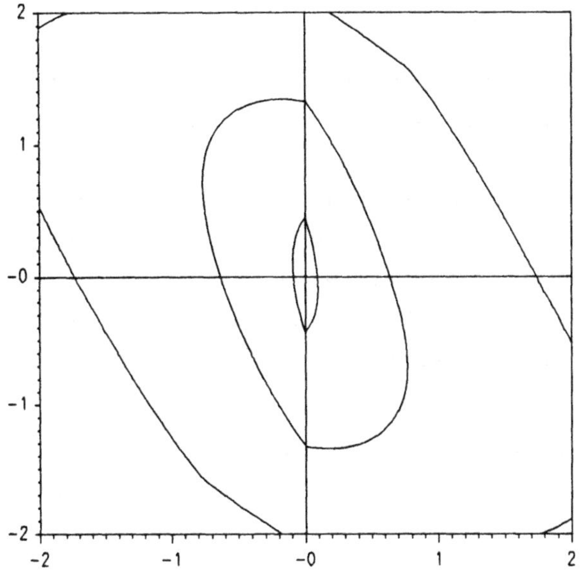

Figure 2.2 Contours of $S_1(\boldsymbol{\theta}) = 1.1,\ 2,\ 5$.

The optimal solution is $\boldsymbol{\theta}^* = (0.0, 0.0)$ with $S_1(\boldsymbol{\theta}^*) = 1$. Thus $A = \{1, 2\}$.

$$\nabla h_1(\boldsymbol{\theta}^*) = \begin{pmatrix} 2\theta_1^* + \theta_2^* \\ \theta_1^* + 2\theta_2^* \end{pmatrix} = \begin{pmatrix} 0 \\ 0 \end{pmatrix}$$

$$\nabla h_2(\boldsymbol{\theta}^*) = \begin{pmatrix} \cos\theta_1^* \\ 0 \end{pmatrix} = \begin{pmatrix} 0 \\ 0 \end{pmatrix}$$

$$\nabla h_3(\boldsymbol{\theta}^*) = \begin{pmatrix} 0 \\ -\sin\theta_2^* \end{pmatrix} = \begin{pmatrix} 0 \\ 0 \end{pmatrix}$$

Thus $\alpha_1 \nabla h_1(\boldsymbol{\theta}^*) + \alpha_2 \nabla h_2(\boldsymbol{\theta}^*) + (1)\nabla h_3(\boldsymbol{\theta}^*) = 0$ holds trivially for all α_1 and α_2. Furthermore $\boldsymbol{d}'\nabla h_i(\boldsymbol{\theta}^*) = 0$ for $i = 1, 2$ and all $\boldsymbol{d} \neq 0$. The projected Hessian is positive definite if $2/3 < \alpha_1 \leq 1$.

2.3. Algorithms for solving the nonlinear L_1-norm estimation problem

Over the past 18 years numerous algorithms have been proposed for solving the nonlinear L_1-norm problem (2.2). These methods vary from being relatively simple to numerically complex and can be classified into the following four classes:

I These methods are of the Gauss-Newton or Levenberg-Marquardt type and use 1st-derivative information only. The original nonlinear problem is reduced to a *sequence of linear L_1-norm problems*, each of which can be solved efficiently as a linear programming problem (sequential linear programming – SLP). In the Gauss-Newton type an *unconstrained* L_1-regression problem is solved — see for example the algorithms of Osborne and Watson (1971) and Anderson and Osborne (1977a). In the Levenberg-Marquardt type a *constrained* L_1-regression problem is solved – see for example Anderson and Osborne (1977b), McLean and Watson (1980) and Schrager and Hill (1980).

II These are single phase methods in which a sequence of quadratic programming problems is solved (also known as the sequential quadratic programming (SQP) approach of NLP). An active set strategy for updating the active set is also used. Curvature effects (2nd-derivative information) are taken into account by the quadratic part of the QP problem whilst linear approximations of the model functions constitute the linear constraints of the QP problem. See for example Murray and Overton (1981), Bartels and Conn (1982), Overton (1982), Busovača (1985) and Gurwitz (1986).

III Two-phase or hybrid methods avoid the need for an active set strategy by attempting to identify the optimal active set in the first phase. In the second phase a system of nonlinear equations is solved using Newton's (or a quasi-Newton) method. These methods use 2nd-derivative information. See for example McLean and Watson (1980) and Hald and Madsen (1985).

IV Approximate methods which use 2nd-derivative information. The objective function's nondifferentiability is overcome by transforming the original L_1-norm problem into a *sequence of unconstrained minimization problems* each requiring solution. See for example El-Attar et al. (1979) and Tishler and Zang (1982), the former use the penalty function method of NLP and the latter a smoothing function approach.

Under the condition of strong uniqueness, which would depend on the data/model combination, some of these classes will be equivalent in a neighbourhood of the solution.

To discuss all the algorithms in great depth is beyond the scope of this book. We shall, however, attempt to give an overview while looking at certain cases in more detail. FORTRAN programs will be provided in Appendix 2H to enable the reader to solve some nonlinear L_1-norm problems. The reader is, however, warned that these algorithms may converge slowly.

Definition 2.2 A *merit function* is a function which measures progress towards the optimal solution.

In the special case of unconstrained minimization the objective function of the problem acts as its own merit function. However, in the case of general NLP problems a merit function must balance the decrease in the objective function with the violation in the constraints. A decrease need therefore not necessarily mean progress.

Most minimization algorithms have the property that at each iteration the merit function decreases. Another desirable property (see Appendix 2E) is that of local superlinear convergence. Most algorithms have the former property. The algorithm of Murray and Overton (1981) has both. Under the condition of strong uniqueness algorithms of Type I, II and III will be locally quadratically convergent.

Before considering specific algorithms for the L_1-norm problem we shall digress for a brief discussion of general unconstrained (differentiable) optimization.

2.3.1. Differentiable unconstrained minimization

Suppose we wish to minimize a function $f(\boldsymbol{\theta})$ where $f : \Re^k \rightarrow \Re$. Given an initial estimate $\boldsymbol{\theta}^1$ of the optimal point $\boldsymbol{\theta}^*$ we attempt to compute a better estimate of $\boldsymbol{\theta}^*$ say

$$\boldsymbol{\theta}^2 = \boldsymbol{\theta}^1 + \gamma_1 \delta\boldsymbol{\theta}^1$$

where γ_1 is a steplength parameter and $\delta\boldsymbol{\theta}^1$ an increment or direction vector. Various approaches exist for calculating $\delta\boldsymbol{\theta}^1$. The so-called gradient procedures for continuously differentiable problems are of the steepest descent, Newton and quasi-Newton type. For example, if we use Newton's method then

$$\delta\boldsymbol{\theta}^1 = -[\nabla^2 f(\boldsymbol{\theta}^1)]^{-1} \nabla f(\boldsymbol{\theta}^1)$$

while for steepest descent [Cauchy (1847)]

$$\delta\boldsymbol{\theta}^1 = -\nabla f(\boldsymbol{\theta}^1)$$

The Gauss-Newton type methods (specific to least squares problems) are discussed in Appendix 2D.

In order to calculate γ_1, the steplength, we attempt to solve

$$\min_{\gamma>0} f(\boldsymbol{\theta}^1 + \gamma \delta\boldsymbol{\theta}^1)$$

Note that this is a one-dimensional minimization or exact line search problem. In practice this is seldom used; instead inexact line search procedures are used. Only a sufficient decrease in $f(\boldsymbol{\theta})$ is therefore sought:

$$f(\boldsymbol{\theta}^1 + \gamma_1 \delta\boldsymbol{\theta}^1) < f(\boldsymbol{\theta}^1) - \epsilon \qquad (\epsilon > 0)$$

In Appendix 2B we discuss some line search procedures.

The abovementioned methods assume that the objective function $f(\boldsymbol{\theta})$ is continuously differentiable. Since the objective functions $S_1(\boldsymbol{\theta})$ and $S_\infty(\boldsymbol{\theta})$ are not differentiable they need special treatment.

2.3.2. Type I Algorithms

We now consider the SLP methods.

2.3.2.1 The Anderson-Osborne-Watson algorithm.

This method is due to Osborne and Watson (1971) and Anderson and Osborne (1977a). The underlying philosophy behind this method is not difficult. At each step a first-order Taylor series (linear approximation) of $f_i(\boldsymbol{x}_i, \boldsymbol{\theta})$ is used. The method is therefore essentially a generalization of the Gauss-Newton method for nonlinear least squares extended to solve L_1-norm estimation problems.

The algorithm

Step 0: Given an initial estimate $\boldsymbol{\theta}^1$ of the optimal $\boldsymbol{\theta}^*$, choose $\mu = 0.1$ and $\lambda = 0.0001$ and set $j = 1$.

Step 1: Calculate $\delta\boldsymbol{\theta}$ to minimize

$$\sum_{i=1}^{n} |y_i - f_i(\boldsymbol{x}_i, \boldsymbol{\theta}^j) - \nabla f_i(\boldsymbol{x}_i, \boldsymbol{\theta}^j)'\delta\boldsymbol{\theta}|$$

Let the minimum be $\hat{S}^j$ with $\delta\boldsymbol{\theta} = \delta\boldsymbol{\theta}^j$.

Step 2: Choose γ_j as the largest number in the set

$$\{1, \mu, \mu^2, \ldots\}$$

$$\text{for which} \qquad \frac{S_1(\boldsymbol{\theta}^j) - S_1(\boldsymbol{\theta}^j + \gamma_j \boldsymbol{d}^j)}{\gamma_j(S_1(\boldsymbol{\theta}^j) - \hat{S}^j)} \geq \lambda$$

Let the minimum be $\bar{S}^{j+1}$ with $\gamma = \gamma_j$.

Step 3: Set $\boldsymbol{\theta}^{j+1} = \boldsymbol{\theta}^j + \gamma_j \delta\boldsymbol{\theta}^j$ and return to Step 1. Repeat until certain convergence criteria are met.

Remarks

(1) The original nonlinear problem is reduced to a sequence of linear L_1-norm regression problems, each of which can be solved efficiently as a standard LP problem using the Barrodale and Roberts (1973,1974) or Armstrong et al. (1979) algorithms.

(2) The inexact line search in Step 2 ensures convergence and is due to Anderson and Osborne (1977a) (see Appendix 2B). Osborne and Watson (1971) originally proposed an exact line search:

Calculate $\gamma > 0$ to minimize $\sum_{i=1}^{n} |y_i - f_i(\boldsymbol{x_i}, \boldsymbol{\theta}^j + \gamma\delta\boldsymbol{\theta}^j)|$.

Example 2.3

We shall illustrate one step of the algorithm on the following simple example. Given the data points (0,0), (1,1), (2,4) we wish to fit the curve $y = e^{\theta x}$ to the three points. The objective function is plotted in Fig. 2.3.

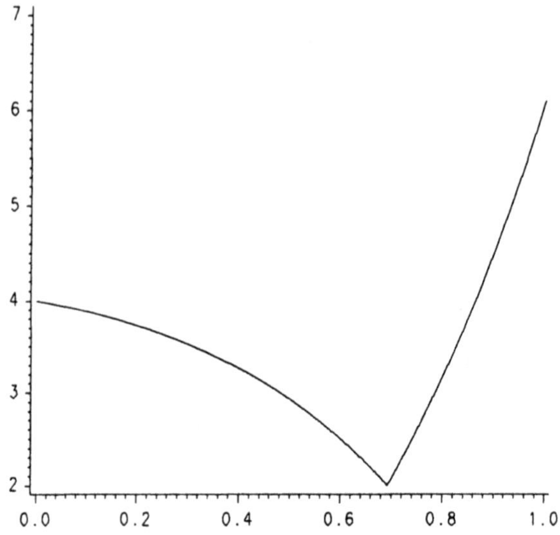

Figure 2.3 $S_1(\theta)$ vs θ.

The solution is $\theta^* = \ln 2 = 0.69315$. Dennis and Schnabel (1983:226) also used this example to illustrate the slow convergence of the Gauss-Newton method when a least squares fit is performed. We start with $\theta^1 = 1$.

Step 1: We have to solve the problem: $\min \sum_{i=1}^{3} |y_i - e^{\theta^1 x_i} - x_i e^{\theta^1 x_i} \delta\theta|$ for $\delta\theta$.

This is equivalent to solving the linear programming problem:

$$\min \sum_{i=1}^{3} (u_i + v_i)$$

subject to the 3 constraints

$$u_1 - v_1 = 1$$

$$u_2 - v_2 - 2.71836\theta = 1.7183$$

$$u_3 - v_3 - 14.77816\theta = 3.3891$$

and $u_1, u_2, u_3, v_1, v_2, v_3 \geq 0$ and $\delta\theta$ unconstrained in sign. The solution to this LP problem yields $\delta\theta^1 = -0.2293, u_1 = 1, u_2 = 1.0949, u_3 = v_1 = v_2 = v_3 = 0$. The minimum $\hat{S}^1 = 2.0949$. The number of simplex iterations using subroutine L1 of Barrodale and Roberts (1974) was 1.

Step 2: The inexact line search proceeds as follows:

Select $\gamma = 1$, calculate

$$\frac{S_1(\theta^1) - S_1(\theta^1 + \gamma\delta\theta^1)}{\gamma(S_1(\theta^1) - \hat{S}^1)} = \frac{6.1073 - 2.8321}{6.1073 - 2.0949}$$
$$= 0.816 > 0.0001$$

Hence $\gamma_1 = 1$ is selected. The approximate value of $\bar{S}^2 = 2.8321$.

Step 3: Set $\theta^2 = \theta^1 + \gamma_1\delta\theta^1 = 0.7707$. The algorithm proceeds in this fashion to the optimal solution $\theta^* = 0.693147$ with $S_1(\theta^*) = 2$. The residuals are -1, -1 and 0. Thus the curve goes through one point exactly as is to be expected. The total number of iterations was 4 and the total number of simplex iterations also equalled 4. The line search step $\gamma = 1$ was chosen at every iteration (quadratic convergence).

Under the usual smoothness assumptions and the assumption that the Jacobian matrix

$$J(\boldsymbol{\theta}) = [\frac{\partial f_i}{\partial \theta_j}]$$

is of full column rank, the following convergence properties were derived:

Convergence properties

Assuming an exact minimum is located in Step 2, then:

(1) $\hat{S}^j \leq \bar{S}^j$ for all j; if $\hat{S}^j < \bar{S}^j$ then $\bar{S}^{j+1} < \bar{S}^j$ for all j.

(2) At a limit point of the algorithm $\hat{S}^j = \bar{S}^j$.

(3) If $\hat{S}^j = \bar{S}^j$ and if $\delta\boldsymbol{\theta}^j$ is an isolated (unique) local minimum in Step 1 then a limit point of the sequence $\{\boldsymbol{\theta}^j\}$ is a stationary point of (2.2).

If the inexact line search is performed in Step 2, then:

(4) If $S_1(\boldsymbol{\theta}^j) \neq \hat{S}^j$ and given that $0 < \lambda < 1$, then there exists a $\gamma_j \in \{1, \mu, \mu^2, \ldots\}$ with $0 < \mu < 1$ such that

$$\frac{S_1(\boldsymbol{\theta}^j) - S_1(\boldsymbol{\theta}^j + \gamma_j \delta\boldsymbol{\theta}^j)}{\gamma_j (S_1(\boldsymbol{\theta}^j) - \hat{S}^j)} \geq \lambda \tag{2.6}$$

(5) If the algorithm does not terminate in a finite number of iterations then the sequence $\{S_1(\boldsymbol{\theta}^j)\}$ converges to $\hat{S}^j$ as $j \to \infty$.

(6) Any limit point of $\{\boldsymbol{\theta}^j\}$ is a stationary point of $S_1(\boldsymbol{\theta})$ — i.e., a point $\boldsymbol{\theta}^*$ such that $S_1(\boldsymbol{\theta}^*) = \hat{S}^*$.

Convergence rate

Anderson and Osborne (1977a) show that if the points $\boldsymbol{\theta}^j$ all lie in a compact (closed and bounded) set and a so-called "*multiplier*" condition holds, then the ultimate convergence rate is quadratic. Cromme (1978) has shown that the condition of *strong uniqueness* is sufficient for ensuring quadratic convergence of Gauss-Newton methods.

Definition 2.3 $\boldsymbol{\theta}^*$ is strongly unique if $\exists$ a $\gamma > 0$ such that

$$S_1(\boldsymbol{\theta}) \geq S_1(\boldsymbol{\theta}^*) + \gamma \|\boldsymbol{\theta} - \boldsymbol{\theta}^*\|$$

for all $\boldsymbol{\theta}$ in a neighbourhood of $\boldsymbol{\theta}^*$. Jittorntrum and Osborne (1978) show that strong uniqueness is implied by the multiplier condition. It is therefore a weaker condition, however, not the weakest condition that ensures quadratic convergence.

- The FORTRAN program ANDOSBWAT with an example problem is listed in Appendix 2H. Convergence occurred at iteration 8 ($\bar{S}^8 = \hat{S}^8 = 0.001563$). The intermediate results are in complete agreement with those found by Anderson and Osborne (1977a).

2.3.2.2 The Anderson-Osborne-Levenberg-Marquardt algorithm.

Anderson and Osborne (1977a,1977b) couched the L_1- and L_∞-norm problems within the framework of so-called polyhedral norms (see Appendix 2F). A discussion of these norms can be found in Fletcher (1981a:Chapter 14). The polyhedral norm approximation problem is defined as:

Find parameters $\boldsymbol{\theta}$ which minimize

$$\|\boldsymbol{f}(\boldsymbol{\theta})\|_B$$

where B is a matrix as defined in Appendix 2F. Step 1 of the previous algorithm can therefore be written in polyhedral norm notation as:

Step 1: Calculate $\delta\boldsymbol{\theta}$ and a scalar u by solving the problem:

$$\text{minimize} \quad u$$

$$\text{subject to} \qquad B(\boldsymbol{y} - \boldsymbol{f}(\boldsymbol{x}, \boldsymbol{\theta}^j) - \nabla_{\boldsymbol{\theta}} \boldsymbol{f}(\boldsymbol{x}, \boldsymbol{\theta}^j)'\delta\boldsymbol{\theta}) \leq u\boldsymbol{e}$$

Let the minimum be $\hat{S}^j$ with $\delta\boldsymbol{\theta} = \delta\boldsymbol{\theta}^j$.

This type of algorithm (Gauss-Newton) will be very effective if the number of active functions at $\boldsymbol{\theta}^*$ is k. In general this will not necessarily be the case and as a consequence SLP methods will converge slowly if at all.

To overcome this problem Anderson and Osborne (1977b) derived a Levenberg-Marquardt algorithm (Appendix 2D). The inexact line search was replaced by a strategy for modifying μ which thereby imposes a bound on the change in $\boldsymbol{\theta}$. The size as well as the direction of $\delta\boldsymbol{\theta}$ is therefore controlled.

The algorithm

Step 0: Given an initial estimate $\boldsymbol{\theta}^1$ of the optimal $\boldsymbol{\theta}^*$, select $\mu_0 = \mu_1 = 1$ and set $j = 1$.

Step 1: Calculate $\delta\boldsymbol{\theta}$ by minimizing

$$\left\| \begin{pmatrix} \boldsymbol{y} - \boldsymbol{f}(\boldsymbol{x}, \boldsymbol{\theta}^j) \\ 0 \end{pmatrix} - \begin{pmatrix} -J(\boldsymbol{\theta}^j) \\ \mu_j I \end{pmatrix} \delta\boldsymbol{\theta} \right\|_1$$

Let the minimum be $\hat{S}^j(\mu_j)$ with $\delta\boldsymbol{\theta} = \delta\boldsymbol{\theta}^j$.

Step 2: Modify $\mu_j > 0$ as follows:

Set $l = 0, T_l = 0, over := false, \mu_j^1 = \mu_{j-1}$ and $\delta = 0.01, \beta = 10, \tau = 0.2$.

2(a): If $l < 1$ or $T_l < 1 - \delta$ go to Step 2(c).

Set $l := l + 1$ and calculate

$$T_l = T(\boldsymbol{\theta}^j, \mu_j^l) = \frac{S_1(\boldsymbol{\theta}^j) - S_1(\boldsymbol{\theta}^j + \delta\boldsymbol{\theta}^j)}{S_1(\boldsymbol{\theta}^j) - \hat{S}^j(\mu_j^l)}$$

2(b): If $T_l < \delta$ and *over* = *true* set $\mu_j^{l+1} = \frac{1}{2}(\tilde{\mu}_j + \mu_j^l)$

If $T_l < \delta$ and *over* = *false* set $\mu_j^{l+1} = \beta\mu_j^l$

If $T_l < \delta$ set $\hat{\mu}_j = \mu_j^l$

If $l > 1$ and $T_l > 1 - \delta$ set *over* := *true*, $\mu_j^{l+1} = \frac{1}{2}(\hat{\mu}_j + \mu_j^l)$ and $\tilde{\mu}_j = \mu_j^l$

2(c): If $T_l < \delta$ or $[l > 1$ and $T_l > 1 - \delta]$ return to Step 2(b).

Otherwise if $l = 1$ set $\mu_{j+1} = \tau\mu_j^l$

If $l \neq 1$ set $\mu_{j+1} = \mu_j^l$

Step 3: set $\boldsymbol{\theta}^{j+1} = \boldsymbol{\theta}^j + \delta\boldsymbol{\theta}^j$ and return to Step 1. Repeat until certain convergence criteria are met.

Remarks

(1) When $\mu_j > 0$ then the matrix $\begin{pmatrix} -J(\boldsymbol{\theta}^j) \\ \mu_j I \end{pmatrix}$ is of rank k.

(2) McLean and Watson (1980) observe that Step 1 can be time consuming since it may involve the solving of many linear programming problems. They suggest an alternative form of the linear subproblem which admits an efficient numerical solution.

(3) Convergence properties analogous to those stated previously can also be found in Anderson and Osborne (1977b). Jittorntrum and Osborne (1980) give necessary and sufficient conditions for quadratic convergence of Gauss-Newton methods when the objective function is a polyhedral norm.

2.3.2.3 The McLean and Watson algorithms.

McLean and Watson (1980) provide two numerically robust and efficient methods that are based on Levenberg-Marquardt like approaches, and are therefore superlinearly convergent. For these algorithms the functions $f_i(\boldsymbol{x}_i, \boldsymbol{\theta})$ are assumed to be at least once continuously differentiable in an arbitrary neighbourhood of a stationary point of (2.2). Their alternative to the previous approach (§2.3.2.2) is similar to Madsen's (1975) proposal for the nonlinear L_∞-norm problem.

In Gauss-Newton methods $f_i(\boldsymbol{x_i}, \boldsymbol{\theta}^j)$ is approximated by a first-order Taylor series $f_i(\boldsymbol{x_i}, \boldsymbol{\theta}^j) + \nabla f_i(\boldsymbol{x_i}, \boldsymbol{\theta}^j)'\boldsymbol{d}$. This approximation will only be accurate for "small" $\boldsymbol{d}$. Hence McLean and Watson (1980) bound the direction by stipulating:

$$\max_{1 \le i \le k} |d_i| \le \lambda$$

where $\boldsymbol{d} = (d_1, \ldots, d_k)'$ and $\lambda > 0$. At each step a constrained linear L_1-norm problem is solved to obtain the direction $\boldsymbol{d}$. This problem can be solved by Barrodale and Roberts' (1978) procedure for constrained linear L_1-norm problems.

In the first algorithm a sequence of LP problems is solved at each step for different values of λ to generate a suitably bounded direction so that $\boldsymbol{\theta}^{j+1} = \boldsymbol{\theta}^j + \boldsymbol{d}^j$.

Algorithm 1

Step 0: Given an initial estimate $\boldsymbol{\theta}^1$ select $\lambda_1, \beta_1, \beta_2, \epsilon_1 > 0$ and $0 < \sigma < \frac{1}{2}$, set $j = 1$.

Step 1: Calculate $\boldsymbol{d}^j(\lambda_j)$ by solving:

$$\min \sum_{i=1}^{n} |y_i - f_i(\boldsymbol{x_i}, \boldsymbol{\theta}^j) - \nabla f_i(\boldsymbol{x_i}, \boldsymbol{\theta}^j)'\boldsymbol{d}|$$

$$\text{subject to} \qquad \max_{1 \le i \le k} |d_i| \le \lambda_j$$

Let the minimum be $\hat{S}^j$.

Step 2: If $\sum_{i=1}^{n} |y_i - f_i(\boldsymbol{x_i}, \boldsymbol{\theta}^j)| - \hat{S}^j \le \epsilon_1$ then stop. Otherwise go to Step 3.

Step 3: Compute

$$T_1(\lambda_j) = \frac{S_1(\boldsymbol{\theta}^j) - S_1[\boldsymbol{\theta}^j + \boldsymbol{d}^j(\lambda_j)]}{S_1(\boldsymbol{\theta}^j) - \hat{S}^j}$$

If $T_1(\lambda_j) < \sigma$ then $\boldsymbol{\theta}^{j+1} = \boldsymbol{\theta}^j$ and $\lambda_{j+1} = \max_{1 \le i \le k} |d_i^j(\lambda_j)|/\beta_1$.

Otherwise $\boldsymbol{\theta}^{j+1} = \boldsymbol{\theta}^j + \boldsymbol{d}^j(\lambda_j)$ and $\lambda_{j+1} = \max_{1 \le i \le k} |d_i^j(\lambda_j)|\beta_2$.

Set $j := j + 1$ and return to Step 1.

The second algorithm utilizes the property that $\boldsymbol{d}^j(\lambda)$ is a descent direction if $\lambda > 0$ when $\boldsymbol{\theta}^j$ is not a stationary point. Then $\boldsymbol{\theta}^{j+1} = \boldsymbol{\theta}^j + \gamma_j \boldsymbol{d}^j(\lambda)$ for some $\gamma_j > 0$. The algorithm proceeds as follows:

Algorithm 2

Step 0: Given an initial estimate $\boldsymbol{\theta}^1$ select $\lambda_1, \beta_1, \epsilon_1, \epsilon_2 > 0$ and $0 < \sigma < \frac{1}{2}$, set $j = 1$.

Step 1: Calculate $\boldsymbol{d}^j(\lambda_j)$ by solving:

$$\min \sum_{i=1}^{n} |y_i - f_i(\boldsymbol{x}_i, \boldsymbol{\theta}^j) - \nabla f_i(\boldsymbol{x}_i, \boldsymbol{\theta}^j)'\boldsymbol{d}|$$

$$\text{subject to} \qquad \max_{1\le i\le k} |d_i| \le \lambda_j$$

Let the minimum be $\hat{S}^j$.

Step 2: If $\sum_{i=1}^{n} |y_i - f_i(\boldsymbol{x}_i, \boldsymbol{\theta}^j)| - \hat{S}^j \le \epsilon_1$ then stop. Otherwise go to Step 3.

Step 3: Determine γ_j so that

$$\sigma \le \frac{S_1(\boldsymbol{\theta}^j) - S_1[\boldsymbol{\theta}^j + \gamma_j \boldsymbol{d}^j(\lambda_j)]}{S_1(\boldsymbol{\theta}^j) - \hat{S}^j} \le 1 - \sigma$$

If $\gamma_j < 1$ then $\lambda_{j+1} = \max\{[\max_{1\le i\le k} |d_i^j(\lambda_j)|]\gamma_j, \epsilon_2\}$.

If $\gamma_j \ge 1$ then $\lambda_{j+1} = \beta_1 \max_{1\le i\le k} |d_i^j(\lambda_j)|$.

Set $\boldsymbol{\theta}^{j+1} = \boldsymbol{\theta}^j + \gamma_j \boldsymbol{d}^j(\lambda_j)$ and return to Step 1.

Remarks

(1) It is shown that

$$\sum_{i=1}^{n} |y_i - f_i(\boldsymbol{x}_i, \boldsymbol{\theta}^j)| = \sum_{i=1}^{n} |y_i - f_i(\boldsymbol{x}_i, \boldsymbol{\theta}^j) - \nabla f_i(\boldsymbol{x}_i, \boldsymbol{\theta}^j)'\boldsymbol{d}(\lambda)|$$

if and only if $\boldsymbol{\theta}^j$ is a stationary point of problem (2.2) for all $\lambda \ge 0$.

(2) If the above equality is not satisfied or $\boldsymbol{\theta}^j$ is not a stationary point then a descent direction $\boldsymbol{d}^j = \boldsymbol{d}(\lambda)$

McLean and Watson (1980) derived the following convergence results.

(3) $S_1(\boldsymbol{\theta}^j) \to \hat{S}^j$ as $j \to \infty$.

(4) The limit points of the iterates $\{\boldsymbol{\theta}^j\}$ are stationary points of $S_1(\boldsymbol{\theta}^j)$.

Program MCLEANWAT which is an implementation of Algorithm 1 is listed in Appendix 2H.

Example 2.4

The following example due to Jennrich and Sampson (1968) was used

$$\min \sum_{l=1}^{10} |y_l - (\exp(l\theta_1) + \exp(l\theta_2)|$$

where $y_l = 2 + 2l$. The contourlines for the objective function are given in Fig. 2.4.

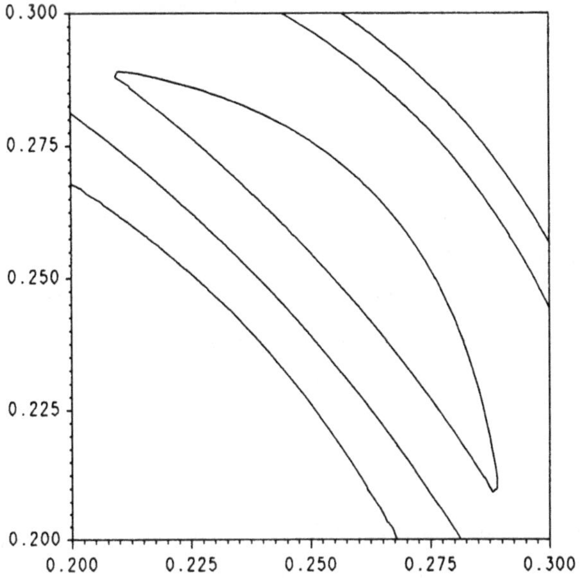

Figure 2.4 Contours of $S_1(\boldsymbol{\theta}) = 33,\ 35,\ 37$.

The optimal solution is $\boldsymbol{\theta}^* = (0.255843, 0.255843)$ and $S_1(\boldsymbol{\theta}^*) = 32.091941$.

General comments: Type I algorithms

- If the conditions for quadratic convergence of GN methods hold then the bound on $\delta\boldsymbol{\theta}^j$ becomes inactive and all Type I methods will be identical
- If $y_i = f_i(\boldsymbol{x_i}, \boldsymbol{\theta})$ for all i then the convergence rate is quadratic for any norm (L_1, L_p and L_∞).
- If $y_i \neq f_i(\boldsymbol{x_i}, \boldsymbol{\theta})$ for some i then convergence is either slow or quadratic for L_1 problems. For L_p-norms the rate will be linear.

2.3.3. Type II Algorithms

In the event that strong uniqueness does not prevail, Gauss-Newton methods may converge slowly and hence 2nd-derivative information has to be used. This is achieved by sequential quadratic programming (SQP). Algorithms of this type are complex, and are therefore stated without worked examples. The interested reader is encouraged to read the relevant papers. Again for notational convenience denote $y_i - f_i(\boldsymbol{x_i}, \boldsymbol{\theta})$ by $h_i(\boldsymbol{\theta})$.

2.3.3.1 The Murray and Overton algorithm.

The Murray and Overton (1981) algorithm is an application and adaptation of the modified Newton algorithm for linearly constrained NLP problems [Gill and Murray (1974)] to solve L_1-norm problems. We give a brief outline of the approach, following the description given in Overton (1982).

In order to obtain a superlinear convergent algorithm, second-order derivative information must be utilized. The method utilizes the structure of (2.3) and essentially reduces the problem back to an optimization problem in k variables (parameters). If we solve problem (2.2) directly, then $S_1(\boldsymbol{\theta})$ may be used as a merit function. Recall (§3.2) that the objective function in (2.3) cannot be used as a merit function. In the algorithm a descent direction is determined from a quadratic program based on a *projected Lagrangian* function. Two types of Lagrange multiplier estimates (1st- and 2nd-order) are calculated to form the Lagrangian function. A line search is then performed using this direction in order to reduce the merit function.

We shall partly adopt the Murray and Overton notation in this section: the constraints are assumed to have been ordered accordingly without loss of generality.

- Active function set

$$A = \{1, \ldots, t\} = \{i | h_i(\boldsymbol{\theta}^*) = 0\}$$

  Thus at any point $\boldsymbol{\theta} \neq \boldsymbol{\theta}^*$ we define A as the set of functions which we think will be zero at the solution $\boldsymbol{\theta}^*$.

- Let

$$\bar{\boldsymbol{h}}(\boldsymbol{\theta}) = [h_{t+1}(\boldsymbol{\theta}), \ldots, h_n(\boldsymbol{\theta})]'$$

$$\hat{\boldsymbol{h}}(\boldsymbol{\theta}) = [h_1(\boldsymbol{\theta}), \ldots, h_t(\boldsymbol{\theta})]'$$

- Active constraints

$$u_i - sign[h_i(\boldsymbol{\theta})]h_i(\boldsymbol{\theta}) \qquad i = t+1, \ldots, n$$

$$\left.\begin{array}{l} u_i - h_i(\boldsymbol{\theta}) \\ u_i + h_i(\boldsymbol{\theta}) \end{array}\right\} \qquad i = 1, \ldots, t$$

- The matrices of gradients

$$\hat{V}_{k \times t} = (\nabla h_1(\boldsymbol{\theta}), \ldots, \nabla h_t(\boldsymbol{\theta}))$$

$$\bar{V}_{k \times n-t} = (\nabla h_{t+1}(\boldsymbol{\theta}), \ldots, \nabla h_n(\boldsymbol{\theta}))$$

$$\hat{\Sigma}_{t \times t} = diag\{ sign[h_i(\boldsymbol{\theta})] : i = 1, \ldots, t\}$$

$$\bar{\Sigma}_{n-t \times n-t} = diag\{ sign[h_i(\boldsymbol{\theta})] : i = t+1, \ldots, n\}$$

- After a further re-ordering of the active constraints so that the active constraints are given in matrix notation

$$\hat{c}(\boldsymbol{\theta}) = \begin{pmatrix} \bar{\boldsymbol{u}} - \bar{\Sigma}\bar{\boldsymbol{h}} \\ \hat{\boldsymbol{u}} - \hat{\boldsymbol{h}} \\ \hat{\boldsymbol{u}} + \hat{\boldsymbol{h}} \end{pmatrix} = \begin{pmatrix} 0 \\ \hat{\Sigma}\hat{\boldsymbol{h}} - \hat{\boldsymbol{h}} \\ \hat{\Sigma}\hat{\boldsymbol{h}} + \hat{\boldsymbol{h}} \end{pmatrix}$$

-

$$\nabla\hat{c}(\boldsymbol{\theta})_{k+n \times t+n} = \begin{pmatrix} -\bar{V}\bar{\Sigma} & -\hat{V} & \hat{V} \\ I_{n-t} & 0 & 0 \\ 0 & I_t & I_t \end{pmatrix}$$

- The gradient of the objective function of (2.3) with respect to $(\boldsymbol{\theta}, \boldsymbol{u})$ is given by:

$$g(\boldsymbol{\theta}) = \begin{pmatrix} 0 \\ \bar{e} \\ \hat{e} \end{pmatrix}$$

where $\bar{e}$ and $\hat{e}$ are $(n-t)$- and t-unit vectors respectively.

- Orthogonal decomposition of $\hat{V}$ (Appendix 2F).

$$\hat{V} = [Y(\boldsymbol{\theta}) \;\; Z(\boldsymbol{\theta})] \begin{pmatrix} R(\boldsymbol{\theta}) \\ \cdots \\ O \end{pmatrix} = Q' \begin{pmatrix} R(\boldsymbol{\theta}) \\ \cdots \\ O \end{pmatrix}$$

where $Y(\boldsymbol{\theta})$ is $k \times t$, $Z(\boldsymbol{\theta})$ is $k \times k-t$ and $R(\boldsymbol{\theta})$ is $t \times t$. The matrices Y and Z are orthogonal i.e., $Y'Y = I_t$ and $Z'Z = I_{k-t}$ and $\hat{V}'Z = Y'Z = O$.

- The Lagrangian function

$$\begin{aligned} L(\boldsymbol{u}, \boldsymbol{\theta}, \boldsymbol{\lambda}) &= \sum_{i=t+1}^{n} u_i + \sum_{i=1}^{t} u_i - \sum_{i=t+1}^{n} \lambda_i [u_i - sign[h_i(\boldsymbol{\theta})] h_i(\boldsymbol{\theta})] \\ &- \sum_{i=1}^{t} \lambda_i^+ [u_i - h_i(\boldsymbol{\theta})] - \sum_{i=1}^{t} \lambda_i^- [u_i + h_i(\boldsymbol{\theta})] \\ &= \bar{e}'\bar{\boldsymbol{u}} + \hat{e}'\hat{\boldsymbol{u}} - [\bar{\boldsymbol{\lambda}}, \hat{\boldsymbol{\lambda}}^+, \hat{\boldsymbol{\lambda}}^-]'\hat{c}(\boldsymbol{\theta}) \end{aligned}$$

- The gradient of $L(\boldsymbol{u}, \boldsymbol{\theta}, \boldsymbol{\lambda})$ with respect to $(\boldsymbol{\theta}, \boldsymbol{u})$ is given by:

$$g(\boldsymbol{\theta}) - \nabla \hat{c}(\boldsymbol{\theta}) \begin{pmatrix} \bar{\boldsymbol{\lambda}} \\ \hat{\boldsymbol{\lambda}}^+ \\ \hat{\boldsymbol{\lambda}}^- \end{pmatrix}$$

- Hessian of the Lagrangian function with respect to $(\boldsymbol{\theta}, \boldsymbol{u})$ is:

$$\nabla_{\boldsymbol{\theta}}^2 L(\boldsymbol{u}, \boldsymbol{\theta}, \boldsymbol{\lambda}) = \begin{pmatrix} \boldsymbol{W}(\boldsymbol{\theta}) & 0 \\ 0 & 0 \end{pmatrix}$$

where

$$\boldsymbol{W}(\boldsymbol{\theta}) = \sum_{i=1}^{t} (\hat{\lambda}_i^+ - \hat{\lambda}_i^-) \nabla^2 h_i(\boldsymbol{\theta}) + \sum_{i=t+1}^{n} \bar{\lambda}_i sign[h_i(\boldsymbol{\theta})] \nabla^2 h_i(\boldsymbol{\theta})$$

- The projected Hessian of $L(\boldsymbol{u}, \boldsymbol{\theta}, \boldsymbol{\lambda})$ onto the null space of $\nabla \hat{c}(\boldsymbol{\theta})'$ is $Z'\boldsymbol{W}(\boldsymbol{\theta})Z$.

Several methods exist for calculating a feasible descent direction. Murray and Overton (1981) use a 2nd-derivative method which essentially solves an equality constrained QP problem [see Gill et al. (1981:159)]. A direct application of this method to the nonlinear L_1-norm problem would be to solve for direction $\boldsymbol{d}$ in:

$$\text{Minimize } \frac{1}{2}\boldsymbol{d}'\boldsymbol{W}(\boldsymbol{\theta})\boldsymbol{d} + \bar{\boldsymbol{e}}'\bar{\Sigma}\bar{V}\,'\boldsymbol{d} - \hat{\boldsymbol{e}}'\hat{\Sigma}\hat{\boldsymbol{h}} \tag{2.7}$$

$$\text{subject to } \quad \hat{V}\,'\boldsymbol{d} + \hat{\boldsymbol{h}}(\boldsymbol{\theta}) = 0$$

The QR factorization of $\hat{V}$ is then used to obtain a solution for $\boldsymbol{d}$.

In order to construct the Hessian $\boldsymbol{W}(\boldsymbol{\theta})$ a reliable estimate of the Lagrange multiplier vector $\boldsymbol{\lambda} = [\bar{\boldsymbol{\lambda}}, \hat{\boldsymbol{\lambda}}^+, \hat{\boldsymbol{\lambda}}^-]$ is needed. This is obtained by solving the linear least squares problem

$$||\boldsymbol{g}(\boldsymbol{\theta}) - \nabla\hat{\boldsymbol{c}}(\boldsymbol{\theta})\boldsymbol{\lambda}||_2$$

The algorithm will now be stated very briefly.

The algorithm

Step 0: Given an initial estimate $\boldsymbol{\theta}^1$ of the optimal $\boldsymbol{\theta}^*$, set $j = 1$.

Step 1: Calculate $\boldsymbol{g}(\boldsymbol{\theta}^j)$ and $\nabla\hat{\boldsymbol{c}}(\boldsymbol{\theta}^j)$. Calculate the Lagrange multiplier vector $\boldsymbol{\lambda}^j$ by solving the least squares problem

$$||\boldsymbol{g}(\boldsymbol{\theta}^j) - \nabla\hat{\boldsymbol{c}}(\boldsymbol{\theta}^j)\boldsymbol{\lambda}||_2$$

Compute $\boldsymbol{W}(\boldsymbol{\theta}^j)$.

Step 2: Calculate $\boldsymbol{d}^j$ by solving the QP problem (2.7) evaluated at $\boldsymbol{\theta}^j$.

Step 3: Set $\boldsymbol{\theta}^{j+1} = \boldsymbol{\theta}^j + \gamma_j\boldsymbol{d}^j$ where γ_j is determined by line search. Return to Step 1. Repeat until certain convergence criteria are met.

Remarks

(1) Murray and Overton (1979) describe a special line search method that can be used in Step 3 of the algorithm.

(2) The choice of a reasonable set of active functions at each step is important. This is known as an "active set strategy". If too few functions are selected ($|h_i(\boldsymbol{\theta})|$ is required to be very small to qualify for inclusion in A) then the iterates $\boldsymbol{\theta}^j$

will follow the constraint boundaries very closely and convergence will therefore be very slow. On the other hand if too many functions are considered active, then the search direction may be poor. Murray and Overton (1981) describe a procedure that strikes a balance between the above two considerations.

2.3.3.2 The Bartels and Conn algorithm.

The Bartels and Conn (1982) algorithm for nonlinear L_1-norm estimation is an extension of their earlier algorithm for solving linear L_1-norm estimation problems [see Bartels and Conn (1980)]. An *active set strategy* is used to determine a good choice of A. The algorithm then iteratively solves the constrained problem:

$$\min \Phi(\boldsymbol{\theta}) = \sum_{i \in I} sign\left[h_i(\boldsymbol{\theta})\right] h_i(\boldsymbol{\theta}) \tag{2.8}$$

$$\text{subject to} \qquad h_i(\boldsymbol{\theta}) = 0 \quad i \in A$$

for the given active set A. It proceeds in this fashion until a convergence criterion is satisfied or until it is evident that the set A is the incorrect active set. It is assumed that the gradients of active functions, $\nabla h_i(\boldsymbol{\theta})$, $i \in A$ are linearly independent for each $\boldsymbol{\theta}$ and A that are considered. We see that this is similar to the constraint qualification discussed in in §2.2.

The Lagrangian function of (2.8) is given by

$$L(\boldsymbol{\theta}, \boldsymbol{\lambda}) = \sum_{i \in I} sign\left[h_i(\boldsymbol{\theta})\right] h_i(\boldsymbol{\theta}) + \sum_{i \in A} \lambda_i h_i(\boldsymbol{\theta})$$

Thus

$$\nabla_{\boldsymbol{\theta}} L(\boldsymbol{\theta}, \boldsymbol{\lambda}) = \sum_{i \in I} sign\left[h_i(\boldsymbol{\theta})\right] \nabla h_i(\boldsymbol{\theta}) + \sum_{i \in A} \lambda_i \nabla h_i(\boldsymbol{\theta})$$

Before describing the algorithm the following concepts are listed. Many are restatements of the necessary and sufficient optimality conditions.

- $\boldsymbol{\theta}^*$ is a *stationary point* of $S_1(\boldsymbol{\theta})$ if there exist multipliers

$$\lambda_i, \quad i \in A \qquad \text{such that } \nabla_{\boldsymbol{\theta}} L(\boldsymbol{\theta}^*, \boldsymbol{\lambda}) = 0$$

- A stationary point $\boldsymbol{\theta}^*$ is a *first-order point* of $S_1(\boldsymbol{\theta})$ if $-1 \leq \lambda_i \leq 1, \quad i \in A$.

- For a given $\boldsymbol{\theta}$ the solution $\boldsymbol{\lambda}$ to the linear least squares problem

$$\min_{\boldsymbol{\lambda}} \left\| \sum_{i \in I} sign[h_i(\boldsymbol{\theta})] \nabla h_i(\boldsymbol{\theta}) + \sum_{i \in A} \lambda_i \nabla h_i(\boldsymbol{\theta}) \right\|_2 \tag{2.9}$$

is the *first-order multiplier vector* at $\boldsymbol{\theta}$.

- A *dropping direction* $\boldsymbol{d}$ is the solution to the set of linear equations

$$\boldsymbol{d}' \nabla h_l(\boldsymbol{\theta}) = -sign(\lambda_l)$$

$$\boldsymbol{d}' \nabla h_i(\boldsymbol{\theta}) = 0 \qquad i \in A - \{l\}$$

where $\boldsymbol{\theta}$ is a stationary point of (2.8) but not a first-order point of $S_1(\boldsymbol{\theta})$, and $|\lambda_l| > 1$ for some $1 \leq l \leq n$. $\boldsymbol{d}$ is a descent direction $\boldsymbol{d}$ for $S_1(\boldsymbol{\theta})$. Note that a constraint is dropped from the active set A.

- A *horizontal direction* $\boldsymbol{d}_{Q_i}$ for (2.8) is a direction $\boldsymbol{d}$ which solves the QP problem:

$$\min_{\boldsymbol{d}} \frac{1}{2} \boldsymbol{d}' Q_i \boldsymbol{d} + \boldsymbol{d}' \left[\sum_{i \in I} sign[h_i(\boldsymbol{\theta})] \nabla h_i(\boldsymbol{\theta}) \right]$$

$$\text{subject to} \qquad \nabla h_i(\boldsymbol{\theta})' \boldsymbol{d} = 0 \qquad i \in A$$

where Q_i is positive definite and may be one of the following:

$$Q_1 = \nabla^2_{\boldsymbol{\theta}} \Phi(\boldsymbol{\theta}) + \mu I = \sum_{i \in I} sign[h_i(\boldsymbol{\theta})] \nabla^2 h_i(\boldsymbol{\theta}) + \mu I$$

where μ is a scalar to ensure that Q_1 is positive definite.

$$Q_2 = \nabla^2_{\boldsymbol{\theta}} \Phi(\boldsymbol{\theta}) + \sum_{i \in A} \lambda_i \nabla^2 h_i(\boldsymbol{\theta})$$

where λ_i are the first-order multiplier estimates at $\boldsymbol{\theta}$.

- Let $\tilde{\boldsymbol{\theta}}$ be termed a reference point. A *vertical direction* $\boldsymbol{d}_v$ at $\boldsymbol{\theta}$ referenced to $\tilde{\boldsymbol{\theta}}$ is a vector which solves the linear least squares problem

$$\min_{\boldsymbol{d}} \left\| \sum_{i \in A} [\boldsymbol{d}' \nabla h_i(\tilde{\boldsymbol{\theta}}) + \nabla h_i(\boldsymbol{\theta})] \right\|_2$$

- The direction $\boldsymbol{\theta}^j + \boldsymbol{d}_{Q_2} + \boldsymbol{d}_v$ is a *Newton step*.

Suppose we are at some point $\boldsymbol{\theta}^j$ then the following outcomes are possible:

- $\boldsymbol{\theta}^j$ is close to a stationary point of (2.8) and A has been correctly chosen. Hence $\boldsymbol{\theta}^j$ is a first-order point of $S_1(\boldsymbol{\theta})$.
- $\boldsymbol{\theta}^j$ is nowhere near a first-order point of $S_1(\boldsymbol{\theta})$ and hence a descent direction is needed.
- $\boldsymbol{\theta}^j$ is in the vicinity of a stationary point of (2.8) but it must be determined whether A has been correctly chosen.

Thus three regions are associated with a given stationary point $\boldsymbol{\theta}^*$ of (2.8). Let ϵ_1 and ϵ_2 be positive constants.

$$R_1 = \left\{\boldsymbol{\theta} : \left\|\sum_{i\in A} \lambda_i \nabla h_i(\boldsymbol{\theta}) + \nabla\Phi(\boldsymbol{\theta})\right\| \geq \epsilon_1\right\}$$

consists of points $\boldsymbol{\theta}$ so distant from $\boldsymbol{\theta}^*$ (i.e., $\|\nabla_{\boldsymbol{\theta}} L(\boldsymbol{\theta}, \boldsymbol{\lambda})\| \gg 0$) that the multiplier estimates will not be required. In any case should an attempt be made to calculate the multipliers, these estimates would be unreliable.

$$R_2 = \left\{\boldsymbol{\theta} \mid \boldsymbol{\theta} \notin R_1 \cup R_3 \text{ and } \left\|\sum_{i\in A} \lambda_i \nabla h_i(\boldsymbol{\theta}) + \nabla\Phi(\boldsymbol{\theta})\right\| < \epsilon_1\right\}$$

and consists of points $\boldsymbol{\theta}$ closer to $\boldsymbol{\theta}^*$ ($\|\nabla L_{\boldsymbol{\theta}}(\boldsymbol{\theta}, \boldsymbol{\lambda})\|$ is small). In this case the associated least squares problem yields first-order multiplier estimates close to the true multipliers.

$$R_3 = \left\{\boldsymbol{\theta} \mid \boldsymbol{\theta} \in N_1 \cap N_2 \text{ and } \left\|\sum_{i\in A} \lambda_i \nabla h_i(\boldsymbol{\theta}) + \nabla\Phi(\boldsymbol{\theta})\right\| < \epsilon_2\right\}$$

with

$$N_1 = \{\boldsymbol{\theta} \mid S_1(\boldsymbol{\theta} + \boldsymbol{d}_v) < S_1(\boldsymbol{\theta})\}$$

$$N_2 = \{\boldsymbol{\theta} \mid S_1(\boldsymbol{\theta} + \boldsymbol{d}_v + \boldsymbol{d}_{Q_2}) < S_1(\boldsymbol{\theta})\}$$

R_3 consists of points $\boldsymbol{\theta}$ close to $\boldsymbol{\theta}^*$ (i.e., $\|\nabla_{\boldsymbol{\theta}} L(\boldsymbol{\theta}, \boldsymbol{\lambda}\|$ is very small).

The authors present their algorithm in the form of a table. We shall, however, state the algorithm in the conventional manner below.

The algorithm

Step 0: Given an initial estimate $\boldsymbol{\theta}^1$ of the optimal $\boldsymbol{\theta}^*$, set $j = 1$, $icount = false$ and $pos\_result = false$. Choose $\epsilon_1, \epsilon_2, \delta_1, \delta_2, \delta_3 > 0$.

Step 1: Select A and determine $I = \{1, \ldots, n\} - A$ and calculate $h_i(\boldsymbol{\theta}^j)$.

Step 2: Set $last\_pos\_result = pos\_result$ and $pos\_result = false$. Determine the multipliers $\boldsymbol{\lambda}^j$ by solving problem (2.9) at $\boldsymbol{\theta}^j$.

Step 3: While $\boldsymbol{\theta}^j \in R_1$ then compute $\boldsymbol{\theta}^{j+1}$ by line search using direction $\boldsymbol{d}_{Q_1}$ and return to Step 2 with $j := j + 1$.

Step 4: If $\boldsymbol{\theta}^j \notin R_2$ go to Step 6.

If $\langle \boldsymbol{\theta}^j \in R_2$ and $\lambda_i^j \leq 1$ for all $i \in A\rangle$ or $\langle(\lambda_l^j > 1$ for some $l \in A)$ and $\nabla\Phi(\boldsymbol{\theta}^j)'\boldsymbol{d}_v \geq -\delta_1$ holds) $\rangle$ then go to Step 5a, otherwise go to Step 5b.

Step 5a: Compute $\boldsymbol{\theta}^{j+1}$ by line search using direction $\boldsymbol{d}_{Q_1}$.

Test if $S_1(\boldsymbol{\theta}^j + \boldsymbol{d}_v) \leq S_1(\boldsymbol{\theta}^j) - \delta_2 \|\sum_{i \in A} h_i(\boldsymbol{\theta}^j)\|$. If true set $icount = true$ and $\boldsymbol{\theta}^{j+1} = \boldsymbol{\theta}^j + \boldsymbol{d}_v$, set $pos\_result = true$.

Return to Step 2 with $j := j + 1$.

Step 5b: Compute $\boldsymbol{\theta}^{j+1}$ by line search using direction $\boldsymbol{d}$. Go to Step 2 with $j := j+1$.

Step 6: If $\boldsymbol{\theta}^j \notin R_3$, or $icount = false$ or $last\_pos\_result = false$, then go to Step 2 with $j := j + 1$.

If

$$S_1(\boldsymbol{\theta}^j + \boldsymbol{d}_v + \boldsymbol{d}_{Q_2}) - \delta_2 \|\sum_{i \in A} h_i(\boldsymbol{\theta}^j)\| < S_1(\boldsymbol{\theta}^j) - \delta_3$$

then go to Step 7a, otherwise go to Step 7b.

Step 7a: Compute $\boldsymbol{\theta}^{j+1} = \boldsymbol{\theta}^j + \boldsymbol{d}_{Q_2} + \boldsymbol{d}_v$, $pos\_result = true$ and go to Step 2 with $j := j + 1$. Repeat until certain convergence criteria are met.

Step 7b: Compute $\boldsymbol{\theta}^{j+1}$ by line search using direction $\boldsymbol{d}_{Q_2}$.

If $S_1(\boldsymbol{\theta}^j + \boldsymbol{d}_v) < S_1(\boldsymbol{\theta}^j) - \delta_2 \|\sum_{i \in A} h_i(\boldsymbol{\theta}^j)\|$ then set $\boldsymbol{\theta}^{j+1} = \boldsymbol{\theta}^j + \boldsymbol{d}_v$.

Go to Step 2 with $j := j + 1$. Repeat until certain convergence criteria are met.

Remarks

(1) The two convergence properties, decreasing merit function at every step and superlinearity, given in §2.3, as noted by Murray and Overton (1981) are also satisfied.

(2) Instead of matrix Q_1 the modified Cholesky method of Gill and Murray (1974) could be used to ensure that $\nabla_{\boldsymbol{\theta}}\Phi(\boldsymbol{\theta})$ is positive definite. A description is given in Appendix 4A of Chapter 4.

(3) **Degeneracy**. Busovača (1985) considers both the Bartels and Conn and Murray and Overton algorithms and amends these approaches to solve degeneracy problems efficiently. These are problems in which many of the model functions closely interpolate the data ($h_i(\boldsymbol{\theta}) \approx 0$). In Busovača's approach degeneracy is treated by placing appropriate bounds on the Lagrange multiplier estimates. Thus the constrained linear least squares problem

$$\min_{\boldsymbol{\lambda}} \left\| \sum_{i \in I} sign[h_i(\boldsymbol{\theta})] \nabla h_i(\boldsymbol{\theta}) + \sum_{i \in A} \lambda_i \nabla h_i(\boldsymbol{\theta}) \right\|_2$$

$$\text{subject to} \quad -1 \le \lambda_i \le 1 \quad \in A$$

is solved. A modification of Lawson and Hanson's (1974) routine NNLS is proposed as a solution procedure for the above constrained least squares problem.

It has been pointed out to us by M.R. Osborne how the degeneracy problem can be resolved by using unbounded directions of recession. The original idea is due to Wolfe (1963). See also Osborne (1985:68-69) and Perold (1980).

2.3.4. Type III methods

If the solution $\boldsymbol{\theta}^*$ lies at an edge then $S_1(\boldsymbol{\theta})$ will be nondifferentiable at $\boldsymbol{\theta}^*$, i.e., no smooth valley traverses the minimum and the minimum point will be well determined (unique). This nondifferentiability can be exploited to obtain a fast rate (quadratic) of convergence. Recall that the strong uniqueness of $\boldsymbol{\theta}^*$ ensures that Gauss-Newton methods converge locally at a quadratic rate. First derivative information will therefore be sufficient and sequential linear programming (SLP) methods will perform well.

However, if the minimum lies in a smooth valley then the minimum point will not be well determined (nonunique) and SLP methods will converge slowly and 2nd-order derivative information has to be invoked to ensure fast convergence (see e.g., Type II or hybrid methods). Type III are hybrid methods and consist of two distinct phases.

2.3.4.1 The Hald and Madsen algorithm.

Hald and Madsen (1985) considered the linearly constrained nonlinear L_1-norm problem. We shall only consider the solution of the unconstrained problem (2.2). In the first phase a Gauss-Newton method (SLP) is used. In the second phase a quasi-Newton method (approximate 2nd-derivative information) is used to solve the system of nonlinear equations arising out of the first-order necessary conditions.

The algorithm always starts in phase 1. Sequential linear programming is used until it seems as if a "good" approximation to the solution has been found. The algorithm then switches over to phase 2 which attempts solve the first-order necessary conditions in conjunction with the active functions, i.e.,

$$\nabla L(\boldsymbol{\theta}, \boldsymbol{\alpha}) = -\sum_{i \in A} \alpha_i \nabla f_i(\boldsymbol{x_i}, \boldsymbol{\theta}) - \sum_{i \notin A} sign[y_i - \nabla f_i(\boldsymbol{x_i} \boldsymbol{\theta})] \nabla f_i(\boldsymbol{x_i}, \boldsymbol{\theta}) = 0 \qquad (2.10)$$

$$y_i - f_i(\boldsymbol{x_i}, \boldsymbol{\theta}) = 0 \qquad i \in A \qquad (2.11)$$

by means of a quasi-Newton (BFGS, Appendix 2C) method. $L(\cdot)$ is a "Lagrangian" function. In the BFGS method the approximation to the inverse Hessian $\nabla^2_{\boldsymbol{\theta}} L(\boldsymbol{\theta}^j, \boldsymbol{\alpha}^j)$ is denoted by H^j. If the phase 2 iteration is unsuccessful, for example if A is incorrectly chosen, then the algorithm reverts back to phase 1. Several such switches (in practice only a few) can occur. Certain criteria for switching back and forth are laid down. It is assumed that the response function is twice continuously differentiable with respect to $\boldsymbol{\theta}$. The algorithm proceeds as follows:

The Algorithm

Step 0: Given an initial estimate $\boldsymbol{\theta}^1$ of the optimal $\boldsymbol{\theta}^*$, an upper bound Λ_1 on the direction length and parameter $\epsilon_1 > 0$. Set $H^j = I_{k \times k}$, _iphase_ = 1 and $j = 1$.

Step 1: Phase 1. Solve the constrained linear L_1-norm problem:

$$\sum_{i=1}^{n} |y_i - f_i(\boldsymbol{x_i}, \boldsymbol{\theta}^j) - \nabla f_i(\boldsymbol{x_i}, \boldsymbol{\theta}^j)'\delta\boldsymbol{\theta}|$$

$$\text{subject to} \qquad \max_{1 \le i \le k} |\delta\theta_i| \le \Lambda_j$$

Let the solution be $\hat{S}^j$ with $\delta\boldsymbol{\theta} = \delta\boldsymbol{\theta}^j$

Step 2: Compute

$$T = \frac{S_1(\boldsymbol{\theta}^j) - S_1(\boldsymbol{\theta}^j + \delta\boldsymbol{\theta}^j)}{S_1(\boldsymbol{\theta}^j) - \hat{S}^j}$$

If $T \geq 0.01$ set $\boldsymbol{\theta}^{j+1} = \boldsymbol{\theta}^j + \delta\boldsymbol{\theta}^j$. Otherwise set $\boldsymbol{\theta}^{j+1} = \boldsymbol{\theta}^j$.

Step 3: If _iphase_ = 1 update the bound as follows:

$$\Lambda_{j+1} = \begin{cases} 2\max_{1 \le i \le k} |\delta\theta_i^j| & \text{if } T \ge 0.75 \\ 0.25\max_{1 \le i \le k} |\delta\theta_i^j| & \text{if } T < 0.25 \\ \max_{1 \le i \le k} |\delta\theta_i^j| & \text{if } 0.25 \le T < 0.75 \end{cases}$$

If _iphase_ = 2 then set $\Lambda_{j+1} = \Lambda_l$ (l the latest phase 1 iteration).

Step 4: Define $A_j = \{i \mid |y_i - f_i(\boldsymbol{x_i}, \boldsymbol{\theta}^j)| \le \epsilon_1\}$.

Calculate the first-order multiplier estimate $\boldsymbol{\alpha}^j$ by solving the linear least squares problem

$$\min_{\boldsymbol{\alpha}} \Big\| \sum_{i \in A_j} \alpha_i \nabla f_i(\boldsymbol{x_i}, \boldsymbol{\theta}^j) + \sum_{i \notin A_j} sign[y_i - f_i(\boldsymbol{x_i}, \boldsymbol{\theta}^j)] \nabla f_i(\boldsymbol{x_i}, \boldsymbol{\theta}^j) \Big\|_2$$

Step 5: If $\boldsymbol{\theta}^{j+1} \neq \boldsymbol{\theta}^j$, calculate

$$\boldsymbol{\delta}^j = \boldsymbol{\theta}^{j+1} - \boldsymbol{\theta}^j$$

$$\boldsymbol{\gamma}^j = \nabla L(\boldsymbol{\theta}^{j+1}, \boldsymbol{\alpha}^j) - \nabla L(\boldsymbol{\theta}^j, \boldsymbol{\alpha}^j)$$

update H^j as follows:

$$H^{j+1} = \left(I_k - \frac{\boldsymbol{\delta}^j(\boldsymbol{\gamma}^j)'}{(\boldsymbol{\gamma}^j)'\boldsymbol{\delta}^j}\right) H^j \left(I_k - \frac{\boldsymbol{\gamma}^j(\boldsymbol{\delta}^j)'}{(\boldsymbol{\gamma}^j)'\boldsymbol{\delta}^j}\right) + \frac{\boldsymbol{\delta}^j(\boldsymbol{\delta}^j)'}{(\boldsymbol{\gamma}^j)'\boldsymbol{\delta}^j}$$

Step 6: Phase 2. Suppose the last 3 different iterates that were calculated in Step 2 are $\boldsymbol{\theta}^l$, $\boldsymbol{\theta}^{l-j_1}$ and $\boldsymbol{\theta}^{l-j_2}$. Suppose the active set has also remained constant:

$$A_l = A_{l-j_1} = A_{l-j_2}$$

and the multiplier estimates are such that

$$|\alpha_i^l| \leq 1 \qquad i \in A_l$$

then go to Step 7. Otherwise return to Step 1 with _iphase_ = 1.

Step 7: Perform one iteration of the BFGS method with H^l, $\boldsymbol{\theta}^l$ and $\boldsymbol{\alpha}^{l-j_1}$ as starting values to solve the system of equations

$$\nabla L(\boldsymbol{\theta}, \boldsymbol{\alpha}) = 0$$

$$y_i - f_i(\boldsymbol{x_i}, \boldsymbol{\theta}) = 0 \qquad i \in A_l$$

Call the next iterate $(\boldsymbol{\theta}^{l+1}, \boldsymbol{\alpha}^{l+1})$.

Step 8: Denote the residual vector of the system of equations in Step 7 by $r(\boldsymbol{\theta}, \boldsymbol{\alpha})$. If the following three conditions hold

$$||r(\boldsymbol{\theta}^{l+1}, \boldsymbol{\alpha}^{l+1})|| \leq \eta ||r(\boldsymbol{\theta}^l, \boldsymbol{\alpha}^{l-j_1}||$$

$$|\alpha_i^{l+1}| \leq 1 \qquad i \in A_j$$

$$sign[y_i - f_i(\boldsymbol{x_i}, \boldsymbol{\theta}^{l+1})] = sign[y_i - f_i(\boldsymbol{x_i}, \boldsymbol{\theta}^l)] \text{ for all } i \notin A_l$$

then return to Step 7. If not then return to Step 1 with _iphase_ = 2.

Remarks:

(1) In Step 8 the residual test ensures that the residuals decrease at every iteration. $\eta = 0.999$ may be used.

(2) In their third algorithm McLean and Watson (1980) also consider the system of nonlinear equations (2.10) and (2.11). They suggested that once the set of active functions, A and the signs of $y_i - f_i(\boldsymbol{x_i}, \boldsymbol{\theta})$ are known and a good estimate of $(\boldsymbol{\theta}^*, \boldsymbol{\alpha}^*)$ is available then one may attempt to solve (2.10) and (2.11) using Newton's method. This will obviously depend on how easily A and the signs of $y_i - f_i(\boldsymbol{x_i}, \boldsymbol{\theta})$, $i \notin A$ can be determined. If Step 1 of the second algorithm of McLean and Watson is performed then the optimal dual variables in the LP

problem are simply the components of $\boldsymbol{\alpha}$. They report that the performance of their algorithm is disappointing. It would appear that unless the number of active functions is small compared to n, the performance of the algorithm will not be satisfactory. This idea was also proposed by Watson (1979, 1980: Chapter 10).

The following convergence results were proved by Hald and Madsen (1985).

Convergence

(a) If the sequence of vectors $\{\boldsymbol{\theta}^j\}$ generated by the phase 1 algorithm converges to a strongly unique local minimum then the rate of convergence is quadratic.

(b) If the sequence of vectors $\{\boldsymbol{\theta}^j\}$ generated by the phase 2 algorithm is convergent then the limit is a stationary point.

2.3.5. Type IV Algorithms

These methods are approximate in the sense that the original problem (2.2) is replaced by a continuously differentiable approximation. This approximation depends on a parameter δ. In the limiting case as $\delta \to 0$ the approximation is exact.

2.3.5.1 The El-Attar-Vidyasagar-Dutta algorithm.

El-Attar et al. (1979) overcome the nondifferentiability of the objective function by transforming the original problem (2.2) into a sequence of unconstrained minimization problems each requiring solution. This is known as the penalty function method of nonlinear programming. These penalty functions are differentiable and of the form:

$$P(\boldsymbol{\theta},\delta) = \sum_{i=1}^{n} \sqrt{[y_i - f_i(\boldsymbol{x}_i,\boldsymbol{\theta})]^2 + \delta} \qquad \delta > 0 \,.$$

The penalty function is differentiable as long as $\delta > 0$. The algorithm proceeds as follows:

The Algorithm

Choose $\delta_1, \epsilon_1, \epsilon_2 > 0$ and set $j = 1$.

Step 1: (Inner iteration). Solve the problem

$$\min_{\boldsymbol{\theta}} P(\boldsymbol{\theta}, \delta_j)$$

let the solution be $\boldsymbol{\theta}^j$.

Step 2: (Outer iteration). Set $\delta_{j+1} = \delta_j/\beta \qquad (\beta = 10)$.

Step 3: If $\delta_{j+1} \leq \epsilon_1$ or $\|\boldsymbol{\theta}^{j+1} - \boldsymbol{\theta}^j\|_2 \leq \epsilon_2$, stop. Otherwise set $j := j + 1$ and return to Step 1.

Example 2.5

We again consider Example 2.3 and apply the algorithm to it.

Step 1: In this step we calculate θ^1 by minimizing

$$\sum_{i=1}^{3}[(y_i - e^{\theta x_i})^2 + \epsilon]^{\frac{1}{2}}$$

we start with $\epsilon = 1$. The minimum point is given as $\theta^1 = 0.67019$

Step 2: Set $\epsilon = \frac{1}{10}$ and repeat.

The same optimal solution $\theta^* = 0.693147$ was found.

Remarks

(1) A test such as

$$\frac{|\,\theta_l^{j+1} - \theta_l^j\,|}{|\,\theta_l^j\,|} \leq \epsilon_2$$

for $l = 1, \ldots, k$ can also be used.

(2) El-Attar et al. (1979) recommend an initial choice of

$$\delta_1 = \max_{1 \leq i \leq n} |y_i - f_i(\boldsymbol{x}_i, \boldsymbol{\theta})|/10$$

δ_1 should not be chosen too small since the unconstrained problem may become highly ill-conditioned and hence the algorithm may not converge at all.

(3) Any unconstrained minimization technique can be used in Step 1. In the listed FORTRAN program ELATTAR (Appendix 2H), Fletcher's (1970) quasi-Newton method (Harwell subroutine VA09AD) was used for the unconstrained

minimizations. (See also the BFGS method in Appendix 2C). It also contains the relevant input data to the extrapolation procedure (Appendix 2G).

(4) El-Attar et al. (1979) prove convergence of their algorithm.

2.3.5.2 The Tishler and Zang algorithm.

Tishler and Zang (1982) smooth out the kinks that occur in $S_1(\boldsymbol{\theta})$. Recall that these derivative discontinuities occur where the model fits the data exactly [i.e., where $y_i = f_i(\boldsymbol{x_i}, \boldsymbol{\theta})$]. Again we let $h_i(\boldsymbol{\theta}) = y_i - f_i(\boldsymbol{x_i}, \boldsymbol{\theta})$. The smoothing function behaves like a spline function since $S_1(\boldsymbol{\theta})$ is approximated by a continuously (up to order 2) differentiable function. It is based on the following observation:

$$|h_i(\boldsymbol{\theta})| = \max[0, h_i(\boldsymbol{\theta})] + \max[0, -h_i(\boldsymbol{\theta})]$$

Hence a continuously differentiable approximation to $|h_i(\boldsymbol{\theta})|$ would be

$$h_i(\boldsymbol{\theta}, \beta) = \begin{cases} -h_i(\boldsymbol{\theta}), & \text{if } h_i(\boldsymbol{\theta}) \leq -\beta; \\ \frac{h_i(\boldsymbol{\theta})^2 + \beta^2}{2\beta}, & \text{if } -\beta \leq h_i(\boldsymbol{\theta}) \leq \beta; \\ h_i(\boldsymbol{\theta}), & \text{if } h_i(\boldsymbol{\theta}) \geq \beta; \end{cases}$$

The smoothing principle is illustrated in Fig. 2.5. Note that a quadratic function is used to approximate the kink.

The accuracy of approximation is therefore determined by controlling the value of β. Care should be taken in the choice of β. If β is too small then the unconstrained minimization procedure may fail to converge because the underlying unconstrained problem will essentially be nondifferentiable and therefore ill-conditioned. Any gradient method may be used to perform the unconstrained minimization. The authors also derive a second-order smooth approximation of $|h_i(\boldsymbol{\theta})|$.

Thus instead of problem (2.2) the following problem

$$\min_{\boldsymbol{\theta}} \sum_{i=1}^{n} h_i(\boldsymbol{\theta}, \beta) \tag{2.12}$$

is considered where $\lim_{\beta \to 0} h_i(\boldsymbol{\theta}, \beta) = |h_i(\boldsymbol{\theta})|$.

The algorithm

Step 0: Given an initial estimate $\boldsymbol{\theta}^1$ and β^1 set $j = 1$.

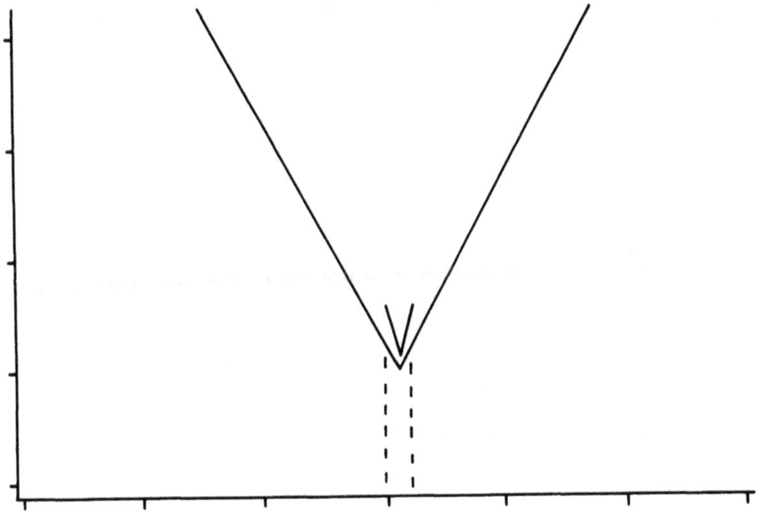

Figure 2.5 $h_i(\boldsymbol{\theta}, \beta)$ smoothing function.

Step 1: Solve $\min_{\boldsymbol{\theta}} \sum_{i=1}^{n} h_i(\boldsymbol{\theta}, \beta)$. Let the solution be $\boldsymbol{\theta}^{j+1}$.

Step 2: Choose $\beta^{j+1} < \beta^j$ set $j := j+1$ and return to Step 1. Repeat until certain convergence criteria are met.

Remarks:

(1) If β is sufficiently large so that $|h_i(\boldsymbol{\theta})| = |y_i - f_i(\boldsymbol{x}_i, \boldsymbol{\theta})| \leq \beta$ for all i then

$$2\beta \sum_{i=1}^{n} h_i(\boldsymbol{\theta}, \beta) - \beta^2 n = \sum_{i=1}^{n} [y_i - f_i(\boldsymbol{x}_i, \boldsymbol{\theta})]^2$$

Thus there exists a threshold value of β at which least squares is performed.

(2) Convergence of the algorithm is proved.

(3) Tishler and Zang (1982) suggest an initial choice of $\beta = 0.1$, but do not mention how β should be decreased. Presumably the scheme of El-Attar et al. (1979) for varying δ can be used here for β.

(4) The FORTRAN program TISHZANG (Appendix 2H) uses Fletcher's (1970) quasi-Newton method (BFGS) (Harwell subroutine VA09AD) for the uncon-

strained minimizations. It also contains the relevant input data to the extrapolation procedure.

Numerical experience with these two algorithms indicates that they are comparable. The example of Madsen (1975) was used. In both algorithms we chose

$$\delta_1 = \max_{1 \le i \le n} |y_i - f_i(\boldsymbol{x}_i, \boldsymbol{\theta})|/100$$

In all the outer iterations $\beta = 10$ was used. The Tishler-Zang algorithm performed 8 outer iterations and 144 function evaluations, whilst the algorithm of El-Attar et al. performed 14 outer iterations and 186 function evaluations. It appears as if the Tishler-Zang algorithm could be less expensive in computation but this would have to be confirmed by more detailed study.

CONCLUSION

L_1-norm estimation has become a very popular tool in recent years. It is a satisfactory robust alternative to least squares. In Chapters 5 and 6 we shall see that it is the method of choice when the underlying data contain outlying observations. It was for this reason that we devoted a fair amount of space to the solution of the L_1-norm problem.

Appendix 2A: Local and global minima

Definition 1 A δ-neigbourhood of a point $\boldsymbol{\theta}^* \in \Re^k$ is the set of points

$$N(\boldsymbol{\theta}^*;\delta) = \{\boldsymbol{\theta} \mid \boldsymbol{\theta} \in \Re^k, ||\boldsymbol{\theta} - \boldsymbol{\theta}^*|| < \delta\}$$

When the *radius* is unimportant we shall write $N(\boldsymbol{\theta}^*)$. A closed δ-neigbourhood of $\boldsymbol{\theta}^* \in \Re^k$ is denoted by $N(\boldsymbol{\theta}^*;\delta) = \{\boldsymbol{\theta} \mid \boldsymbol{\theta} \in \Re^k, ||\boldsymbol{\theta} - \boldsymbol{\theta}^*|| \leq \delta\}$.

Definition 2 $\boldsymbol{\theta}^*$ is a local minimum point of a function $f(\boldsymbol{\theta})$, $f : \Re^k \rightarrow \Re$, if there exists a neigbourhood $N(\boldsymbol{\theta}^*)$ of $\boldsymbol{\theta}^*$ such that $\boldsymbol{\theta} \in N(\boldsymbol{\theta}^*) \Rightarrow f(\boldsymbol{\theta}^*) \leq f(\boldsymbol{\theta})$.

$\boldsymbol{\theta}^*$ is an isolated (strong) local minimum point of a function $f(\boldsymbol{\theta})$ if $\boldsymbol{\theta} \in N(\boldsymbol{\theta}^*)$, $\boldsymbol{\theta} \neq \boldsymbol{\theta}^* \Rightarrow f(\boldsymbol{\theta}^*) < f(\boldsymbol{\theta})$. $\boldsymbol{\theta}^*$ will be a global minimum point of a function $f(\boldsymbol{\theta})$ with respect to $X \subset \Re^k$ if $\boldsymbol{\theta} \in X$ and $N(\boldsymbol{\theta}^*) = X$.

Quadratic behaviour of a nonlinear function near a local minimum

Consider the quadratic function $f(\boldsymbol{x}) = \frac{1}{2}\boldsymbol{x}'Q\boldsymbol{x} + \boldsymbol{b}'\boldsymbol{x} + c$ where Q is an $n \times n$ matrix, $\boldsymbol{x}$ and $\boldsymbol{b}$ are n-vectors and c a scalar. Given two points $\boldsymbol{x}^1$ and $\boldsymbol{x}^2$ then

$$\nabla f(\boldsymbol{x}^1) - \nabla f(\boldsymbol{x}^2) = Q(\boldsymbol{x}^1 - \boldsymbol{x}^2) \qquad (A)$$

Consider a general function $f(\boldsymbol{x})$, then by a first-order Taylor series expansion:

$$\nabla f(\boldsymbol{x}^* + \boldsymbol{h}) = \nabla f(\boldsymbol{x}^*) + \nabla^2 f(\boldsymbol{x}^*)\boldsymbol{h}$$

or

$$\nabla^2 f(\boldsymbol{x}^*)\boldsymbol{h} \approx \nabla f(\boldsymbol{x}^* + \boldsymbol{h}) - \nabla f(\boldsymbol{x}^*)$$

Relation (A) is also known as the *quasi-Newton* or *secant* relation for nonlinear functions where $Q = \nabla^2 f$. It illustrates the quadratic behaviour of the nonlinear function in the neighbourhood of an optimal point $\boldsymbol{x}^*$.

Definition 3 A set $F \subset \Re^k$ is convex if $\boldsymbol{x}, \boldsymbol{y} \in F \Rightarrow \boldsymbol{w} = \lambda\boldsymbol{x} + (1-\lambda)\boldsymbol{y} \in F$ whenever $0 \leq \lambda \leq 1$.

Definition 4 A function $f(\boldsymbol{x})$ is convex on a convex set F if $\boldsymbol{x}, \boldsymbol{y} \in F \Rightarrow$

$$f(\lambda\boldsymbol{x} + (1-\lambda)\boldsymbol{y}) \leq \lambda f(\boldsymbol{x}) + (1-\lambda)f(\boldsymbol{y}) \quad \text{for} \quad 0 \leq \lambda \leq 1$$

A function will be strictly convex if the strict inequality holds for $0 < \lambda < 1$.

The following theorems characterise convexity. Proofs may be found in Mangasarian (1969).

Theorem 1 Let $f(\boldsymbol{x})$ be twice continuously differentiable then the Hessian matrix of $f(\boldsymbol{x})$ is positive semi-definite if and only if $f(\boldsymbol{x})$ is convex. It will be positive definite if and only if $f(\boldsymbol{x})$ is strictly convex.

Theorem 2 Let $f(\boldsymbol{x})$ be a convex function on a convex set F then any local minimum point of $f(\boldsymbol{x})$ in F will also be a global minimum point of $f(\boldsymbol{x})$ over F. If $f(\boldsymbol{x})$ is strictly convex on F then there exists at the most one global minimum point of $f(\boldsymbol{x})$ over F.

Appendix 2B: One-dimensional line search algorithms

These methods are generally known as steplength algorithms. Recall that we wish to find

$$\boldsymbol{\theta}^{j+1} = \boldsymbol{\theta}^j + \gamma_j \boldsymbol{d}^j$$

where the vectors $\boldsymbol{\theta}^j$, $\boldsymbol{d}^j \in \Re^k$ are known. Given $\boldsymbol{\theta}^j$ and the search direction $\boldsymbol{d}^j$ then all we need to determine is the value of the scalar γ_j. We need to consider the search direction first.

Search directions

We characterize descent directions by means of the following useful theorems which may be found in standard texts in optimization [see e.g., Zangwill (1969)].

Definition 5 Suppose $f : \Re^k \to \Re$. The directional derivative of a continuous function $f(\boldsymbol{\theta})$ at $\boldsymbol{\theta}$ is defined as:

$$\lim_{\gamma \to 0} \frac{f(\boldsymbol{\theta} + \gamma \boldsymbol{d}) - f(\boldsymbol{\theta})}{\gamma} = \nabla f(\boldsymbol{\theta})'\boldsymbol{d}$$

Theorem 3 Suppose the continuous function $f(\boldsymbol{\theta})$ is differentiable (i.e., its first-order partial derivatives exist) at $\boldsymbol{\theta}$ and for a given direction $\boldsymbol{d} \neq \mathbf{o}$ the directional derivative

$$\nabla f(\boldsymbol{\theta})'\boldsymbol{d} < 0$$

then there exists a sufficiently small constant δ so that for all $0 < \gamma < \delta$

$$f(\boldsymbol{\theta} + \gamma \boldsymbol{d}) < f(\boldsymbol{\theta})$$

Proof The directional derivative,

$$\lim_{\gamma \to 0} \frac{f(\boldsymbol{\theta} + \gamma \boldsymbol{d}) - f(\boldsymbol{\theta})}{\gamma} = \nabla f(\boldsymbol{\theta})'\boldsymbol{d}$$

is negative by assumption. By definition of the limit there exists a constant $\delta > 0$ such that for all $\gamma \neq 0$ and $-\delta < \gamma < \delta$ it follows that

$$\frac{f(\boldsymbol{\theta} + \gamma \boldsymbol{d}) - f(\boldsymbol{\theta})}{\gamma} > 0$$

Choose γ to preserve this inequality and the theorem is proved. ∎

Definition 6 A direction $\boldsymbol{d}$ such that the directional derivative $\nabla f(\boldsymbol{\theta})'\boldsymbol{d} < 0$ is termed a descent direction.

Theorem 4 The direction $\boldsymbol{d} = -\nabla^2 f(\boldsymbol{\theta})^{-1}\nabla f(\boldsymbol{\theta})$ is a descent direction if $\nabla^2 f(\boldsymbol{\theta})$ is a positive definite matrix.

Proof If $\nabla^2 f(\boldsymbol{\theta})$ is a positive definite matrix then its inverse will also be positive definite. Thus

$$\nabla f(\boldsymbol{\theta})'\boldsymbol{d} = -\nabla f(\boldsymbol{\theta})'\nabla^2 f(\boldsymbol{\theta})^{-1}\nabla f(\boldsymbol{\theta}) < 0$$

since $\nabla^2 f(\boldsymbol{\theta})$ is positive definite. From Theorem 3 it follows that $\boldsymbol{d}$ is a descent direction and the theorem is proved. ∎

Remark

Theorem 3 shows that a decrease can be obtained by taking a sufficiently small step in the direction $\boldsymbol{d}$ provided that $\nabla f(\boldsymbol{\theta})'\boldsymbol{d} < 0$.

Exact line search

In this case the maximum possible decrease in $f(\boldsymbol{\theta})$ occurs for a given $\boldsymbol{d}^j$ if we solve

$$f(\boldsymbol{\theta}^j + \gamma_j \boldsymbol{d}^j) = \min_{\gamma > 0} f(\boldsymbol{\theta}^j + \gamma \boldsymbol{d}^j)$$

where $\boldsymbol{d}^j$ is a descent direction i.e., $f(\boldsymbol{\theta}^j + \gamma_j \boldsymbol{d}^j) < f(\boldsymbol{\theta}^j)$. The main drawback is the number of function evaluations required in the minimization process. Through the years various line search procedures have been proposed e.g., Fibonacci search, quadratic and cubic line search etc. A thorough discussion of these methods as well as other heuristic line search methods may be found in Avriel (1976:215–242).

Inexact line search.

Anderson-Osborne (1977a) line search strategy.

Assume that the model functions are sufficiently smooth to allow a Taylor series of order 2. Thus

$$h_i(\boldsymbol{\theta} + \gamma \boldsymbol{d}) = h_i(\boldsymbol{\theta}) + \gamma \nabla h_i(\boldsymbol{\theta})'\boldsymbol{d} + \gamma^2 \boldsymbol{d}' w_i(\xi)\boldsymbol{d}$$

where $\boldsymbol{\theta} < \xi < \boldsymbol{\theta} + \gamma \boldsymbol{d}$. See Dennis and Schnabel (1983:72). In general we can write $w_i(\xi) = \frac{1}{2}\nabla^2 h_i(\xi)$. Assume further that $\|w(\xi)\| \leq W$ i.e., the Hessian is bounded, $0 < \gamma \leq 1$ in a neighbourhood of $\boldsymbol{\theta}$. The $\|.\|$ can be any appropriate vector norm so we assume it is the L_1-norm. Suppose that $\boldsymbol{h}(\boldsymbol{\theta}) = [h_1(\boldsymbol{\theta}), \ldots, h_n(\boldsymbol{\theta})]$. Then it follows that

$$\boldsymbol{h}(\boldsymbol{\theta}^j + \gamma_j \boldsymbol{d}^j) = \boldsymbol{h}(\boldsymbol{\theta}^j) + \gamma_j \nabla h(\boldsymbol{\theta}^j)'\boldsymbol{d}^j + \gamma_j^2 ||\boldsymbol{d}||^2 w_i(\xi)$$

Define $\hat{S}^j = \boldsymbol{h}(\boldsymbol{\theta}^j) + \nabla h(\boldsymbol{\theta}^j)'\boldsymbol{d}^j$ then

$$\boldsymbol{h}(\boldsymbol{\theta}^j + \gamma_j \boldsymbol{d}^j) = \boldsymbol{h}(\boldsymbol{\theta}^j) + \gamma_j [\hat{S}^j - h(\boldsymbol{\theta}^j)] + \gamma_j^2 ||\boldsymbol{d}||^2 w_i(\xi)$$

Since $0 < \gamma_j \leq 1$ and $||w_i(\xi)|| \leq W$ we have

$$||\boldsymbol{h}(\boldsymbol{\theta}^j + \gamma_j \boldsymbol{d}^j)|| \leq (1 - \gamma_j)||\boldsymbol{h}(\boldsymbol{\theta}^j)|| + \gamma_j ||\hat{S}^j|| + \gamma_j^2 \lambda_j^2 W$$

Where $\max_{1 \leq i \leq k} |d_i^j| \leq \lambda_j$. Thus we can write (the L_1-norm is implied).

$$\begin{aligned} S_1(\boldsymbol{\theta}^j + \gamma_j \boldsymbol{d}^j) &\leq (1 - \gamma_j) S_1(\boldsymbol{\theta}^j) + \gamma_j ||\hat{S}^j|| + \gamma_j^2 \lambda_j^2 W \\ &= S_1(\boldsymbol{\theta}^j) - \gamma_j (S_1(\boldsymbol{\theta}^j) - ||\hat{S}^j||) + \gamma_j^2 \lambda_j^2 W \end{aligned}$$

Thus

$$\frac{S_1(\boldsymbol{\theta}^j) - S_1(\boldsymbol{\theta}^j + \gamma_j \boldsymbol{d}^j)}{\gamma_j (S_1(\boldsymbol{\theta}^j) - ||\hat{S}^j||)} \geq 1 - \frac{\gamma_j \lambda_j^2 W}{S_1(\boldsymbol{\theta}^j) - ||\hat{S}^j||}$$

The left hand side of the inequality can be denoted by $T_1(\gamma_j, \lambda_j)$. Thus for fixed j

$$T_1(\gamma_j, \lambda_j) \geq \sigma$$

can be satisfied:

(i) for fixed λ_j let $\gamma_j \to 0$.

(ii) for fixed $\gamma_j = 1$ let $\lambda_j \to 0$.

In the first algorithm by McLean and Watson (1980) we need $T_1(1, \lambda_j) < \sigma$. In their second algorithm, however, we need $\sigma \leq T_1(\gamma_j, \lambda_j) < 1 - \sigma$. The authors provide a simple strategy for doing this. (See also Anderson and Osborne (1977b) for a similar strategy).

The line search algorithm proceeds as follows: Choose γ_j as the largest number in the set

$$\{1, \mu, \mu^2, \ldots\}, \quad 0 < \mu < 1$$

for which

$$\frac{S_1(\boldsymbol{\theta}^j) - S_1(\boldsymbol{\theta}^j + \gamma_j \boldsymbol{d}^j)}{\gamma_j (S_1(\boldsymbol{\theta}^j) - \hat{S}^j)} \geq \lambda$$

where $\hat{S}^j$ is the objective function value of the linearized L_1-norm problem solved in Step 1 of the Anderson-Osborne-Watson algorithm (§2.3.2.1). Typical values of

$\mu = 0.1$ and $\lambda = 0.0001$ are used. This line search has been incorporated in the FORTRAN program ANDOSBWAT.

Armijo-Goldstein line search

This method [Armijo (1966)] is based on the Goldstein-Armijo principle:

A sufficient decrease in $f(\boldsymbol{\theta})$ is obtained if γ_j satisfies

$$\mu_2 \gamma_j \nabla f(\boldsymbol{\theta}^j)'\boldsymbol{d}^j \leq f(\boldsymbol{\theta}^j + \gamma_j \boldsymbol{d}^j) - f(\boldsymbol{\theta}^j) \leq \mu_1 \gamma_j \nabla f(\boldsymbol{\theta}^j)'\boldsymbol{d}^j$$

where μ_1 and μ_2 are scalars and $0 < \mu_1 \leq \mu_2 < 1$. The bounds ensure that γ_j is neither too small nor too large. The algorithm proceeds as follows:

Step 0: Select a trial value $\gamma_1 (= 1)$ set $j = 1$.

Step 1: If $f(\boldsymbol{\theta}^j + \gamma_j \boldsymbol{d}^j) - f(\boldsymbol{\theta}^j) \leq \mu_1 \gamma_j \nabla f(\boldsymbol{\theta}^j)'\boldsymbol{d}^j$, stop.

Step 2: Set $\gamma_j := \omega \gamma_j$, $j := j + 1$ and return to Step 1.

Typical choices are $\mu_1 = \mu_2 = 0.0001$ and $\omega = 0.1$. The procedure works well when $f(\boldsymbol{\theta})$ is differentiable.

Appendix 2C: The BFGS approach in unconstrained minimization

The BFGS [Broyden (1970), Fletcher (1970), Goldfarb (1970) and Shanno (1970)] is a special case of the general quasi-Newton approach to unconstrained minimization (see e.g., Luenberger (1984) and Dennis and Moré (1977) for more detailed discussions of quasi-Newton methods).

The basic idea is to provide a scheme for approximating the Hessian (or its inverse) of the multivariable function $f(\boldsymbol{\theta})$ to be minimized. As the algorithm progresses this approximation is updated so that it eventually is accurate (hopefully) at convergence to a local minimum.

Given that H^j is an approximation of the inverse Hessian at iteration j. We want to choose H^{j+1} so that the *secant* relationship

$$H^{j+1}[\nabla f(\boldsymbol{\theta}^{j+1}) - \nabla f(\boldsymbol{\theta}^j)] = \boldsymbol{\theta}^{j+1} - \boldsymbol{\theta}^j$$

is satisfied. It is furthermore assumed that H^j is positive definite for all j. A conceptual algorithm for minimizing a continuously differentiable function $f(\boldsymbol{\theta})$ is the following:

The Algorithm

Given an initial estimate $\boldsymbol{\theta}^1$ of the optimal $\boldsymbol{\theta}^*$, select a positive definite matrix $H^j(= I_{k\times k})$, set $j = 1$.

Step 1: Set direction $\boldsymbol{d}^j = -H^j\nabla f(\boldsymbol{\theta}^j)$.

Step 2: Determine $\boldsymbol{\theta}^{j+1} = \boldsymbol{\theta}^j + \alpha_j \boldsymbol{d}^j$ (α_j is determined by line search).

Step 3: Calculate

$$\delta^j = \boldsymbol{\theta}^{j+1} - \boldsymbol{\theta}^j$$

$$\gamma^j = \nabla f(\boldsymbol{\theta}^{j+1}) - \nabla f(\boldsymbol{\theta}^j)$$

Update the approximation of the inverse Hessian:

$$H^{j+1} = \left(I_k - \frac{\delta^j(\gamma^j)'}{(\gamma^j)'\delta^j}\right) H^j \left(I_k - \frac{\gamma^j(\delta^j)'}{(\gamma^j)'\delta^j}\right) + \frac{\delta^j(\delta^j)'}{(\gamma^j)'\delta^j}$$

Return to Step 1. Repeat until certain convergence criteria are met.

Appendix 2D: Levenberg-Marquardt approach in nonlinear estimation

It can be shown [see Dennis and Schnabel (1983)] that Gauss-Newton methods converge quadratically on linear or zero-residual least squares problems. If the second-order terms are small compared to the first-order terms then these methods will only be linearly convergent. If the second-order terms dominate then these methods may not converge at all. One way to overcome the latter problem is the Levenberg-Marquardt approach.

This method was developed to solve nonlinear least squares problems [see for example Levenberg (1944) and Marquardt (1963)]. More recently it has been applied to the nondifferentiable L_1 and L_∞-norm problems. We shall consider the formulation of the Gauss-Newton and Levenberg-Marquardt iteration for both the differentiable L_p-norm and nondifferentiable L_1 and L_∞-norm problems.

Consider the least squares problem:

$$\min_{\boldsymbol{\theta}} \sum_{i=1}^{n} [y_i - f_i(\boldsymbol{x}_i, \boldsymbol{\theta})]^2 = \|\boldsymbol{y} - \boldsymbol{f}(\boldsymbol{x}, \boldsymbol{\theta})\|^2$$

Given a point $\boldsymbol{\theta}^j$ then the Gauss-Newton method solves the *linear least squares* problem:

$$\min_{\boldsymbol{\theta}} \|\boldsymbol{y} - \boldsymbol{f}(\boldsymbol{x}, \boldsymbol{\theta}^j) - J(\boldsymbol{\theta}^j)(\boldsymbol{\theta} - \boldsymbol{\theta}^j)\|^2$$

Vector differentiation with respect to $\boldsymbol{\theta}$, setting the result equal to zero and further simplification yields the Gauss-Newton iteration:

$$J(\boldsymbol{\theta}^j)'J(\boldsymbol{\theta}^j)(\boldsymbol{\theta} - \boldsymbol{\theta}^j) = J(\boldsymbol{\theta}^j)'[\boldsymbol{y} - \boldsymbol{f}(\boldsymbol{x}, \boldsymbol{\theta}^j)]$$

The Levenberg-Marquardt rule limits the length of direction $\boldsymbol{d}^j = \boldsymbol{\theta} - \boldsymbol{\theta}^j$. We therefore solve the problem

$$\min_{\boldsymbol{\theta}} \|\boldsymbol{y} - \boldsymbol{f}(\boldsymbol{x}, \boldsymbol{\theta}^j) - J(\boldsymbol{\theta}^j)(\boldsymbol{\theta} - \boldsymbol{\theta}^j)\|^2 \tag{1}$$

$$\text{subject to} \qquad \|\boldsymbol{\theta} - \boldsymbol{\theta}^j\| \leq \delta$$

The Lagrangian function for this constrained problem is

$$L(\boldsymbol{\theta}, \lambda) = \|\boldsymbol{y} - \boldsymbol{f}(\boldsymbol{x}, \boldsymbol{\theta}^j) - J(\boldsymbol{\theta}^j)(\boldsymbol{\theta} - \boldsymbol{\theta}^j)\|^2 + \lambda[\|\boldsymbol{\theta} - \boldsymbol{\theta}^j)\|^2 - \delta^2]$$

The Karush-Kuhn-Tucker (KKT) conditions are:

(a) $(\boldsymbol{\theta} - \boldsymbol{\theta}^j)'(\boldsymbol{\theta} - \boldsymbol{\theta}^j) \leq \delta^2$

(b) $\lambda[(\boldsymbol{\theta} - \boldsymbol{\theta}^j)'(\boldsymbol{\theta} - \boldsymbol{\theta}^j) - \delta^2] = 0$

(c) $\nabla_{\boldsymbol{\theta}} L = -2J(\boldsymbol{\theta}^j)'[\boldsymbol{y} - \boldsymbol{f}(\boldsymbol{x}, \boldsymbol{\theta}^j) - J(\boldsymbol{\theta}^j)(\boldsymbol{\theta} - \boldsymbol{\theta}^j)] + 2\lambda(\boldsymbol{\theta} - \boldsymbol{\theta}^j) = 0$

KKT condition (c) can be written in the more familiar form

$$[J(\boldsymbol{\theta}^j)'J(\boldsymbol{\theta}^j) + \lambda I](\boldsymbol{\theta} - \boldsymbol{\theta}^j) = J(\boldsymbol{\theta}^j)[\boldsymbol{y} - \boldsymbol{f}(\boldsymbol{x}, \boldsymbol{\theta}^j)]$$

KKT condition (a) can be used to provide information about the Lagrange multiplier or the Levenberg-Marquardt parameter λ. Condition (a) is equivalent to

$$||\boldsymbol{\theta} - \boldsymbol{\theta}^j|| \leq \delta$$

Now

$$\begin{aligned} ||\boldsymbol{\theta} - \boldsymbol{\theta}^j|| &= ||[J(\boldsymbol{\theta}^j)'J(\boldsymbol{\theta}^j) + \lambda I]^{-1}J(\boldsymbol{\theta}^j)'[\boldsymbol{y} - \boldsymbol{f}(\boldsymbol{x}, \boldsymbol{\theta}^j)]|| \\ &\leq ||[J(\boldsymbol{\theta}^j)'J(\boldsymbol{\theta}^j)]^{-1}J(\boldsymbol{\theta}^j)[\boldsymbol{y} - \boldsymbol{f}(\boldsymbol{x}, \boldsymbol{\theta}^j)]|| \end{aligned}$$

Now if

$$\delta \geq ||[J(\boldsymbol{\theta}^j)'J(\boldsymbol{\theta}^j)]^{-1}J(\boldsymbol{\theta}^j)[\boldsymbol{y} - \boldsymbol{f}(\boldsymbol{x}, \boldsymbol{\theta}^j)]||$$

then λ plays no role and may be chosen as 0.

Problem (1) is therefore equivalent to

$$\boldsymbol{\theta}^{j+1} = \boldsymbol{\theta}^j - [J(\boldsymbol{\theta}^j)'J(\boldsymbol{\theta}^j) + \lambda I]^{-1}J(\boldsymbol{\theta}^j)[\boldsymbol{y} - \boldsymbol{f}(\boldsymbol{x}, \boldsymbol{\theta}^j)]$$

with $\lambda > 0$ if $\delta < ||[J(\boldsymbol{\theta}^j)'J(\boldsymbol{\theta}^j)]^{-1}J(\boldsymbol{\theta}^j)[\boldsymbol{y} - \boldsymbol{f}(\boldsymbol{x}, \boldsymbol{\theta}^j)]||$.

Some authors use an alternative formulation to derive the above strategy [also known as a trust region approach, see Dennis and Schnabel (1983:227)]. Consider the linear least squares problem

$$\min_{\boldsymbol{\theta}} \left\| \begin{pmatrix} \boldsymbol{y} - \boldsymbol{f}(\boldsymbol{x}, \boldsymbol{\theta}^j) \\ 0 \end{pmatrix} - \begin{pmatrix} -J(\boldsymbol{\theta}^j) \\ \mu I \end{pmatrix} (\boldsymbol{\theta} - \boldsymbol{\theta}^j) \right\|_2^2$$

This problem is equivalent to

$$\min_{\boldsymbol{d}} ||\boldsymbol{y} - \boldsymbol{f}(\boldsymbol{x}, \boldsymbol{\theta}^j) - J(\boldsymbol{\theta}^j)\boldsymbol{d}||^2 + \mu^2||\boldsymbol{d}||^2$$

Expanding and solving by vector differentiation yields:

$$[J(\boldsymbol{\theta}^j)'J(\boldsymbol{\theta}^j) + \mu^2 I]\boldsymbol{d}^j = J(\boldsymbol{\theta}^j)[\boldsymbol{y} - \boldsymbol{f}(\boldsymbol{x}, \boldsymbol{\theta}^j)]$$

Thus $\mu^2 = \lambda \geq 0$, the Lagrange multiplier of problem (1). The two approaches are therefore equivalent.

For the nondifferentiable problems such equivalences are not possible. The L_1-norm problem is

$$\min_{\boldsymbol{\theta}} \sum_{i=1}^{n} |y_i - f_i(\boldsymbol{x}_i, \boldsymbol{\theta})| \tag{2}$$

The Gauss-Newton iteration can only be written as a *linear L_1-norm problem*:

$$\min_{\boldsymbol{d}} \sum_{i=1}^{n} |y_i - f_i(\boldsymbol{x}_i, \boldsymbol{\theta}^j) - \nabla_{\boldsymbol{\theta}} f_i(\boldsymbol{x}_i, \boldsymbol{\theta}^j)'\boldsymbol{d}|$$

and similarly the Levenberg-Marquardt iteration is the *constrained linear L_1-norm problem*:

$$\min_{\boldsymbol{d}} \sum_{i=1}^{n} |y_i - f_i(\boldsymbol{x}_i, \boldsymbol{\theta}^j) - \nabla_{\boldsymbol{\theta}} f_i(\boldsymbol{x}_i, \boldsymbol{\theta}^j)'\boldsymbol{d}| \qquad \text{subject to} \qquad \|\boldsymbol{d}\| \le \delta$$

The equivalent formulation is

$$\min_{\boldsymbol{d}} \left\| \begin{pmatrix} \boldsymbol{y} - \boldsymbol{f}(\boldsymbol{x}, \boldsymbol{\theta}^j) \\ 0 \end{pmatrix} - \begin{pmatrix} -J(\boldsymbol{\theta}^j) \\ \mu I \end{pmatrix} \boldsymbol{d} \right\|_1$$

In Chapter 3 the L_∞-norm problem is discussed and in Chapter 4 the general L_p-norm problem. We shall simply state the results for these two problems.

L_∞-norm or Chebychev

GN

$$\min_{\boldsymbol{d}} \max_{1 \le i \le n} |y_i - f_i(\boldsymbol{x}_i, \boldsymbol{\theta}^j) - \nabla_{\boldsymbol{\theta}} f_i(\boldsymbol{x}_i, \boldsymbol{\theta}^j)'\boldsymbol{d}|$$

Levenberg-Marquardt

$$\min_{\boldsymbol{d}} \max_{1 \le i \le n} |y_i - f_i(\boldsymbol{x}_i, \boldsymbol{\theta}^j) - \nabla_{\boldsymbol{\theta}} f_i(\boldsymbol{x}_i, \boldsymbol{\theta}^j)'\boldsymbol{d}|$$
$$\text{subject to} \qquad \|\boldsymbol{d}\|_\infty = \max_{1 \le i \le k} |\boldsymbol{d}_i| \le \delta$$

L_p-norm $(1 < p < \infty)$

GN

$$(p-1)J_p(\boldsymbol{\theta}^j)'J_p(\boldsymbol{\theta}^j)\boldsymbol{d}^j = J_p(\boldsymbol{\theta}^j)'[\boldsymbol{y} - \boldsymbol{f}(\boldsymbol{x}, \boldsymbol{\theta})]_p$$

Levenberg-Marquardt

$$(p-1)[J_p(\boldsymbol{\theta}^j)'J_p(\boldsymbol{\theta}^j) + \lambda I]\boldsymbol{d}^j = J_p(\boldsymbol{\theta}^j)'[\boldsymbol{y} - \boldsymbol{f}(\boldsymbol{x}, \boldsymbol{\theta})]_p$$

Appendix 2E: Rates of convergence

Let $\theta^1, \theta^2, \ldots \in \Re^k$ be a sequence of vectors that converges to θ^*. The *order* of convergence of the sequence is the supremum of nonnegative numbers q so that

$$\overline{\lim}_{j\to\infty} \frac{\|\theta^{j+1} - \theta^*\|}{\|\theta^j - \theta^*\|^q} = \ell$$

where $0 \leq \ell < \infty$ is the *root convergence factor* and $\|\theta^n - \theta^*\|$ is the *error* of the nth approximant. If $q = 1$ we say the convergence rate is *linear*. If $q = 1$ and $\ell = 0$ or if $q > 1$ then we say the convergence rate is *superlinear*. If $q = 2$ then the convergence rate is *quadratic*.

It can be shown that the steepest descent method converges at a linear rate and Newton's method is quadratic [see e.g., Luenberger (1984)]. Gill and Murray (1978) remark that in the case of small residual problems " *Gauss-Newton methods will ultimately converge at the same rate as Newton's method despite the fact that only first derivatives are required* ". Obviously the higher the order of convergence the quicker convergence of $\theta^j \to \theta^*$ occurs. Powell (1971) shows that if the objective function to be minimized is convex and if exact line searches are used then a class of quasi-Newton or variable metric methods for unconstrained minimization is superlinearly convergent.

Appendix 2F: Linear Algebra

A matrix $J_{n\times k}$ is *rank-deficient* if its rank$(J) < \min(n,k)$. It is of *full rank* if rank$(J) = min(n,k)$. Matrix J will be termed of *full column rank* if rank $(J) = k$. A square $n \times n$ matrix B is *nonsingular* if rank$(B) = n$ and singular if rank$(B) < n$. Matrix B will be termed *orthogonal* if $B'B = BB' = I$. Matrix B is termed *ill-conditioned* if B is nearly singular. In this case its *condition number* (defined as the ratio of the largest to the smallest eigenvalue) will be very large.

The Euclidean norm of a vector $\boldsymbol{x} \in \Re^k$, denoted by $||\boldsymbol{x}||$, is defined as

$$||\boldsymbol{x}|| = \sqrt{\sum_{i=1}^{n} x_i^2}$$

The Euclidean (Frobenius) norm of a square $n \times n$ matrix B, denoted by $||B||$, is defined as

$$||B|| = \sqrt{\sum_{i=1}^{n}\sum_{j=1}^{n} b_{ij}^2}$$

Definition 7 Polyhedral norms - Let B, $\boldsymbol{x}$ and $\boldsymbol{e}$ be an $m \times n$ matrix and n- and m-vectors respectively with $m > n$ and $\boldsymbol{e}$ is a unit vector. Consider the set of inequalities $B\boldsymbol{x} \leq \boldsymbol{e}$ given the following conditions:

(i) The feasible region $F = \{\boldsymbol{x} \mid B\boldsymbol{x} \leq \boldsymbol{e}\}$, is bounded and nonempty $(F \neq \emptyset)$.

(ii) $\boldsymbol{x} \in F \Rightarrow -\boldsymbol{x} \in F$.

For computational efficiency, Anderson and Osborne (1977a) stipulate the stronger condition $\boldsymbol{x} \in F \iff |\boldsymbol{x}| \in F$. The polyhedral norm of $\boldsymbol{x}$ is written as

$$||\boldsymbol{x}||_B = \min\{\, \eta \mid B\boldsymbol{x} \leq \eta \boldsymbol{e}\}, \qquad (\eta \text{ is a scalar number})$$

Examples

(1) To obtain the L_1-norm, matrix B is chosen so that its rows correspond to the 2^n different ways of filling n locations with either $+1$ or -1. For example consider $n = 2$.

$$||\boldsymbol{x}||_1 = \sum_{i=1}^{2} |x_i|$$

Then

$$Bx = \begin{pmatrix} 1 & 1 \\ 1 & -1 \\ -1 & 1 \\ -1 & -1 \end{pmatrix} \begin{pmatrix} x_1 \\ x_2 \end{pmatrix} \leq \eta \begin{pmatrix} 1 \\ 1 \\ 1 \\ 1 \end{pmatrix}$$

These inequalities simply state that the minimum η is the same as

$$\|x\|_1 = |x_1| + |x_2|$$

(2) To obtain the L_∞-norm, $B_{2n\times n}$ is chosen as

$$B = \begin{pmatrix} I_{n\times n} \\ -I_{n\times n} \end{pmatrix}$$

Orthogonal decomposition of a matrix

Let $A_{n\times m}$ be a given real matrix of rank m. The QR decomposition of A is of the form:

$$A = Q'R = Q' \begin{pmatrix} \tilde{R} \\ \cdots \\ O \end{pmatrix}$$

where $Q_{n\times n}$ is an orthogonal matrix $\tilde{R}_{m\times m}$ is an upper triangular matrix and $O_{(n-m)\times m}$ is matrix of zeroes. This numerically stable approach is used in solving linear least squares problems [see e.g., Businger and Golub (1965)]. An effective way of carrying out the QR decomposition is by means of Householder transformations:

$$P = I - \beta \boldsymbol{u}\boldsymbol{u}'$$

where P is orthogonal and symmetric. Denote all matrices at iteration k by the superscript (k). Let $A^{(1)} = A$ and define $A^{(2)}, \ldots, A^{(m+1)}$ by $A^{(k+1)} = P^{(k)}A^{(k)}$ for $k = 1, \ldots, m$. The elements of $P^{(k)}$ are such that $a_{ik}^{(k+1)} = 0$ for $i = k+1, \ldots, n$.

Householder (1958) shows that $P^{(k)}$ is generated by choosing

$$u_i^{(k)} = \begin{cases} 0 & \text{if } i < k \\ sign(a_{kk}^{(k)})(\sigma_k + |a_{kk}^{(k)}|) & \text{if } i = k \\ a_{ik}^{(k)} & \text{if } i > k \end{cases}$$

$$\beta_k = \frac{1}{\sigma_k(\sigma_k + |a_{kk}^{(k)}|)}$$
$$\text{where } \sigma_k = \Big[\sum_{i=k}^{n} (a_{ik}^{(k)})^2\Big]^{1/2}$$

The matrix $P^{(k)} = I - \beta_{(k)} \boldsymbol{u}^{(k)} \boldsymbol{u}^{(k)\prime}$ is not computed explicitly since

$$\begin{aligned} A^{(k+1)} &= \left[I - \beta_k \boldsymbol{u}^{(k)} \boldsymbol{u}^{(k)\prime}\right] A^{(k)} \\ &= A^{(k)} - \boldsymbol{u}^{(k)} \boldsymbol{y}_k' \end{aligned}$$

with

$$\boldsymbol{y}_k' = \beta_k \boldsymbol{u}^{(k)\prime} A^{(k)}$$

In the computation of $\boldsymbol{y}_k$ and $A^{(k+1)}$ we utilize the fact that the first $(k-1)$ components of $\boldsymbol{u}^{(k)}$ are zero. Then finally $A^{(m+1)} = R$ and $Q = P^{(m)} \dots P^{(2)} P^{(1)}$.

Suppose we want to solve the system of linear equations $A\boldsymbol{x} = \boldsymbol{b}$. We can write

$$Q\boldsymbol{b} = \begin{pmatrix} e_m \\ f_{n-m} \end{pmatrix}$$

Since Q is orthogonal

$$\begin{aligned} \|A\boldsymbol{x} - \boldsymbol{b}\|_2^2 &= \|QA\boldsymbol{x} - Q\boldsymbol{b}\|_2^2 \\ &= \|\tilde{R}\boldsymbol{x} - \boldsymbol{e}\|_2^2 + \|\boldsymbol{f}\|_2^2 \end{aligned}$$

Furthermore $\|A\boldsymbol{x} - \boldsymbol{b}\|_2$ is minimized if $\tilde{R}\boldsymbol{x} = \boldsymbol{e}$. Hence given the orthogonal factors of A we can solve the system of linear equations $A\boldsymbol{x} = \boldsymbol{b}$ by obtaining $\boldsymbol{x}$ by back substitution from $\tilde{R}\boldsymbol{x} = \boldsymbol{e}$. The reader is also referred to Ansley (1985).

Example

Compute the QR decomposition of

$$A = \begin{pmatrix} 1 & 1 \\ 0 & 1 \\ 0 & 0 \end{pmatrix} = A^{(1)}$$

For $k = 1$ it follows that $\sigma_1 = 1, \beta_1 = \frac{1}{2}$. Thus $u_1^{(1)} = 2,\ u_2^{(1)} = 0$ and $u_3^{(1)} = 0$ or $\boldsymbol{u}^{(1)} = (2 \quad 0 \quad 0)'$.

$$A^{(2)} = \begin{pmatrix} 1 & 1 \\ 0 & 1 \\ 0 & 0 \end{pmatrix} - \frac{1}{2} \begin{pmatrix} 2 \\ 0 \\ 0 \end{pmatrix} (2 \quad 0 \quad 0) \begin{pmatrix} 1 & 1 \\ 0 & 1 \\ 0 & 0 \end{pmatrix} = \begin{pmatrix} -1 & -1 \\ 0 & 1 \\ 0 & 0 \end{pmatrix}$$

For $k = 2$ it follows that $\sigma_2 = 1$ and $\beta_2 = \frac{1}{2}$. Thus $u_1^{(2)} = 0$, $u_2^{(2)} = 2$ and $u_3^{(2)} = 0$ or $\mathbf{u}^{(2)} = (0 \quad 2 \quad 0)$.

$$A^{(3)} = \begin{pmatrix} -1 & -1 \\ 0 & 1 \\ 0 & 0 \end{pmatrix} - \frac{1}{2}\begin{pmatrix} 0 & 0 & 0 \\ 0 & 4 & 0 \\ 0 & 0 & 0 \end{pmatrix}\begin{pmatrix} -1 & -1 \\ 0 & 1 \\ 0 & 0 \end{pmatrix} = \begin{pmatrix} -1 & -1 \\ 0 & -1 \\ 0 & 0 \end{pmatrix} = R$$

To obtain Q we need $P^{(1)}$ and $P^{(2)}$

$$P^{(1)} = \begin{pmatrix} 1 & 0 & 0 \\ 0 & 1 & 0 \\ 0 & 0 & 1 \end{pmatrix} - \frac{1}{2}\begin{pmatrix} 4 & 0 & 0 \\ 0 & 0 & 0 \\ 0 & 0 & 0 \end{pmatrix} = \begin{pmatrix} -1 & 0 & 0 \\ 0 & 1 & 0 \\ 0 & 0 & 1 \end{pmatrix}$$

$$P^{(2)} = \begin{pmatrix} 1 & 0 & 0 \\ 0 & 1 & 0 \\ 0 & 0 & 1 \end{pmatrix} - \frac{1}{2}\begin{pmatrix} 0 & 0 & 0 \\ 0 & 4 & 0 \\ 0 & 0 & 0 \end{pmatrix} = \begin{pmatrix} 1 & 0 & 0 \\ 0 & -1 & 0 \\ 0 & 0 & 1 \end{pmatrix}$$

It follows that

$$Q = P^{(2)}P^{(1)} = \begin{pmatrix} 1 & 0 & 0 \\ 0 & -1 & 0 \\ 0 & 0 & 1 \end{pmatrix}\begin{pmatrix} -1 & 0 & 0 \\ 0 & 1 & 0 \\ 0 & 0 & 1 \end{pmatrix} = \begin{pmatrix} -1 & 0 & 0 \\ 0 & -1 & 0 \\ 0 & 0 & 1 \end{pmatrix}$$

and $Q'R = A$ by inspection.

Appendix 2G: Extrapolation procedures to enhance convergence

El-Attar et al. (1979) also suggest the use of extrapolation procedures to enhance convergence of the penalty function method. This will overcome the ill-conditioning in the problem. Extrapolation procedures are often used in numerical analysis and the interested reader is referred to Henrici (1964:Chapter 12) for a thorough discussion.

A brief outline of the extrapolation procedure will now be given. Suppose $P(\boldsymbol{\theta}, \delta)$ has been minimized for the penalty constants $\delta_1 > \delta_2 > \ldots \delta_r > 0$ yielding solutions $\boldsymbol{\theta}^1(\delta_1) \ldots \boldsymbol{\theta}^r(\delta_r)$. Let $\boldsymbol{\theta}^*$ be the true optimal solution and let

$$\boldsymbol{\theta}_{ij} = j^{th}\text{-order estimate of } \boldsymbol{\theta}^* = \boldsymbol{\theta}(0)$$

after i minima have been calculated. Fiacco and McCormick (1968:189) give the following recursive relationships:

$$\boldsymbol{\theta}_{i0} = \boldsymbol{\theta}\left(\frac{\delta_1}{\beta^{i-1}}\right) = \boldsymbol{\theta}(\delta_i) \qquad i = 1, \ldots, r$$

$$\boldsymbol{\theta}_{ij} = \frac{\beta^j \boldsymbol{\theta}_{i,j-1} - \boldsymbol{\theta}_{i-1,j-1}}{\beta^j - 1} \qquad i = 2, \ldots, r, \quad j = 1, \ldots, i-1$$

where δ_1 and β are defined as in §2.3.5.1. Then the best estimate of $\boldsymbol{\theta}^*$ is given by $\boldsymbol{\theta}_{r,r-1}$. Schematically we shall calculate the estimates as follows:

Minima	order j			
i	0	1	2	3
δ_1	$\boldsymbol{\theta}_{10}$			
δ_2	$\boldsymbol{\theta}_{20}$	$\boldsymbol{\theta}_{21}$		
δ_3	$\boldsymbol{\theta}_{30}$	$\boldsymbol{\theta}_{31}$	$\boldsymbol{\theta}_{32}$	
δ_4	$\boldsymbol{\theta}_{40}$	$\boldsymbol{\theta}_{41}$	$\boldsymbol{\theta}_{42}$	$\boldsymbol{\theta}_{43}$
$\vdots$	$\vdots$	$\vdots$	$\vdots$	$\vdots$

We consider the second Example in Madsen (1975). We shall use β and δ as before. We found:

$$\boldsymbol{\theta}_{10} = \begin{pmatrix} 0.1271246 \\ -0.63873 \end{pmatrix}$$

$$\boldsymbol{\theta}_{20} = \begin{pmatrix} 0.0758286 \\ -0.514263 \end{pmatrix}$$

$$\boldsymbol{\theta}_{30} = \begin{pmatrix} 0.029752 \\ -0.353664 \end{pmatrix}$$

Then

$$\boldsymbol{\theta}_{21} = \frac{\beta\boldsymbol{\theta}_{20} - \boldsymbol{\theta}_{10}}{\beta - 1} = \begin{pmatrix} 0.070129 \\ -0.500433 \end{pmatrix}$$

$$\boldsymbol{\theta}_{31} = \begin{pmatrix} 0.0246324 \\ -0.33582 \end{pmatrix}$$

$$\boldsymbol{\theta}_{32} = \frac{\beta^2\boldsymbol{\theta}_{31} - \boldsymbol{\theta}_{21}}{\beta^2 - 1} = \begin{pmatrix} 0.0241728 \\ -.334157 \end{pmatrix}$$

Appendix 2H: FORTRAN Programs

Program ANDOSBWAT

```
C
C         PROGRAM ANDOSBWAT IS BASED ON THE ALGORITHM BY OSBORNE & WA
C         (1971), COMPUTER J., 14, PP 184-188.
C
C         IT USES SUBROUTINE L1 DUE TO BARRODALE AND ROBERTS (1974)
C         COMMUN. ASSOC. COMPUT. MACH., 17, PP 319-320
C
C         SUBROUTINE LINAND PERORMS THE ANDERSON AND OSBORNE LINE SEA
C         RULE: ANDERSON D.H. AND OSBORNE, M.R. (1977A). DISCRETE,
C         NONLINEAR APPROXIMATION PROBLEMS IN POLYHEDRAL NORMS.
C         NUMER. MATH. 28, PP 143-156.
C
C         OPTIMIZATION PROBLEM DATA
C         N = NUMBER OF PARAMETERS
C         M = NUMBER OF OF PROBLEM FUNCTIONS (OR OBSERVATIONS)
C         ITER = MAXIMUM NUMBER OF ITERATIONS
C         THETA(.)  = PARAMETER VECTOR
C
C         THIS EXAMPLE IS DUE TO :
C
C         MADSEN, K. (1975).  AN ALGORITHM FOR MINIMAX SOLUTION TO OV
C         DETERMINED SYSTEMS OF NON-LINEAR EQUATIONS. J. INST. MATH.
C         APPLIC. 16, PP 321-328.
C
          IMPLICIT REAL*8 (A-H,O-Z)
          DIMENSION A(23,7),B(51),F(51),THETA(20),DTHETA(20),T(5),E(5
          INTEGER S(51)
          COMMON/DATA/N,M,X(30),Y(30)
          M=21
          N=5
          ITER = 20
          DO 93 I=1,M
          XI1 = I-1
          X(I)=XI1*0.1D0 -1D0
93        Y(I)=DEXP(X(I))
          THETA(1)=0.5D0
          THETA(2)=0.5D0
          THETA(3)= 0D0
          THETA(4)= 0D0
          THETA(5)= 0D0
          TOLER =1.E-11
          M2=M+2
          N2=N+2
          SJ =1D10
10        CONTINUE
          KOUNT=0
          IOPT = 0
```

```
              CALL FUNC(IOPT,THETA,A,B,F,F1)
              WRITE(6,98)(THETA(I),I=1,N)
98            FORMAT(1HO,' INITIAL THETA PARAMETERS: ',5F8.3)
15            CONTINUE
              KOUNT=KOUNT+1
              CALL L1(M,N,M2,N2,A,B,TOLER,T,E,S)
              IF(DABS((A(M+1,N+1)-SJ)/SJ).LT.1D-10) GO TO 195
              SJ=A(M+1,N+1)
75            CONTINUE
              IF(KOUNT.EQ.ITER) GO TO 195
              CALL LINAND(T,THETA,A,B,F,SJ,FNEW,F1)
              WRITE(6,149)KOUNT,(THETA(I),I=1,N)
              WRITE(6,150)FNEW,SJ
149           FORMAT(1HO,'ITERATION NUMBER ',I3/' THETA PARAMETERS',5F11.6)
150           FORMAT(' SBARJ+1 = ',F12.6,' SHATJ = ',F12.6)
              IOPT = 0
              CALL FUNC(IOPT,THETA,A,B,F,F1)
              GO TO 15
195           STOP
              END
              SUBROUTINE FUNC(IOPT,THETA,A,B,F,F1)
              IMPLICIT REAL*8 (A-H,O-Z)
              DIMENSION A(23,7),B(1),F(1),THETA(1)
              COMMON/DATA/N,M,X(30),Y(30)
C
C             A(I,J) IS THE DERIVATIVE OF THE I-TH MODEL FUNCTION F(I) W.R.T.
C             THE J-TH PARAMETER THETA(J)
C
              F1=0.
              IF (IOPT.NE.0) GO TO 3
              DO 2 I=1,M
              FA=THETA(1)+THETA(2)*X(I)
              FB=1DO+THETA(3)*X(I)+THETA(4)*X(I)**2+THETA(5)*X(I)**3
              A(I,1)=1DO/FB
              A(I,2)=X(I)/FB
              A(I,3)=-FA*X(I)/FB**2
              A(I,4)=-FA*X(I)**2/FB**2
              A(I,5)=-FA*X(I)**3/FB**2
              F(I)=FA/FB
              B(I)=Y(I)-F(I)
              F1=F1+DABS(B(I))
2             CONTINUE
              RETURN
3             F1 = 0DO
              DO 4 I=1,M
              FA=THETA(1)+THETA(2)*X(I)
              FB=1DO+THETA(3)*X(I)+THETA(4)*X(I)**2+THETA(5)*X(I)**3
              F(I)=FA/FB
              B(I)=Y(I)-F(I)
4             F1=F1+DABS(B(I))
              RETURN
              END
```

```
C
C          THIS IS AN INACCURATE LINE SEARCH DUE TO ANDERSON AND OSBORNE
C          (1977a)
C
           SUBROUTINE LINAND(T,THETA,A,B,F,SJ,FNEW,F1)
           IMPLICIT REAL*8 (A-H,O-Z)
           DIMENSION THETA(1),T(1),F(1),DTHETA(20),A(23,7),B(1)
           COMMON/DATA/N,M,X(30),Y(30)
           DELTA = 1D-4
           GAM=1DO
1          CONTINUE
           DO 10 I=1,N
10         DTHETA(I)=THETA(I) + GAM*T(I)
           IOPT = 1
           CALL FUNC(IOPT,DTHETA,A,B,F,FNEW)
           PSI=(F1 - FNEW)/(GAM*(F1-SJ))
           IF(PSI.GE.DELTA) GO TO 30
           GAM = GAM*0.1DO
           GO TO 1
30         DO 35 I=1,N
35         THETA(I)=DTHETA(I)
           IOPT = 1
           CALL FUNC(IOPT,THETA,A,B,F,FNEW)
           RETURN
           END
```

```
INITIAL THETA PARAMETERS: 0.500 0.500 0.000 0.000 0.000
ITERATION NUMBER 1
THETA PARAMETERS 0.550000 0.513108 -0.073438 -0.026505 -0.008632
SBARJ+1 = 13.048823 SHATJ = 0.006731
ITERATION NUMBER 2
THETA PARAMETERS 0.595000 0.521498 -0.133736 -0.038682 -0.005605
SBARJ+1 = 11.705455 SHATJ = 0.008688
ITERATION NUMBER 3
THETA PARAMETERS 0.999759 0.569733 -0.635913 -0.086177 0.053566
SBARJ+1 = 6.493958 SHATJ = 0.008142
ITERATION NUMBER 4
THETA PARAMETERS 1.001554 0.246945 -0.751834 0.295571 -0.138880
SBARJ+1 = 0.823983 SHATJ = 0.028848
ITERATION NUMBER 5
THETA PARAMETERS 1.000512 0.194418 -0.806054 0.312597 -0.071069
SBARJ+1 = 0.045525 SHATJ = 0.012197
ITERATION NUMBER 6
THETA PARAMETERS 0.999936 0.253018 -0.747053 0.246180 -0.038430
SBARJ+1 = 0.002762 SHATJ = 0.000959
```

```
ITERATION NUMBER 7
THETA PARAMETERS 0.999904 0.256836 -0.743214 0.241940 -0.036398
SBARJ+1 = 0.001563 SHATJ = 0.001560
ITERATION NUMBER 8
THETA PARAMETERS 0.999904 0.256848 -0.743202 0.241927 -0.036391
SBARJ+1 = 0.001563 SHATJ = 0.001563
ITERATION NUMBER 9
THETA PARAMETERS 0.999904 0.256848 -0.743202 0.241927 -0.036391
SBARJ+1 = 0.001563 SHATJ = 0.001563
```

Program MCLEANWAT

```
C
C         THIS PROGRAM IS BASED ON THE FIRST ALGORITHM BY
C         MCLEAN AND WATSON (1980)
C
C         IT USES SUBROUTINE CL1 DUE TO BARRODALE AND ROBERTS (1978).
C
C         OPTIMIZATION PROBLEM DATA
C         NPAR = NUMBER OF PARAMETERS
C         NOBS = NUMBER OF OF PROBLEM FUNCTIONS (OR OBSERVATIONS)
C         M = NUMBER OF OF INEQUALITY CONSTRAINTS
C         L = NUMBER OF OF EQUALITY CONSTRAINTS
C         KLMD, KLM2D, NKLMD, N2D, KODE,TOLER,ITER, ARE INPUT PARAMETERS
C         TO SUBROUTINE CL1
C         NITER = MAXIMUM NUMBER OF ITERATIONS REQUIRED
C         THETA(.)  = PARAMETER VECTOR
C         MATRIX Q CONTAINS ALL THE DATA OF THE CONSTRAINED L1-NORM
C         PROBLEM
C
C         THIS EXAMPLE IS DUE TO : JENNRICH AND SAMPSON (1969)
C
          IMPLICIT REAL*8 (A-H,O-Z)
          INTEGER IU(2,110), S(100)
          DIMENSION A(40,40),F(40),THETA(10)
          DIMENSION Q(102,12), X(12), RES(100), CU(2,110)
          DATA KLMD, KLM2D, NKLMD, N2D /100,102,110,12/
          NOBS=10
          NPAR=2
          K=NOBS
          L=0
          N=NPAR
          M = 2*NPAR
          N1= NPAR +1
          KLM = K + L + M
```

```
      KODE = 0
      TOLER = 1.D-30
      NITER = 100
      THETA(1)=.3
      THETA(2)=.4
      XLAM=1.0
      KOUNT=0
      SJ=1E10
15    CONTINUE
      CALL FUNC(NPAR,NOBS,FOLD,F,THETA,A)
      IF (KOUNT.NE.0) GO TO 16
      WRITE(6,98)(THETA(I),I=1,NPAR),FOLD,XLAM
98    FORMAT(' INITIAL THETA:',2F8.3,' S(THETA)',F10.5,
     * ' LAMBDA =',F12.4)
16    KOUNT = KOUNT + 1
      DO 999 I=1,NOBS
      Q(I,N1) =-F(I)
      DO 999 J=1,NPAR
999   Q(I,J) = A(I,J)
      DO 17 I=1,M
      DO 17 J=1,N
17    Q(NOBS+I,J) = 0.0
      DO 18 I =1,N
      Q(NOBS+I,I) = 1.0
18    Q(NOBS+N+I,I) =-1.0
      DO 19 I=1,M
19    Q(NOBS+I,N1) = XLAM
      ITER =NITER
      CALL CL1(K,L,M,N,KLMD,KLM2D,NKLMD,N2D,Q,KODE,TOLER,ITER,
     * X,RES,ERROR,CU,IU,S)
      WRITE(6,776) (X(I),I=1,NPAR)
776   FORMAT(1X,'DELTA THETA=',2F12.6)
      SJ=ERROR
      IF(KOUNT.EQ.NITER) GO TO 200
      CALL MCLEAN (NPAR,NOBS,X,THETA,SJ,FOLD,FNEW,XLAM)
      WRITE(6,150)KOUNT,(THETA(I),I=1,NPAR),SJ,FNEW,XLAM
150   FORMAT(' ITERATION',I4,
     * ' THETA ' ,2G13.6/' DHATJ',F13.7,' SBARJ+1 ',F13.7
     * ,' LAMBDA =',F10.6)
      IF(DABS(FNEW-SJ).LT.1E-9)GO TO 200
      FOLD=FNEW
      GO TO 15
200   STOP
      END
      SUBROUTINE FUNC(NPAR,NOBS,FOBJ,F,THETA,A)
      IMPLICIT REAL*8 (A-H,O-Z)
      DIMENSION A(40,40),F(1),THETA(1)
C
C     A(I,J) IS THE DERIVATIVE OF THE I-TH MODEL FUNCTION F(I) W.R.T.
C     THE J-TH PARAMETER THETA(J)
C
      IF(THETA(1).LE.1D-30) THETA(1)=0D0
```

```
      FOBJ=0.
      DO 1 I=1,NOBS
      XI =I
      YI = 2.0 + 2.0*XI
      F(I) = DEXP(THETA(1)*XI) + DEXP(THETA(2)*XI) -YI
      FOBJ = FOBJ + DABS(F(I))
      A(I,1) = XI*DEXP(THETA(1)*XI)
1     A(I,2) = XI*DEXP(THETA(2)*XI)
      RETURN
      END
      SUBROUTINE FUNL1(NOBS,THETA,FL1)
      IMPLICIT REAL*8 (A-H,O-Z)
      DIMENSION F(30),THETA(1)
      IF(THETA(1).LE.1D-30) THETA(1)=0D0
      FL1=0.
      DO 1 I=1,NOBS
      XI =I
      YI = 2.0 + 2.0*XI
      F(I) = DEXP(THETA(1)*XI) + DEXP(THETA(2)*XI) -YI
1     FL1 = FL1 + DABS(F(I))
      RETURN
      END
C
C     SUBROUTINE MCLEAN(NPAR,NOBS,X,THETA,SJ,FOLD,FNEW,XLAM)
C
      SUBROUTINE MCLEAN(NPAR,NOBS,X,THETA,SJ,FOLD,FNEW,XLAM)
      IMPLICIT REAL*8 (A-H,O-Z)
      DIMENSION A(40,40),F(1),THETA(1),X(1)
      SIGMA = .0001
      BETA1=.1
      BETA2=5.
      DO 10 I=1,NPAR
10    THETA(I)=THETA(I) + X(I)
      CALL FUNL1(NOBS,THETA,FL1)
      PSI=(FOLD - FL1)/(FOLD-SJ)
      WRITE(6,78) PSI
78    FORMAT(1X,'PSI=',F12.6)
      IF(PSI.GE.SIGMA) GO TO 30
      XLAM=BETA1*DMAX1(DABS(X(1)),DABS(X(2)))
      DO 35 I=1,NPAR
35    THETA(I)=THETA(I) -X(I)
      FNEW = FOLD
      RETURN
30    XLAM=BETA2*DMAX1(DABS(X(1)),DABS(X(2)))
      FNEW=FL1
      RETURN
      END
```

```
INITIAL THETA: 0.300 0.400 FNORM 120.58802 LAMBDA = 1.0000

DELTA THETA= 0.168524 -0.158490
```

```
PSI= -0.901962
ITERATION 1 THETA 0.300000 0.400000
DHATJ 15.0499146 SBARJ+1 120.5880199 LAMBDA = 0.016852
DELTA THETA= -0.016852 -0.016852
PSI= 0.926013
ITERATION 2 THETA 0.283148 0.383148
DHATJ 92.8055173 SBARJ+1 94.8610740 LAMBDA = 0.084262
DELTA THETA= -0.003344 -0.084262
PSI= 0.766418
ITERATION 3 THETA 0.279804 0.298885
DHATJ 24.9432565 SBARJ+1 41.2748199 LAMBDA = 0.421311
DELTA THETA= 0.421311 -0.408668
PSI= -133.105667
ITERATION 4 THETA 0.279804 0.298885
DHATJ 25.9687643 SBARJ+1 41.2748199 LAMBDA = 0.042131
DELTA THETA= -0.013900 -0.042131
PSI= 0.861799
ITERATION 5 THETA 0.265904 0.256754
DHATJ 31.2811704 SBARJ+1 32.6623057 LAMBDA = 0.210656
DELTA THETA= -0.204459 0.210656
PSI= -91.767310
ITERATION 35 THETA 0.255842 0.255843
DHATJ 32.0919411 SBARJ+1 32.0919411 LAMBDA = 0.000006
DELTA THETA= 0.000006 -0.000006
PSI= -8.560843
ITERATION 36 THETA 0.255842 0.255843
DHATJ 32.0919411 SBARJ+1 32.0919411 LAMBDA = 0.000001
DELTA THETA= 0.000001 -0.000001
PSI= 0.615560
ITERATION 37 THETA 0.255843 0.255843
DHATJ 32.0919411 SBARJ+1 32.0919411 LAMBDA = 0.000003
```

Program ELATTAR

```
C
C           PROGRAM ELATTAR USES FLETCHER'S ALGORITHM (1970) - HARWELL
C           SUBROUTINE VA09AD (ANALYTICAL DERIVATIVES)
C           TO SOLVE THE GENERAL NONLINEAR L1-NORM
```

```
C           PROBLEM AS SUGGESTED BY EL-ATTAR ET AL. (1979).
C
            IMPLICIT REAL*8 (A-H,O-Z)
            DIMENSION X(100),G(100),H(210),W(100),EPS(100),ALP(100),THETA(30,5)
            COMMON/DATA/T(30),Y(30),FF(30),DELTA,M
C
C           OPTIMIZATION PROBLEM DATA
C           N = NUMBER OF PARAMETERS
C           M = NUMBER OF OF PROBLEM FUNCTIONS (OR OBSERVATIONS)
C           X(.)  = PARAMETER VECTOR
C           THETA(K,L) CONTAINS THE INPUT DATA FOR EXTRAPOLATION
C           DFN, MAXFN, EPS(1) ..., EPS(N), IPRINT, MODE ARE CONTROL
C           PARAMETERS OF SUBROUTINE VA09AD.
C
C           THIS EXAMPLE IS DUE TO :
C
C           MADSEN, K. (1975).  AN ALGORITHM FOR MINIMAX SOLUTION TO OVER-
C           DETERMINED SYSTEMS OF NON-LINEAR EQUATIONS. J. INST. MATH.
C           APPLIC. 16, PP 321-328.
C
            M=21
            N=5
            DO 93 I=1,M
            TI1 = I-1
            T(I)=TI1*0.1D0 -1D0
 93         Y(I)=DEXP(T(I))
            X(1)=0.5D0
            X(2)=0.5D0
            X(3)= 0D0
            X(4)= 0D0
            X(5)= 0D0
            DO 20 I=1,N
            EPS(I)=1.D-10
 20         ALP(I)=X(I)
            DFN=0.
            MAXFN=600
            IPRINT=1
            MODE=1
            CALL FUNCT(N,X,F,G)
            FMAX=-1D10
            DO 5 I=1,M
            DEL =DMAX1(FMAX,DABS(FF(I)))
 5          FMAX =DEL
            DELTA = DEL/1D2
            LTHET =1
 30         CONTINUE
            WRITE(6,999) DELTA
 999        FORMAT(' DELTA ' = F10.6)
            WRITE(6,1000)
 1000       FORMAT(1H0,' ENTRY TO VA09AD'/)
            IPRINT=MAXFN + 1
            CALL VA09AD(N,X,F,G,H,W,DFN,EPS,MODE,MAXFN,IPRINT,IEXIT,JX)
```

```
      DO 777 I = 1,N
777   THETA(LTHET,I) = X(I)
      LTHET= LTHET + 1
      WRITE(6,1002)F
1002  FORMAT(1HO,'P(THETA,DELTA) =',F12.8)
      WRITE(6,1003) (X(I),I=1,N)
1003  FORMAT(1HO,'THETA =',5F12.8)
      DO 25 I=1,N
      IF(DABS((X(I)-ALP(I))/ALP(I)).GT.1D-10) GO TO 26
25    CONTINUE
      GO TO 500
26    DO 27 I=1,N
27    ALP(I)=X(I)
      DELTA = DELTA/1D1
      GO TO 30
500   WRITE(6,888)((THETA(K,L),L=1,N),K=1,30)
888   FORMAT(1X,5F14.8)
      STOP
      END
      SUBROUTINE FUNCT(N,X,F,G)
      IMPLICIT REAL*8 (A-H,O-Z)
      DIMENSION A(21,5),X(1),G(1)
      COMMON/DATA/T(30),Y(30),FF(30),DELTA,M
      F=ODO
      DO 2 I =1,N
2     G(I)=0.
      DO 10 I =1,M
      FA=X(1)+X(2)*T(I)
      FB=1DO+X(3)*T(I)+X(4)*T(I)**2+X(5)*T(I)**3
      FF(I)=Y(I)-FA/FB
      A(I,1)=1DO/FB
      A(I,2)=T(I)/FB
      A(I,3)=-FA*T(I)/FB**2
      A(I,4)=-FA*T(I)**2/FB**2
      A(I,5)=-FA*T(I)**3/FB**2
      TERM =DSQRT(FF(I)**2 + DELTA)
      F = F + TERM
      DO 3 J =1,N
      G(J) = G(J) - FF(I)/TERM*A(I,J)
3     CONTINUE
10    CONTINUE
      RETURN
      END
```

```
DELTA = 0.017183
ENTRY TO VA09AD
P(THETA,DELTA) = 2.75274888
THETA = 0.99989763 0.25461104 -0.74552382 0.24418741 -0.03717220
DELTA = 0.001718
```

```
ENTRY TO VA09AD

P(THETA,DELTA) = 0.87049745

THETA = 0.99989763 0.25461104 -0.74552382 0.24418741 -0.03717220
```

Program TISHZANG

```
C
C          PROGRAM TISHZANG USES FLETCHER'S ALGORITHM (1970) - HARWELL
C          SUBROUTINE VA09AD (ANALYTICAL DERIVATIVES)
C          TO SOLVE THE GENERAL NONLINEAR L1-NORM
C          PROBLEM AS SUGGESTED BY TISHLER AND ZANG (1982).
C
           IMPLICIT REAL*8(A-H,O-Z)
           DIMENSION X(100),G(100),H(210),W(100),EPS(100),ALP(100),THETA(30,2)
           COMMON/DATA/FF(6),DELTA,M
C
C          OPTIMIZATION PROBLEM DATA
C          N = NUMBER OF PARAMETERS
C          M = NUMBER OF OF PROBLEM FUNCTIONS (OR OBSERVATIONS)
C          X(.)  = PARAMETER VECTOR
C          THETA(K,L) CONTAINS THE INPUT DATA FOR EXTRAPOLATION
C          DFN, MAXFN, EPS(1) ..., EPS(N), IPRINT, MODE ARE CONTROL
C          PARAMETERS OF SUBROUTINE VA09AD.
C
C          THIS EXAMPLE IS DUE TO :
C
C          MADSEN, K. (1975).  AN ALGORITHM FOR MINIMAX SOLUTION TO OVER-
C          DETERMINED SYSTEMS OF NON-LINEAR EQUATIONS. J. INST. MATH.
C          APPLIC. 16, PP 321-328.
C
           M=3
           N=2
           X(1)=3D0
           X(2)=1D0
           DO 20 I=1,N
20         ALP(I)=X(I)
           DFN=0.
           MAXFN=100
           EPS(1)=1.D-10
           EPS(2)=1.D-10
           MODE=1
           CALL FUNCT(N,X,F,G)
           FMAX=-1D10
           DO 6 I=1,M
           DEL =DMAX1(FMAX,DABS(FF(I)))
6          FMAX = DEL
```

```
      DELTA = DEL/1D1
      LTHET =1
30    CONTINUE
      WRITE(6,999) DELTA
999   FORMAT(1X, ' DELTA =',F12.8)
      IPRINT=MAXFN + 1
      WRITE(6,998)
998   FORMAT(' ENTRY TO VA09AD')
      CALL VA09AD(N,X,F,G,H,W,DFN,EPS,MODE,MAXFN,IPRINT,IEXIT,JX)
      DO 777 I = 1,N
777   THETA(LTHET,I) = X(I)
      LTHET= LTHET + 1
      WRITE(6,1002)F
1002  FORMAT(1X,'H(THETA,BETA) =',F12.8)
      WRITE(6,1003) (X(I),I=1,N)
1003  FORMAT(1HO,'THETA =',5F20.8)
      WRITE(6,1004)(G(I),I=1,N)
1004  FORMAT(1HO,'G =',5F20.8)
      DO 25 I=1,N
      IF(DABS((ALP(I)-X(I))/ALP(I)).GT.1D-10) GO TO 26
25    CONTINUE
      GO TO 500
26    DO 27 I=1,N
27    ALP(I)=X(I)
      FMIN=1D10
      DELTA = DELTA/1D1
      GO TO 30
500   WRITE(6,888)((THETA(K,L),L=1,N),K=1,30)
888   FORMAT(1X,2F15.8)
      STOP
      END
      SUBROUTINE FUNCT(N,X,F,G)
      IMPLICIT REAL*8 (A-H,O-Z)
      DIMENSION X(1),G(1),HF(3),A(3,2)
      COMMON/DATA/FF(6),DELTA,M
      IF(X(1).LE.1D-20)X(1)=0D0
      FF(1) = X(1)**2 + X(2)**2 +X(1)*X(2)
      FF(2) = DSIN(X(1))
      FF(3) = DCOS(X(2))
      A(1,1) = 2D0*X(1) +X(2)
      A(1,2) = 2D0*X(2) +X(1)
      A(2,1) = DCOS(X(1))
      A(2,2) = 0D0
      A(3,1) = 0D0
      A(3,2) = - DSIN(X(2))
      DO 2 I=1,N
2     G(I) =0D0
      F=0D0
      DO 10 I =1,M
      IF(FF(I).LT.-DELTA) GO TO 20
      IF(FF(I).GT.DELTA) GO TO 30
      IF(FF(I).GE.-DELTA .AND.FF(I).LE.DELTA) GO TO 40
```

```
40          HF(I)=(FF(I)**2+DELTA**2)/(2DO*DELTA)
            G(1) = G(1) +FF(I)*A(I,1)/DELTA
            G(2) = G(2) +FF(I)*A(I,2)/DELTA
            GO TO 10
20          HF(I)=-FF(I)
            G(1) = G(1) - A(I,1)
            G(2) = G(2) - A(I,2)
            GO TO 10
30          HF(I)= FF(I)
            G(1) = G(1) + A(I,1)
            G(2) = G(2) + A(I,2)
10          F = F + HF(I)
            RETURN
            END
```

```
DELTA = 1.30000000
ENTRY TO VA09AD
H(THETA,BETA) = 2.24738425
THETA = 0.15543723 -0.69456377
DELTA = 0.13000000
ENTRY TO VA09AD
H(THETA,BETA) = 1.11295230
THETA = 0.01546825 -0.26463607
DELTA = 0.01300000
ENTRY TO VA09AD
H(THETA,BETA) = 1.01136628
THETA = 0.00052063 -0.08096743
DELTA = 0.00130000
ENTRY TO VA09AD
H(THETA,BETA) = 1.00113741
THETA = 0.00001656 -0.02550613
DELTA = 0.00013000
ENTRY TO VA09AD
H(THETA,BETA) = 1.00011375
THETA = 0.00000052 -0.00806261
DELTA = 0.00001300
ENTRY TO VA09AD
H(THETA,BETA) = 1.00001137
THETA = 0.00000002 -0.00254952
DELTA = 0.00000130
ENTRY TO VA09AD
H(THETA,BETA) = 1.00000114
```

```
THETA = 0.00000000 -0.00080623
DELTA = 0.00000013
ENTRY TO VA09AD
H(THETA,BETA) = 1.00000011
THETA = 0.00000000 -0.00025495
DELTA = 0.00000001
ENTRY TO VA09AD
H(THETA,BETA) = 1.00000004
THETA = 0.00000000 -0.00025495
```

Chapter 2: Additional notes

§2.1

There are efficient numerical methods for solving the general nonlinearly constrained NLP problem. See for example: Buys (1972), Bertsekas (1976), Powell (1978), Gill et al. (1981), McCormick (1983) and Luenberger (1984).

§2.2

1 For a proof of the Karush-Kuhn-Tucker (KKT) theorem and a discussion of various constraint qualifications see McCormick (1983:214) and Sposito (1975).

2 El-Attar et al. (1979) and Murray and Overton (1981) and Busovăca (1985) state the first-order optimality conditions in terms of alternative constraint qualifications, for example the Kuhn-Tucker constraint qualification.

3 The set of conditions in Theorem 2.3 was derived by Charalambous (1979) and El-Attar et al. (1979). Their equivalence to those of Theorem 2.2 is shown in these two papers. These conditions are in terms of directional derivatives. (see Appendix 2B).

4 Charalambous (1979) derived the second-order sufficiency conditions for optimality of $\boldsymbol{\theta}^*$ in Theorem 2.4. These conditions are a special case of sufficiency conditions for the general NLP problem, see for example Avriel (1976) or McCormick (1967). Murray and Overton (1981) and Overton (1982) provide similar conditions.

5 Optimality conditions for more general nondifferentiable problems may be found in Hiriart-Urruty (1978), Fletcher and Watson (1980), Womersley (1985) and Wright (1987). Ben-Tal and Zowe (1982) also point out that the derivation by Charalambous is incorrect.

§2.3

1 Much work has also been done in the more general field of nonsmooth optimization and the interested reader is referred to Fletcher (1981a:Chapter 14), Lemaréchal (1975,1982), Lemaréchal et al. (1981), Clark and Osborne (1983), Gill et al. (1981:Chapter 4), Osborne (1981,1985), Overton (1982), Watson (1978,1984,1986), Fletcher (1981b), Powell (1983) and Gurwitz (1986).

2 A broad class of problems, composite nonsmooth optimization, which subsumes L_1 and L_∞, is considered by Womersley and Fletcher (1982,1986),

3 An overview of methods for solving the nonlinear L_1- and L_∞-norm problem as well as regression problems with errors in all variables is given by Watson

(1986). The classification of methods is in terms of single phase or two-phase methods.

§2.3.1

A discussion of line search procedures may be found in Avriel (1976).

§2.3.2.1

A discussion of the Gauss-Newton method for nonlinear least squares may be found in Draper and Smith (1981:459-465).

§2.3.2.2

1 The Levenberg-Marquardt rule is analogous to ridge regression. This method has proved useful in nonlinear least squares when $J(\boldsymbol{\theta})'J(\boldsymbol{\theta})$ is ill-conditioned. Adding a term μI to $J(\boldsymbol{\theta})'J(\boldsymbol{\theta})$ results in a matrix which is positive definite. When $\mu = 0$ the usual Gauss-Newton method results.

2 No examples of other polyhedral norms are, however, given.

3 Shrager and Hill (1980) also considered a Levenberg-Marquardt algorithm for solving the nonlinear L_1- and L_∞-norm estimation problems. Since linear programming is involved in calculating successive estimates of the parameters, they suggested that a good starting estimate would reduce the overall computational burden. See also McCormick and Sposito (1976), Hand and Sposito (1980) and Sklar and Armstrong (1982) for similar suggestions. A second difficulty, the nonuniqueness of the parameters, is also overcome by using the linear programming procedure to select a single solution. The algorithm by Barrodale and Roberts (1974) has this capability.

4 Jittorntrum and Osborne (1980) show that if $\boldsymbol{\theta}^*$ is an isolated local minimum then ultimately the steplength $\gamma_j = 1$ and the convergence rate will be quadratic.

5 Algorithm 563 of Bartels and Conn (1980) can also be used to solve the constrained linear L_1-norm problems.

§2.3.3.1

1 There are two subspaces in question:

   $t-$ dimensional subspace defined by the first t columns of $\hat{V}$. Thus Y forms a basis for the range (column) space of $\hat{V}$.

   $k-t-$ dimensional subspace (orthogonal complement). Its vectors are orthogonal to the rows of $\hat{V}$ or columns of $\hat{V}'$. Thus Z forms a basis of the null space of $\hat{V}'$.

2 Powell and Yuan (1984) derive general conditions for superlinear convergence of SQP methods for L_1- (and L_∞-) norm problems.

§2.3.4.1

If $\boldsymbol{h}(\boldsymbol{\theta})$ has fewer than k zeros ($k < n$) then information about all the nonzero functions need also be taken into account since these terms also contribute to $S_1(\boldsymbol{\theta})$.

§2.3.5.1

1 The Anderson-Osborne-Watson algorithm is a Gauss-Newton type method applied to nonlinear L_1-norm problems. Whilst El-Attar et al. (1979) call it a steepest descent method, it in fact uses $\nabla_{\boldsymbol{\theta}} f_i(\boldsymbol{x}_i, \boldsymbol{\theta})$ instead of $\nabla S_1(\boldsymbol{\theta})$ as a direction and is therefore strictly speaking a Gauss-Newton method.

2 A discussion of penalty function methods in nonlinear programming may be found in Fiacco and McCormick (1968).

3 The HARWELL FORTRAN subroutine VA09AD is an implementation of the BFGS method.

§2.3.5.2

The algorithm is based on a smoothing concept due to Bertsekas (1975). See also Zang (1980).

Bibliography Chapter 2

Anderson, D.H. and Osborne, M.R. (1977a). Discrete, nonlinear approximation problems in polyhedral norms. *Numer. Math.* **28**, pp 143–156.

Anderson, D.H. and Osborne, M.R. (1977b). Discrete, nonlinear approximation problems in polyhedral norms. A Levenberg-like algorithm. *Numer. Math.* **28**, pp 157–170.

Ansley, C.F. (1985). Quick proofs of some regression theorems via the QR algorithm. *The American Statistician* **39**, pp 55–59.

Armijo, L. (1966). Minimization of functions having Lipschitz continuous partial derivatives. *Pacific J. Math.* **16**, pp 1–3.

Armstrong, R.D., Frome, E.L. and Kung, D.S. (1979). A revised simplex algorithm for the absolute deviation curve fitting problem. *Commun. Statist.–Simula. Computa.* **8**, pp 175–190.

Avriel, M. (1976). Nonlinear programming, analysis and methods. *Prentice-Hall, Englewood Cliffs, New Jersey.*

Barrodale, I. and Roberts, F.D.K. (1973). An improved algorithm for discrete ℓ_1 linear approximation. *SIAM J. Numer. Anal.* **10**, pp 839–848.

Barrodale, I. and Roberts, F.D.K. (1974). Algorithm 478: Solution of an overdetermined system of equations in the ℓ_1-norm. *Commun. Assoc. Comput. Mach.* **17**, pp 319–320.

Barrodale, I. and Roberts, F.D.K. (1978). An efficient algorithm for discrete ℓ_1 linear approximation with linear constraints. *SIAM J. Numer. Anal.* **15**, pp 603–611.

Bartels, R.H. and Conn, A.R. (1980): Linearly constrained discrete ℓ_1 problems. *ACM Trans. Math. Software* **6**, pp 594–608.

Bartels, R.H. and Conn, A.R. (1982). An approach to nonlinear ℓ_1 data fitting. In: J.P. Hennart, ed., *Numerical analysis: Lecture notes in mathematics, Springer Verlag, New York,* pp 48–58.

Ben–Tal, A and Zowe, J. (1982). Necessary and sufficient optimality conditions for a class of nonsmooth minimization problems. *Math. Programming* **24**, pp 70–88.

Bertsekas, D.P. (1975). Nondifferentiable optimization via approximation. *Math. Programming Study* **3**, pp 1–25.

Bertsekas, D.P. (1976). Multiplier methods: A survey. *Automatica* **12**, pp 133–145.

Broyden, C.G. (1970). The convergence of a class of double rank minimization algorithms: Parts I and II. *J. Inst. Math. Applic.* **6**, pp 76–90, 222-231.

Businger, P. and Golub, G.H. (1965). Linear least squares solutions by Householder transformations. *Numer. Math.* **7**, pp 269–276.

Buys, J.D. (1972). Dual algorithms for constrained optimization problems. *PhD Thesis. Leiden University, Leiden, Netherlands.*

Busovača, S. (1985). Handling degeneracy in a nonlinear ℓ_1 algorithm. *PhD Thesis University of Waterloo, Waterloo, Ontario.*

Cauchy, A-L. (1847). Méthode générale pour la résolution des systèmes d'équations simultanées, *C.R. Acad. Sci. A. Ser.*, **25**, pp 536–538.

Charalambous, C. (1979). On conditions for optimality of the nonlinear L_1 problem. *Math. Programming* **17**, pp 123–135.

Clark, D.I. and Osborne, M.R. (1983). A descent algorithm for minimizing polyhedral convex functions. *SIAM J. Sci. Stat. Comput.* **4**, pp 757–786.

Cromme, L. (1978). Strong uniqueness: A far reaching criterion for the convergence analysis of iterative procedures. *Numer. Math.* **29**, pp 179–194.

Dennis, J.E. and Moré, J.J. (1977). Quasi-Newton methods, motivation and theory. *SIAM Rev.* **19**, pp 46–89.

Dennis, J.E. and Schnabel, R.B. (1983). Numerical methods for unconstrained optimization and nonlinear equations. *Prentice-Hall, Englewood Cliff, New Jersey.*

Draper, N. and Smith, H. (1981). Applied regression analysis, *2nd edition, John Wiley, New York.*

El-Attar, R.A. Vidyasagar, M. and Dutta, S.R.K. (1979). An algorithm for L_1-norm minimization with application to nonlinear L_1-approximation. *SIAM J. Numer. Anal.* **16**, pp 70–86.

Fiacco, A.V. and McCormick, G.P. (1968). Nonlinear programming: Sequential unconstrained minimization techniques, *John Wiley, New York.*

Fletcher, R. (1970). A new approach to variable metric algorithms. *Comput. J.* **13**, pp 317–322.

Fletcher, R. (1981a). Practical methods of optimization, Volume 2: Constrained optimization. *John Wiley, Chichester.*

Fletcher, R. (1981b). Numerical experiments with an exact L_1 penalty function. In: O.L. Mangasarian, R.R. Meyer, and S. Robinson, eds. *Nonlinear Programming, Academic Press, New York.*

Fletcher, R. and Watson, G.A. (1980). First and second order conditions for a class of nondifferentiable optimization problems. *Math. Programming* **18**, pp 291–307.

Gill, P.E. and Murray, W. (1974). Newton-type methods for unconstrained and linearly constrained optimization. *Math. Programming* **7**, pp 311–350.

Gill, P.E. and Murray, W. (1978). Algorithms for the solution of the nonlinear least-squares problem. *SIAM J. Numer. Anal.* **15**, pp 977–992.

Gill, P.E. , Murray, W. and Wright, M. (1981). Practical optimization. *Academic Press, London.*

Goldfarb, D. (1970). A family of variable metric methods derived by variational means. *Math. Comput.* **24**, pp 23–26.

Gurwitz, C.B. (1986). Sequential quadratic programming methods based on approximating a projected Hessian matrix. *Technical Report #219, Computer Science Department, Courant Institute of Mathematical Sciences, New York University, New York.*

Hald, J. and Madsen, K. (1985). Combined LP and Quasi-Newton methods for nonlinear l_1 optimization. *SIAM J. Numer. Anal.* **22**, pp 68–80.

Hand, M. and Sposito, V.A. (1980). Using the least squares estimator in Chebychev estimation. *Commun. Statist.-Simula. Computa.*, **9**, pp 43–49.

Henrici, P. (1964). Elements of numerical analysis. *John Wiley, New York.*

Hiriart-Urruty, J.B. (1978). On optimality conditions in nondifferentiable programming. *Math. Programming* **14**, pp 73–86.

Householder, A.S. (1958). Unitary triangularization of a nonsymmetric matrix. *JACM* **5**, pp 339–342.

Jennrich, R.I. and Sampson, P.F. (1968). Application of stepwise regression to non-linear estimation. *Technometrics* **10**, pp 63–71.

Jittorntrum, K. and Osborne, M.R. (1980). Strong uniqueness and second order convergence in non-linear discrete approximation. *Numer. Math.* **34**, pp 439–455.

Karush, W. (1939). Minima of functions of several variables with inequalities as side conditions. *MS Thesis, University of Chicago, Chicago, Illinois.*

Kuhn, H.W. and Tucker, A.W. (1951). Nonlinear programming. In: J. Neyman, ed. *Proceedings of the second Berkeley symposium on mathematical statistics and probability, UCLA Press, Berkeley, California*, pp 481–492.

Lawson, C.L. and Hanson, R.J. (1974). Solving least squares problems. *Prentice-Hall, New Jersey.*

Lemaréchal, C. (1975). An extension of Davidon methods to non-differentiable problems. *Math. Programming Study* **3**, pp 95–109.

Lemaréchal, C. (1982). Non-differentiable optimization. In: M.J.D. Powell, ed. *Nonlinear Optimization, Academic Press, London.*

Lemaréchal, C., Strodiot, J.J. and Bihain, A. (1981). On a bundle algorithm for non-smooth optimization. In: O.L. Mangasarian, R.R. Meyer and S.M. Robinson, eds. *Nonlinear Programming 4, Academic Press, New York.*

Levenberg, K. (1944). A method for the solution of certain problems in least squares. *Quart. Appl. Math.* **2**, pp 164–168.

Luenberger, D.G. (1984). Linear and nonlinear programming. *2nd edition, Addison-Wesley, Reading, Massachusetts.*

Madsen, K. (1975). An algorithm for minimax solution of overdetermined systems of non–linear equations. *J. Inst. Math. Applic.* **16**, pp 321–328.

Mangasarian, O.L. (1969). Nonlinear programming. *McGraw–Hill, New York.*

Marquardt, D. (1963). An algorithm for least-squares estimation of nonlinear parameters. *SIAM J. Appl. Math.* **11**, pp 431–441.

McCormick, G.F. and Sposito, V.A. (1976). Using the L_2 estimator in L_1 estimation. *SIAM J. Numer. Anal.* **13**, pp 337–343.

McCormick, G.P. (1967). Second order conditions for constrained minima. *SIAM J. Appl. Math.* **15**, pp 37–47.

McCormick, G.P. (1983). Nonlinear programming: Theory, algorithms and applications. *John Wiley, New York.*

McLean, R.A. and Watson, G.A. (1980). Numerical methods for nonlinear discrete L_1 approximation problems. In: L. Collatz, G. Meinardus and H. Werner, eds. *Numerical methods of approximation theory, Birkhäuser Verlag, Basel.*

Murray, W. and Overton, M (1979). Steplength algorithms for minimizing a class of nondifferentiable functions. *Computing* **23**, pp 309–331.

Murray, W. and Overton, M (1981). A projected Lagrangian algorithm for nonlinear ℓ_1 optimization. *SIAM J. Sci. Stat. Comput.* **2**, pp 207–224.

Osborne, M.R. (1981). Algorithms for nonlinear discrete approximation. In: C.T.H. Baker and C. Phillips, eds. *The numerical solution of nonlinear problems, Clarenden Press*, pp 270–286.

Osborne, M.R. (1985). Finite algorithms in optimization and data analysis. *John Wiley, Chichester.*

Osborne, M.R. and Watson, G.A. (1971). On an algorithm for discrete nonlinear L_1 approximation. *Comput. J.* **14**, pp 184–188.

Overton, M. (1982). Algorithms for nonlinear l_1 and l_∞ fitting. In: M.J.D. Powell, ed. *Nonlinear optimization, Academic Press, London*, pp 91–101.

Perold, A.F. (1980). A degeneracy exploiting LU factorization for the simplex method. *Math. Programming* **19**, pp 239–254.

Powell, M.J.D. (1971). On the convergence of the variable metric algorithm. *J. Inst. Math. Applic.* **7**, pp 21–36.

Powell, M.J. (1978). Algorithms for nonlinear constraints that use Lagrangian functions. *Math. Programming* **14** pp 224–248.

Powell, M.J.D. (1983). General algorithms for discrete nonlinear approximation calculations. In: C.K. Chui, L.L. Schumaker and J.D. Ward, eds. *Approximation theory IV, Academic Press, New York.*

Powell, M.J.D. and Yuan, Y. (1984). Conditions for superlinear convergence in ℓ_1 and ℓ_∞ solutions of overdetermined non-linear equations. *IMA J. Numer. Anal.* **4**, pp 241–251.

Shanno, D.F. (1970). Conditioning of quasi-Newton methods for function minimization. *Math. Comput.* **24** pp 647–656.

Shrager, R.I. and Hill, E. (1980). Nonlinear curve–fitting in the L_1 and L_∞ norms. *Math. Comput.* **34**, pp 529–541.

Sklar, M.G. and Armstrong, R.D. (1982). Least absolute value and Chebychev estimation utilizing least squares results. *Math. Programming* **24**, pp 346–352.

Sposito, V.A. (1975). Linear and nonlinear programming. *Iowa State Press, Ames, Iowa.*

Tishler, A. and Zang, I. (1982). An absolute deviations curve- fitting algorithm for nonlinear models. In: S.H. Zanakis and J.S. Rustagi, eds. *Optimization in Statistics, TIMS Studies in Management Science* **19**, *North Holland.*

Watson, G.A. (1978). A class of programming problems whose objective function contains a norm. *J. Approx. Theory* **23**, pp 401–411.

Watson, G.A. (1979). Dual methods for nonlinear best approximation problems. *J. Approx. Theory* **26**, pp 142–150.

Watson, G.A. (1980). Approximation theory and numerical methods. *John Wiley, New York.*

Watson, G.A. (1984). Discrete ℓ_1 approximation by rational functions. *IMA J. Numer. Anal.* **4**, pp 275–288.

Watson G.A. (1986). Methods for best approximation and regression problems. *Numerical analysis Report NA/96, University of Dundee, Scotland.*

Wolfe, P. (1963). A technique for resolving degeneracy in linear programming. *SIAM J.* **11**, pp 205–211.

Womersley, R.S. (1985). Local properties of algorithms for minimizing nonsmooth composite functions. *Math. Programmming* **32**, pp 69–89.

Womersley, R.S. and Fletcher, R. (1982). An algorithm for composite nonsmooth optimization problems. *Mathematics Department Report NA/60, University of Dundee, Scotland.*

Womersley, R.S. and Fletcher, R. (1986). An algorithm for composite nonsmooth optimization problems. *J. Optim. Theory Appl.* **48** pp 493–523.

Wright, S.J. (1987). Local properties of inexact methods for minimizing nonsmooth composite functions. *Math. Programmming* **37**, pp 232–252.

Zang, I. (1980). A smoothing-out technique for min-max optimization. *Math. Programmming* **19**, pp 61–77.

Zangwill, W.I. (1969). Nonlinear Programming: A unified approach. *Prentice–Hall, New Jersey*.

3

The Nonlinear L_∞-Norm Estimation Problem

In this Chapter we consider the other end of the spectrum where $p \to \infty$. The nondifferentiable L_∞-norm estimation problem more frequently referred to as Chebychev estimation is of importance when one's objective is to minimize the maximum absolute error. It is also a special case of the minimax problem which is encountered in Game Theory. The minimax problem, although more general, may not always have a solution whilst the L_∞-norm problem will always have a solution.

The aim of this Chapter is to define the L_∞-norm estimation problem, to give the conditions under which a solution will be optimal and to discuss algorithms which can be used to solve this problem.

3.1. The nonlinear L_∞-norm estimation problem

The L_∞-norm estimation problem for the nonlinear model

$$y_i = f_i(\boldsymbol{x_i}, \boldsymbol{\theta}) + e_i, \qquad i = 1, \ldots, n \quad (n \geq k) \tag{3.1}$$

is defined as:

Find parameters $\boldsymbol{\theta}$ which minimize

$$S_\infty(\boldsymbol{\theta}) = \max_{1 \leq i \leq n} |y_i - f_i(\boldsymbol{x_i}, \boldsymbol{\theta})| \tag{3.2}$$

Note that $S_\infty(\boldsymbol{\theta})$ is a nondifferentiable function. Problem (3.2) is equivalent to the nonlinear programming (NLP) problem:

$$\begin{aligned} &\min d_\infty \\ \text{subject to} \quad & y_i - d_\infty \leq f_i(\boldsymbol{x_i}, \boldsymbol{\theta}) \leq y_i + d_\infty \qquad i = 1, \ldots, n \\ & d_\infty \geq 0 \end{aligned} \tag{3.3}$$

$d_\infty \geq 0$ (actually $d_\infty > 0$) ensures that the feasible region is connected. Observe that the constraints of (3.3) can be stated as $|y_i - f_i(\boldsymbol{x_i}, \boldsymbol{\theta})| \leq d_\infty$ for all i and thus it follows that (3.3) is equivalent to (3.2). The NLP formulation (3.3) has $2n$ nonlinear constraints and $k+1$ variables.

3.2. L_∞-norm optimality conditions

Suppose $(d_\infty^*, \boldsymbol{\theta}^*)$ is a local minimum point of problem (3.3). Define

- Active function set

$$A = \{i \mid |y_i - f_i(\boldsymbol{x_i}, \boldsymbol{\theta})| = S_\infty(\boldsymbol{\theta})\}$$

- Inactive function set

$$I = \{i \mid |y_i - f_i(\boldsymbol{x_i}, \boldsymbol{\theta})| \neq S_\infty(\boldsymbol{\theta})\} = \{1, \ldots, n\} - A$$

- Active constraint sets

$$K_1 = \{i \mid y_i - f_i(\boldsymbol{x_i}, \boldsymbol{\theta}) = d_\infty\}$$

$$K_2 = \{i \mid y_i - f_i(\boldsymbol{x_i}, \boldsymbol{\theta}) = -d_\infty\}$$

$$K_1 \cup K_2 = A$$

- Feasible region

$$F = \{\, (d_\infty, \boldsymbol{\theta}) \in \Re \times \Re^k \mid y_i - f_i(\boldsymbol{x_i}, \boldsymbol{\theta}) \leq d_\infty, \ -y_i + f_i(\boldsymbol{x_i}, \boldsymbol{\theta}) \leq d_\infty \}$$

- The Lagrangian function for problem (3.3)

$$L(d_\infty, \boldsymbol{\theta}, \boldsymbol{\lambda}, \boldsymbol{\mu}) = d_\infty + \sum_{i=1}^{n} \lambda_i (y_i - f_i(\boldsymbol{x_i}, \boldsymbol{\theta}) - d_\infty) - \sum_{i=1}^{n} \mu_i (y_i - f_i(\boldsymbol{x_i}, \boldsymbol{\theta}) + d_\infty)$$

- Assume that $(d_\infty^*, \boldsymbol{\theta}^*)$ is a *regular* minimum point of (3.3). Then the active constraint gradients evaluated at $(d_\infty^*, \boldsymbol{\theta}^*)$

$$\begin{pmatrix} -1 \\ -\nabla f_i(\boldsymbol{x_i}, \boldsymbol{\theta}^*) \end{pmatrix} \quad i \in K_1 \qquad \begin{pmatrix} 1 \\ -\nabla f_i(\boldsymbol{x_i}, \boldsymbol{\theta}^*) \end{pmatrix} \quad i \in K_2$$

  are linearly independent.

The Karush-Kuhn-Tucker (KKT) conditions for problem (3.3) are stated in the following theorem:

Theorem 3.1 Suppose the functions $f_i(\boldsymbol{x}_i, \boldsymbol{\theta})$ are continuously differentiable with respect to $\boldsymbol{\theta}$ and that $(d^*_\infty, \boldsymbol{\theta}^*)$ is a regular local minimum point of problem (3.3). Then the following conditions are necessary for such a local minimum point:

(a) (Feasibility). $(d^*_\infty, \boldsymbol{\theta}^*) \in F$.

(b) (Complementary slackness). There exist multipliers $\lambda^*_i, \mu^*_i \geq 0,\ i = 1, \ldots, n$ such that:

$$\lambda^*_i (y_i - f_i(\boldsymbol{x}_i, \boldsymbol{\theta}^*) - d^*_\infty) = 0$$

$$\mu^*_i (y_i - f_i(\boldsymbol{x}_i, \boldsymbol{\theta}^*) + d^*_\infty) = 0$$

(c) (Stationarity).

$$\sum_{i=1}^{n} (\lambda^*_i + \mu^*_i) = 1$$

$$-\sum_{i=1}^{n} (\lambda^*_i - \mu^*_i) \nabla f_i(\boldsymbol{x}_i, \boldsymbol{\theta}^*) = 0$$

For notational convenience set $h_i(\boldsymbol{\theta}) = y_i - f_i(\boldsymbol{x}_i, \boldsymbol{\theta})$ then the KKT conditions can be stated as the following **Necessity Theorem**:

Theorem 3.2 Suppose $h_i(\boldsymbol{\theta})$ is continuously differentiable with respect to $\boldsymbol{\theta}$, that $(d^*_\infty, \boldsymbol{\theta}^*)$ is a regular minimum point and $d^*_\infty > 0$, then a necessary condition for $\boldsymbol{\theta}^*$ to be a local minimum point of (3.3) is the existence of *positive* multipliers α_i for all $i \in A$ such that

$$\sum_{i \in A} \alpha_i = 1$$

$$\sum_{i \in A} \alpha_i sign[h_i(\boldsymbol{\theta}^*)] \nabla h_i(\boldsymbol{\theta}^*) = 0$$

$$\alpha_i = 0, \qquad i \in I$$

Proof The complementary slackness and stationarity conditions (b) and (c) can be written as:

$$\lambda^*_i (h_i(\boldsymbol{\theta}^*) - d^*_\infty) = 0$$

$$\mu^*_i (h_i(\boldsymbol{\theta}^*) + d^*_\infty) = 0$$

$$\sum_{i=1}^{n} (\lambda^*_i + \mu^*_i) = 1$$

$$\sum_{i=1}^{n}(\lambda_i^* - \mu_i^*)\nabla h_i(\boldsymbol{\theta}^*) = 0$$

If $\lambda_i^* > 0$ then $h_i(\boldsymbol{\theta}^*) = d_\infty^*$ hence $\mu_i^* = 0$ and vice versa. Thus either λ_i^* or μ_i^* can be unequal to zero but not simultaneously. However, if $\lambda_i^* > 0$ then $sign[h_i(\boldsymbol{\theta}^*)] = 1$; if $\mu_i^* > 0$ then $sign[h_i(\boldsymbol{\theta}^*)] = -1$. Set

$$\alpha_i = \begin{cases} \lambda_i & \text{if } i \in K_1 \\ \mu_i & \text{if } i \in K_2 \end{cases}$$

We see that if $i \in I$ then $\mu_i^* = \lambda_i^* = 0$ or $\alpha_i = 0$. We can therefore write

$$\sum_{i=1}^{n}(\lambda_i^* + \mu_i^*) = \sum_{i \in K_1} \lambda_i^* + \sum_{i \in K_2} \mu_i^* = \sum_{i \in A} \alpha_i^* = 1$$

Similarly

$$\begin{aligned}\sum_{i=1}^{n}(\lambda_i^* - \mu_i^*)\nabla h_i(\boldsymbol{\theta}^*) &= \sum_{i \in K_1} \lambda_i^* sign[h_i(\boldsymbol{\theta}^*)]\nabla h_i(\boldsymbol{\theta}^*) + \sum_{i \in K_2} \mu_i^* sign[h_i(\boldsymbol{\theta}^*)]\nabla h_i(\boldsymbol{\theta}^*) \\ &= \sum_{i \in A} \alpha_i sign[h_i(\boldsymbol{\theta}^*)]\nabla h_i(\boldsymbol{\theta}^*) \\ &= 0\end{aligned}$$

Thus these conditions become:

$$\alpha_i = 0, \quad \text{for} \quad i \in I \qquad \sum_{i \in A} \alpha_i = 1 \qquad \sum_{i \in A} \alpha_i sign[(h_i(\boldsymbol{\theta}^*)]\nabla h_i(\boldsymbol{\theta}^*) = 0$$

and the proof is complete ∎

Corollary 3.1 In linear L_∞-norm estimation a necessary and sufficient condition for $\boldsymbol{\theta}^*$ to be a global L_∞-norm solution is the existence of *nonnegative* multipliers α_i such that

$$\sum_{i=1}^{n} \alpha_i = 1$$

$$\sum_{i \in A} \alpha_i sign[y_i - \boldsymbol{x_i}\boldsymbol{\theta}]\boldsymbol{x_i} = 0$$

Second-order sufficiency conditions for (3.2) are stated as a **Sufficiency Theorem**.

Theorem 3.3 Suppose the functions $h_i(\boldsymbol{\theta}^*)$ are twice continuously differentiable. Then $\boldsymbol{\theta}^*$ is an isolated local minimum point of the L_∞-norm estimation problem if

(a) the necessary condition for a local minimum holds (see Theorem 3.2) and

(b) if for every $i \in A$ and $\boldsymbol{d} \neq 0$ satisfying $\boldsymbol{d}'\nabla h_i(\boldsymbol{\theta}^*) = 0$ it follows that

$$\boldsymbol{d}'\left[\sum_{i \in A} \alpha_i sign[h_i(\boldsymbol{\theta}^*)]\nabla^2 h_i(\boldsymbol{\theta}^*)\right]\boldsymbol{d} > 0 \tag{3.4}$$

3.3. The nonlinear minimax problem

Since L_∞-norm estimation is a special case of the minimax problem it will be appropriate to consider it separately. The nonlinear minimax problem is defined as: Find parameters $\boldsymbol{\theta}$ which minimize

$$h_{max}(\boldsymbol{\theta}) = \max_{1 \le i \le n} \{y_i - f_i(\boldsymbol{x_i}, \boldsymbol{\theta})\} \tag{3.5}$$

We observe that $h_{max}(\boldsymbol{\theta})$ is a nondifferentiable function. Similarly problem (3.5) can be reformulated as the NLP problem:

$$\min \tilde{d}_\infty$$

$$\text{subject to} \qquad y_i - f_i(\boldsymbol{x_i}, \boldsymbol{\theta}) \le \tilde{d}_\infty, \qquad i = 1, \ldots, n \tag{3.6}$$

$\tilde{d}_\infty$ is assumed nonzero to ensure that the feasible region is connected. Problem (3.6) has n nonlinear constraints and $k+1$ variables. The optimality conditions for problem (3.6) are stated in the following Corollary (see Theorem 3.2).

Corollary 3.2 If $h_i(\boldsymbol{\theta}^*)$ is continuously differentiable, and $(\tilde{d}^*_\infty, \boldsymbol{\theta}^*)$ is a regular minimum point of (3.6) and $\tilde{d}^*_\infty \neq 0$, then a necessary condition for $\boldsymbol{\theta}^*$ to be a local minimum point of the minimax problem is the existence of *nonnegative* multipliers α_i such that

$$\sum_{i \in A} \alpha_i = 1$$

$$\sum_{i \in A} \alpha_i \nabla h_i(\boldsymbol{\theta}^*) = 0$$

$$\alpha_i = 0, \quad \text{for} \quad i \in I$$

We see that $0 \in$ convex hull of the gradients of the active functions. Second-order sufficiency conditions for problem (3.6) are stated as follows:

Theorem 3.4 Suppose the functions $h_i(\boldsymbol{\theta}^*)$ are twice continuously differentiable. Then $\boldsymbol{\theta}^*$ is an isolated local minimum point of the minimax problem if:

(a) the necessary condition for a local minimum holds (see Corollary 3.2) and

(b) for every $i \in A$ and $\boldsymbol{d} \neq 0$ satisfying $\boldsymbol{d}'\nabla h_i(\boldsymbol{\theta}^*) = 0$, it follows that

$$\boldsymbol{d}'\left[\sum_{i\in A} \alpha_i \nabla^2 h_i(\boldsymbol{\theta}^*)\right]\boldsymbol{d} > 0 \tag{3.7}$$

3.4. Examples

Example 3.1 This example is reported in Fletcher and Watson (1980).

$$\min_{\boldsymbol{\theta}} \max_{1\le i\le 2} |\, h_i(\boldsymbol{\theta}) \,|$$

$$\text{where} \quad h_1(\boldsymbol{\theta}) = \theta_1 - (\theta_2)^3 + 5(\theta_2)^2 - 2\theta_2 - 13$$

$$h_2(\boldsymbol{\theta}) = \theta_1 + (\theta_2)^3 + (\theta_2)^2 - 14\theta_2 - 29$$

The contours of the objective function are given in Fig. 3.1.

Observe the long narrow valleys of $S_\infty(\boldsymbol{\theta})$. The equations are satisfied at $\boldsymbol{\theta} = (5, 4)$. A local optimal L_∞ (and L_1) solution is

$$\boldsymbol{\theta}^* = (11.4128, -0.8968)$$

$$S_\infty(\boldsymbol{\theta}^*) = 4.9490$$

$$h_1(\boldsymbol{\theta}^*) = 4.9490$$

$$h_2(\boldsymbol{\theta}^*) = -4.9490$$

$$\nabla h_1(\boldsymbol{\theta}^*) = \begin{pmatrix} 1 \\ -13.3808 \end{pmatrix}$$

$$\nabla h_2(\boldsymbol{\theta}^*) = \begin{pmatrix} 1 \\ -13.3808 \end{pmatrix}$$

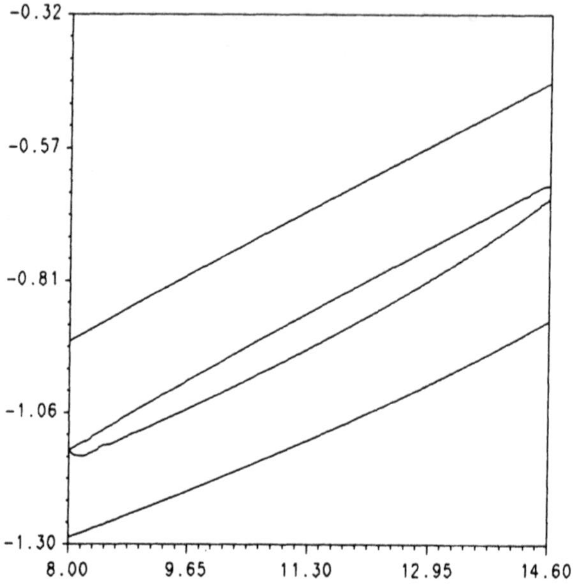

Figure 3.1 Contours of $S_\infty(\boldsymbol{\theta}) = 5.4$ and 8.

Since both constraints are active we need to solve the equations

$$\alpha_1 + \alpha_2 = 1$$
$$\alpha_1 - \alpha_2 = 0$$
$$-13.3808\alpha_1 + 13.3808\alpha_2 = 0$$

Thus $\alpha_1 = \alpha_2 = \frac{1}{2}$ and hence the necessary optimality conditions are satisfied. The 2nd-order sufficiency conditions are also satisfied, since the projected Hessian of the Lagrangian function in (3.4) is

$$\boldsymbol{d}' \begin{pmatrix} 0 & 0 \\ 0 & 9.3808 \end{pmatrix} \boldsymbol{d} = 9.3808b^2 > 0 \qquad \text{for} \quad \boldsymbol{d} = [a, b] \neq 0$$

where $\nabla h_1(\boldsymbol{\theta}^*)'\boldsymbol{d} = \nabla h_2(\boldsymbol{\theta}^*)'\boldsymbol{d} = 0$ yields $\boldsymbol{d} = (13.3808b, b)'$ for $b \neq 0$.

Note that $\boldsymbol{\theta}^* = (11.4128, -0.8968)$ is not a minimax solution.

Example 3.2 This minimax example is due to Charalambous and Bandler (1976).

$$\min_{\boldsymbol{\theta}} \max_{1 \leq i \leq 3} h_i(\boldsymbol{\theta})$$

where

$$h_1(\boldsymbol{\theta}) = (\theta_1)^4 + (\theta_2)^2$$

$$h_2(\boldsymbol{\theta}) = (2 - \theta_1)^2 + (2 - \theta_2)^2$$

$$h_3(\boldsymbol{\theta}) = 2\exp(-\theta_1 + \theta_2)$$

The contours of the objective function are given in Fig. 3.2.

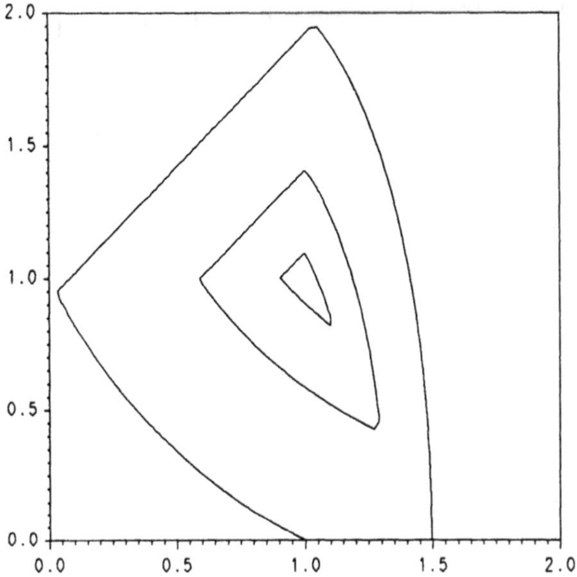

Figure 3.2 Contours of $h_{max}(\boldsymbol{\theta}) = 2.2,\ 3,\ 5.$

$\boldsymbol{\theta}^* = (1, 1)$ and $A = \{1, 2, 3\}$, $\quad h_1(\boldsymbol{\theta}^*) = h_2(\boldsymbol{\theta}^*) = h_3(\boldsymbol{\theta}^*) = 2.$

$$\nabla h_1(\boldsymbol{\theta}^*) = \begin{pmatrix} 4 \\ 2 \end{pmatrix}$$

$$\nabla h_2(\boldsymbol{\theta}^*) = \begin{pmatrix} -2 \\ -2 \end{pmatrix}$$

$$\nabla h_3(\boldsymbol{\theta}^*) = \begin{pmatrix} -2 \\ 2 \end{pmatrix}$$

Thus for

$$\alpha_1 + \alpha_2 + \alpha_3 = 1$$

$$4\alpha_1 - 2\alpha_2 - 2\alpha_3 = 0$$

$$2\alpha_1 - 2\alpha_2 + 2\alpha_3 = 0$$

The unique solution is $\alpha_1 = \frac{1}{3}$, $\alpha_2 = \frac{1}{2}$, $\alpha_3 = \frac{1}{6}$, and the necessary optimality conditions are satisfied. The projected Hessian in (3.7) is:

$$H = \frac{1}{3}\begin{pmatrix} 12 & 0 \\ 0 & 2 \end{pmatrix} + \frac{1}{2}\begin{pmatrix} 2 & 0 \\ 0 & 2 \end{pmatrix} + \frac{1}{6}\begin{pmatrix} 2 & 2 \\ 2 & 2 \end{pmatrix} = \begin{pmatrix} \frac{16}{3} & \frac{1}{3} \\ \frac{1}{3} & 2 \end{pmatrix}$$

which is clearly positive definite irrespective of $\boldsymbol{d}$. Note that only $\boldsymbol{d} = 0$ satisfies $\boldsymbol{d}'\nabla h_i(\boldsymbol{\theta}^*) = 0$ for $i = 1, 2, 3$. Hence the 2nd-order sufficiency conditions are satisfied. Note that $\boldsymbol{\theta}^* = (1, 1)$ is also the L_∞-norm solution of the problem.

3.5. Algorithms for solving nonlinear L_∞-norm estimation problems

Over the past 25 years numerous algorithms have been proposed for solving the L_∞-norm problem and minimax problems. These methods can be classified into the following four classes:

I These methods are of the Gauss-Newton or Levenberg-Marquardt type and only use 1st-derivative information. The original nonlinear problem is reduced to a *sequence of linear L_∞-norm problems*, each of which can be solved efficiently as a linear programming problem.

- In the Gauss-Newton type an *unconstrained* L_∞-regression problem is solved — see for example the algorithms of Osborne and Watson (1969) and Anderson and Osborne (1977a).

- In the Levenberg-Marquardt type a *constrained* L_1-regression problem is solved - see for example Madsen (1975), Anderson and Osborne (1977b) and Schrager and Hill (1980).

- Gradient projection methods for the minimax problem were suggested by Zangwill (1967) and Charalambous and Conn (1978).

II These are single phase methods in which a sequence of quadratic programming problems (SQP) is solved. An active set strategy for updating the active set is also used. Curvature effects (2nd-derivative information) are taken into account by the quadratic part of the QP problem whilst linear approximations of the model functions constitute the linear constraints of the QP problem. For the Chebychev problem see Overton (1982) and for the minimax problem see Murray and Overton (1980) and Han (1981).

III Two-phase or hybrid methods avoid the need for an active set strategy by attempting to identify the optimal active set in the first phase and in the second phase a system of nonlinear equations is solved using Newton's (or a quasi-Newton) method. These methods therefore use 2nd-derivative information. See for example Watson (1979) for the L_∞-norm case and Hald and Madsen (1978,1981) for the minimax problem.

IV Approximate methods which use 2nd-derivative information. The objective function's nondifferentiability is overcome by transforming the original L_∞-norm problem into a *sequence of unconstrained minimization problems* each requiring solution. For the minimax problem Charalambous and Bandler (1976) considered a least p^{th} optimization approach, Zang (1980) a penalty function and Charalambous (1979) an augmented Lagrangian function approach.

3.5.1. Type I Algorithms

In this class we consider the sequential linear programming methods.

3.5.1.1 The Anderson-Osborne-Watson algorithm.

This algorithm is due to Osborne and Watson (1969) and Anderson and Osborne (1977a). It is an analogue of the algorithm presented in §2.3.2.1.

The Algorithm

Step 0: Given an initial estimate $\boldsymbol{\theta}^1$ of the optimal $\boldsymbol{\theta}^*$, set $j = 1$.

Step 1: Calculate the solution $\delta\boldsymbol{\theta}$ to the following *linear* L_∞-norm problem:

$$\min d_\infty$$

$$\text{subject to} \quad \left.\begin{aligned} y_i - f_i(\boldsymbol{x_i}, \boldsymbol{\theta}^j) - \nabla f_i(\boldsymbol{x_i}, \boldsymbol{\theta}^j)'\delta\boldsymbol{\theta} \le d_\infty \\ -y_i + f_i(\boldsymbol{x_i}, \boldsymbol{\theta}^j) + \nabla f_i(\boldsymbol{x_i}, \boldsymbol{\theta}^j)'\delta\boldsymbol{\theta} \le d_\infty \end{aligned}\right\} \quad i = 1, \ldots, n$$

Let the minimum be $\hat{d}^j_\infty$ with $\delta\boldsymbol{\theta} = \delta\boldsymbol{\theta}^j$.

Step 2: Choose γ_j as the largest number in the set $\{1, \mu, \mu^2, \ldots\}$ with $0 < \mu < 1$ for which

$$\frac{S_\infty(\boldsymbol{\theta}^j) - S_\infty(\boldsymbol{\theta}^j + \gamma\delta\boldsymbol{\theta}^j)}{\gamma_j(S_\infty(\boldsymbol{\theta}^j) - \hat{d}^j_\infty)} \ge 0.0001$$

Set $\bar{d}^{j+1}_\infty = S_\infty(\boldsymbol{\theta}^j + \gamma\delta\boldsymbol{\theta}^j)$ with $\gamma = \gamma_j$.

Step 3: Set $\boldsymbol{\theta}^{j+1} = \boldsymbol{\theta}^j + \gamma_j\delta\boldsymbol{\theta}^j$ and return to Step 1. Repeat until certain convergence criteria are met.

Remarks

(1) The original nonlinear problem is reduced to a sequence of linear L_∞-norm problems (Step 1), each of which can be solved efficiently as a standard LP problem using the Barrodale and Phillips (1975) subroutine CHEB.

(2) The inexact line search in Step 2 ensures convergence and is due to Anderson and Osborne (1977a) (see Appendix 2B). Osborne and Watson (1969) originally proposed an exact line search:

$$\text{Calculate} \quad \gamma > 0 \quad \text{to minimize} \quad \max_{1 \le i \le n} |y_i - f_i(\boldsymbol{x_i}, \boldsymbol{\theta}^j + \gamma\delta\boldsymbol{\theta}^j)|$$

Example 3.3 Suppose we wish to minimize the maximum absolute error when fitting the curve $y = e^{\theta x}$ to the three data points (0,0),(1,1), (2,4), see Fig. 3.3.

This problem does not have a unique solution. By inspection $S_\infty(\theta^*) = 1$ for all $\frac{1}{2}\ln 3 \le \theta^* \le \ln 2$. Since $A = \{1\}$ and $I = \{2, 3\}$ it follows that $\alpha_1 = 1$ and

$$\alpha_1 sign[(h_1(\theta^*)]\nabla h_1(\theta^*) = \alpha_1(-1) \times 0 = 0$$

The optimality conditions are therefore satisfied.

We shall apply one iteration of the algorithm starting with $\theta^1 = 1$.

Step 1: We have to solve the problem:

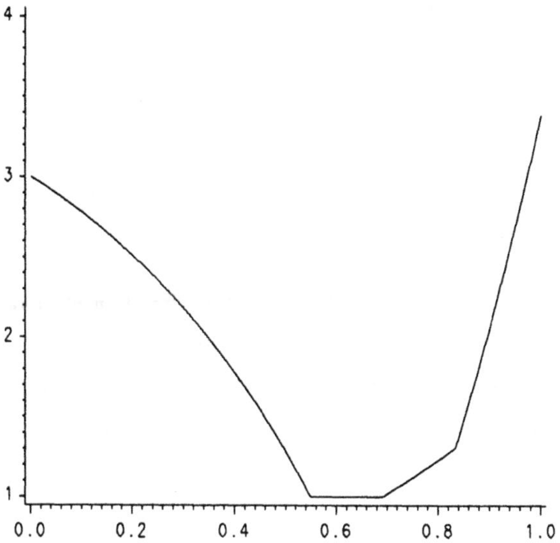

Figure 3.3 $S_\infty(\theta)$ vs θ.

$$\min \max_{1\le i\le 3} |y_i - e^{\theta^1 x_i} - x_i e^{\theta^1 x_i}\delta\theta|$$

for $\delta\theta$. This is equivalent to solving the linear programming problem:

$$\min \quad d_\infty$$

$$\begin{aligned} \text{subject to} \quad -1 - d_\infty &\le 0 \\ 1 - d_\infty &\le 0 \\ -1.7183 - 2.7183\delta\theta - d_\infty &\le 0 \\ 1.7183 + 2.7183\delta\theta - d_\infty &\le 0 \\ -3.3891 - 14.7781\delta\theta - d_\infty &\le 0 \\ 3.3891 + 14.7781\delta\theta - d_\infty &\le 0 \end{aligned}$$

and $d_\infty \ge 0$ and $\delta\theta$ unconstrained in sign. The solution to this LP problem yields $\delta\theta^1 = -0.2642$, and the minimum $\hat{d}^1_\infty = 1$. The number of simplex iterations using subroutine CHEB of Barrodale and Phillips (1975) was 3. By inspection we can see that the solution is not unique (as is to be expected) since $\hat{d}^1_\infty = 1$ for all $-0.2970 \le \delta\theta \le -0.2642$.

Step 2: Select $\gamma = 1$ and calculate

$$\frac{S_\infty(\theta^1) - S_\infty(\theta^1 + \gamma\delta\theta^1)}{\gamma(S_\infty(\theta^1) - \hat{d}^1_\infty)} = \frac{3.3891 - 1.0871}{3.3891 - 1}$$
$$= 0.964 > 0.0001$$

Hence $\gamma_1 = 1$ is selected. The approximate value of $\bar{S}^2 = 1.0871$.

Step 3: Set $\theta^2 = \theta^1 + \gamma_1\delta\theta^1 = 0.7358$.

The algorithm proceeds in this fashion to the optimal solution $\theta^* = 0.5801$ with $S_\infty(\theta^*) = 1$. The residuals are -1, -0.7863 and 0.8093. The total number of iterations was 2 and the total number of simplex iterations equalled 6. The line search step $\gamma = 1$ was chosen at every iteration (quadratic convergence).

Definition 3.1 Haar condition: Every $r \times r$ submatrix of J (Jacobian matrix) is nonsingular.

Suppose the Jacobian matrix J is of full column rank and that no exact solution exists ($d_\infty > 0$), then the following properties can be derived.

Convergence properties

Assuming an exact minimum is located in Step 2, then:

(1) $\hat{d}^j_\infty \leq \bar{d}^j_\infty$ for all j; if $\hat{d}^j_\infty < \bar{d}^j_\infty$ then $\bar{d}^{j+1}_\infty < \bar{d}^j_\infty$ for all j.

(2) At a limit point of the algorithm $\hat{d}^j_\infty = \bar{d}^j_\infty$.

(3) If the Jacobian J satisfies the Haar condition then $\{\boldsymbol{\theta}^j\}$ converges as $j \to \infty$ and the minimum will be isolated.

For inexact line searches in Step 2:

(4) If $S_\infty(\boldsymbol{\theta}^j) \neq \hat{d}^j_\infty$ and given that $0 < \lambda < 1$ then there exists a $\gamma_j \in \{1, \mu, \mu^2, \ldots\}$ with $0 < \mu < 1$ such that

$$\frac{S_\infty(\boldsymbol{\theta}^j) - S_\infty(\boldsymbol{\theta}^j + \gamma_j\delta\boldsymbol{\theta}^j)}{\gamma_j(S_\infty(\boldsymbol{\theta}^j) - \hat{d}^j_\infty)} \geq \lambda$$

(5) If the algorithm does not terminate in a finite number of iterations then the sequence $\{S_\infty(\boldsymbol{\theta}^j)\}$ converges to $\hat{d}^j_\infty$ as $j \to \infty$.

(6) Any limit point of $\{\boldsymbol{\theta}^j\}$ is a stationary point of $S_\infty(\boldsymbol{\theta})$ — i.e., a point $\boldsymbol{\theta}^*$ such that $S_\infty(\boldsymbol{\theta}^*) = \hat{S}^*$.

Convergence rate

Cromme (1978) shows that the condition of strong uniqueness (§2.3.2.1), which is weaker than the Haar condition, ensures quadratic convergence. Strong uniqueness implies that $k+1$ functions are active at $\boldsymbol{\theta}^*$. A necessary condition for the quadratic convergence of the Gauss-Newton method is that the number of active functions in A should be greater than or equal to $k+1$.

Example 3.4 This is the example due to Madsen (1975) considered in §2.2.2. Instead of the L_1-norm we shall consider the L_∞-norm:

$$\min_{\boldsymbol{\theta}} \max_{1 \le i \le 3} |h_i(\boldsymbol{\theta})|$$

The contours of the objective function are given in Fig. 3.4.

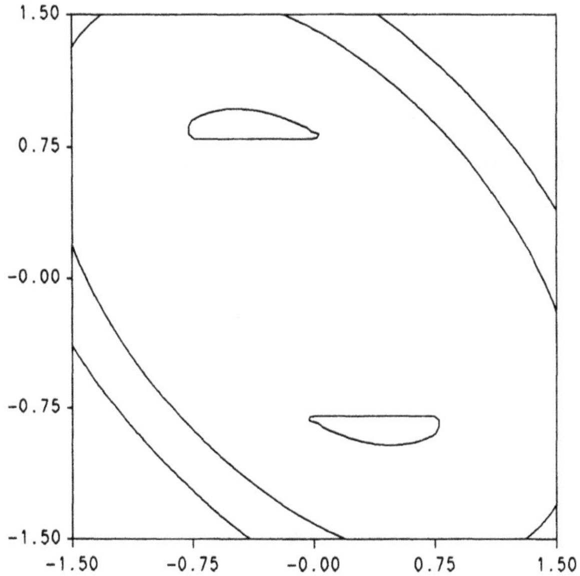

Figure 3.4 Contours of $S_\infty(\boldsymbol{\theta}) = 0.7,\ 2,\ 3.$

The optimal solution is $\boldsymbol{\theta}^* = (\pm 0.4533, \mp 0.9066)$. The FORTRAN program MINMAXABS listed in Appendix 3A uses the linear L_∞-norm subroutine CHEB of

Barrodale and Phillips (1975). Note the slow convergence of the algorithm on this example. The intermediate results yielded by MINMAXABS are in complete agreement with those reported by Anderson and Osborne (1977a).

3.5.1.2 The Madsen algorithm.

Madsen (1975) proposed a Levenberg-Marquardt (Appendix 2D) approach in which the length of the direction $\delta\boldsymbol{\theta}$ is restricted. The algorithm is in other respects similar to the McLean and Watson (1980) algorithm (L_1).

The Algorithm

Step 0: Choose $(\rho_1, \rho_2, \rho_3, \sigma_1, \sigma_2) = (0.01, 0.25, 0.25, 0.25, 2)$. Select the initial estimate $\boldsymbol{\theta}^1$ and an upper bound λ_1 on the solution length; set $j = 1$.

Step 1: Solve the constrained linear L_∞-norm problem:

$$\min_{\delta\boldsymbol{\theta}} \max_{1\le i\le n} \{|y_i - f_i(\boldsymbol{x}_i, \boldsymbol{\theta}^j) - \nabla f_i(\boldsymbol{x}_i, \boldsymbol{\theta}^j)'\delta\boldsymbol{\theta}|\}$$

$$\text{subject to} \quad \max_{1\le i\le k} |\delta\theta_i| \le \lambda_j$$

Call the solution $\delta\boldsymbol{\theta}^j$ with objective function value S_∞^j.

Step 2: If

$$\frac{S_\infty(\boldsymbol{\theta}^j) - S_\infty(\boldsymbol{\theta}^j + \delta\boldsymbol{\theta}^j)}{S_\infty(\boldsymbol{\theta}^j) - S_\infty^j} \ge \rho_1$$

set $\boldsymbol{\theta}^{j+1} := \boldsymbol{\theta}^j + \delta\boldsymbol{\theta}^j$, otherwise set $\boldsymbol{\theta}^{j+1} := \boldsymbol{\theta}^j$.

Step 3: If

$$\frac{S_\infty(\boldsymbol{\theta}^j) - S_\infty(\boldsymbol{\theta}^j + \delta\boldsymbol{\theta}^j)}{S_\infty(\boldsymbol{\theta}^j) - S_\infty^j} \le \rho_2$$

$$\text{set} \quad \lambda_{j+1} = \sigma_1 \max_{1\le i\le k} |\delta\theta_i^j|$$

otherwise if

$$\max_{1\le i\le n} \frac{\{|f_i(\boldsymbol{x}_i, \boldsymbol{\theta}^j) - f_i(\boldsymbol{x}_i, \boldsymbol{\theta}^{j+1}) + \nabla f_i(\boldsymbol{x}_i, \boldsymbol{\theta}^j)'\delta\boldsymbol{\theta}|\}}{S_\infty(\boldsymbol{\theta}^j) - S_\infty(\boldsymbol{\theta}^j + \delta\boldsymbol{\theta}^j)} \le \rho_3$$

then set

$$\lambda_{j+1} = \sigma_2 \max_{1\le i\le k} |\delta\theta_i^j|$$

else set

$$\lambda_{j+1} = \max_{1 \le i \le k} |\delta\theta_i^j|$$

Set $j := j+1$ and return to Step 1. Continue until certain convergence criteria are met.

Remarks

(1) A constrained linear L_∞-norm problem is solved in Step 1. The Joe and Bartels (1983) procedure can be used in this step. Note that in Step 1 the problem can be written as

$$\min d_\infty$$

$$\text{subject to} \quad \left.\begin{array}{r} y_i - f_i(\boldsymbol{x_i}, \boldsymbol{\theta}^j) - \nabla f_i(\boldsymbol{x_i}, \boldsymbol{\theta}^j)'\delta\boldsymbol{\theta} \le d_\infty \\ -y_i + f_i(\boldsymbol{x_i}, \boldsymbol{\theta}^j) + \nabla f_i(\boldsymbol{x_i}, \boldsymbol{\theta}^j)'\delta\boldsymbol{\theta} \le d_\infty \end{array}\right\} \quad i = 1, \ldots, n$$

$$-\lambda_j \le \delta\theta_i \le \lambda_j \qquad i = 1, \ldots, k$$

(2) λ_j controls the accuracy of the linear approximation of $f_i(\boldsymbol{x_i}, \boldsymbol{\theta})$ at $\boldsymbol{\theta}^j$ and is chosen as large as possible whilst maintaining a fair agreement between $f_i(\boldsymbol{x_i}, \boldsymbol{\theta})$ and its linear approximation.

Example 3.5 We shall consider the example due to Madsen (1975) and demonstrate one iteration. Choose $\boldsymbol{\theta}^1 = (3, 1)$ and $\lambda_1 = 1.1$.

Step 1: The problem:

$$\min_{\delta\boldsymbol{\theta}} \max\{|13 + 7\delta\theta_1 + 5\delta\theta_2|, |0.14112 - 0.9899\delta\theta_1|, |0.5403 - 0.84147\delta\theta_2|\}$$

$$\text{subject to} \quad \max_{1 \le i \le 3} |\delta\theta_i| \le 1.1$$

has to be solved. The solution is $\delta\boldsymbol{\theta}^1 = (-1.0979, -0.8173)$ and $S_\infty^1 = 1.2281$.

Step 2:

$$\frac{S_\infty(\boldsymbol{\theta}^1) - S_\infty(\boldsymbol{\theta}^1 + \delta\boldsymbol{\theta}^1)}{S_\infty(\boldsymbol{\theta}^1) - S_\infty^1} = 0.7646 > \rho_1 = 0.01$$

Thus $\boldsymbol{\theta}^2 = \boldsymbol{\theta}^1 + \delta\boldsymbol{\theta}^1 = (1.9021, 0.1827)$ and $S_\infty(\boldsymbol{\theta}^2) = 3.9988$.

Step 3:

$$\frac{S_\infty(\boldsymbol{\theta}^1) - S_\infty(\boldsymbol{\theta}^1 + \delta\boldsymbol{\theta}^1)}{S_\infty(\boldsymbol{\theta}^1) - S_\infty^1} = 0.7646 > \rho_2 = 0.25$$

$$h_1(\boldsymbol{\theta}^2) - h_1(\boldsymbol{\theta}^1) - \nabla h_1(\boldsymbol{\theta}^1)'\delta\boldsymbol{\theta}^1 = 2.7707$$

$$h_2(\boldsymbol{\theta}^2) - h_2(\boldsymbol{\theta}^1) - \nabla h_2(\boldsymbol{\theta}^1)'\delta\boldsymbol{\theta}^1 = 0.2824$$

$$h_3(\boldsymbol{\theta}^2) - h_3(\boldsymbol{\theta}^1) - \nabla h_3(\boldsymbol{\theta}^1)'\delta\boldsymbol{\theta}^1 = 0.2447$$

$$S_\infty(\boldsymbol{\theta}^1) - S_\infty(\boldsymbol{\theta}^2) = 9.0012$$

Thus

$$\max_{1\le i\le 3} \frac{\{|h_i(\boldsymbol{\theta}^j + \delta\boldsymbol{\theta}^j) - h_i(\boldsymbol{\theta}^j) - \nabla h_i(\boldsymbol{\theta}^j)'\delta\boldsymbol{\theta}^j|\}}{S_\infty(\boldsymbol{\theta}^j) - S_\infty(\boldsymbol{\theta}^j + \delta\boldsymbol{\theta}^j)} = \frac{2.7707}{9.0012} = 0.3078 > \rho_3 = 0.25$$

Thus we set $\lambda_2 = \max\{1.0979, 0.8173\} = 1.0979$. The algorithm proceeds in this fashion to the optimal point $\boldsymbol{\theta}^* = (0.4533, -0.9066)$.

If the problem is well-scaled (θ_i's are of a similar order of magnitude) and $f_i(\boldsymbol{x}_i, \boldsymbol{\theta})$ is continuously differentiable then the following can be proved:

Convergence properties

(1) If the sequence $\{\boldsymbol{\theta}^j\}$ generated by the algorithm has a limit $\boldsymbol{\theta}^*$ then $\boldsymbol{\theta}^*$ is a stationary point of $S_\infty(\boldsymbol{\theta})$ i.e.,

$$S_\infty(\boldsymbol{\theta}^*) = \min_{\delta\boldsymbol{\theta}} \max_{1\le i\le n} \{|y_i - f_i(\boldsymbol{x}_i, \boldsymbol{\theta}^j) - \nabla f_i(\boldsymbol{x}_i, \boldsymbol{\theta}^j)'\delta\boldsymbol{\theta}|\}$$

(2) Let $\{\boldsymbol{\theta}^j\}$ be the sequence generated by the algorithm. Let L be the set of stationary points of $S_\infty(\boldsymbol{\theta})$ and let $\boldsymbol{d}(\boldsymbol{\theta}) = \inf_{\boldsymbol{y}\in \mathrm{L}} \|\boldsymbol{y} - \boldsymbol{\theta}\|$. If $\{\boldsymbol{\theta}^j\}$ stays in a finite region then $\boldsymbol{d}(\boldsymbol{\theta}^j) \to 0$ as $j \to \infty$.

3.5.1.3 The Anderson-Osborne-Levenberg-Marquardt algorithm.

The Anderson and Osborne (1977b) algorithm is similar to the Madsen (1975) algorithm in the sense that the length of the direction vector $\delta\boldsymbol{\theta}$ is also bounded.

The Algorithm

Step 0: Given an initial estimate $\boldsymbol{\theta}^1$ of the optimal $\boldsymbol{\theta}^*$, select $\mu_0 = \mu_1 = 1$ and set $j = 1$.

Step 1: Calculate the solution $\delta\boldsymbol{\theta}$ to the following *constrained linear* L_∞-norm problem:

$$\min d_\infty$$

$$\text{subject to}\quad \left.\begin{aligned} y_i - f_i(\boldsymbol{x_i},\boldsymbol{\theta}^j) - \nabla f_i(\boldsymbol{x_i},\boldsymbol{\theta}^j)'\delta\boldsymbol{\theta} &\le d_\infty \\ -y_i + f_i(\boldsymbol{x_i},\boldsymbol{\theta}^j) + \nabla f_i(\boldsymbol{x_i},\boldsymbol{\theta}^j)'\delta\boldsymbol{\theta} &\le d_\infty \\ \mu_j\delta\boldsymbol{\theta} &\le d_\infty\mathbf{1}_k \\ -\mu_j\delta\boldsymbol{\theta} &\le d_\infty\mathbf{1}_k \end{aligned}\right\}\quad i=1,\ldots,n$$

Let the minimum be $\hat{d}^j_\infty(\mu_j)$ with $\delta\boldsymbol{\theta} = \delta\boldsymbol{\theta}^j$.

Step 2: Modify $\mu_j > 0$ as follows:

Set $l = 0, T_l = 0, over := false$, $\mu^1_j = \mu_{j-1}$ and $\delta = 0.01, \beta = 10, \tau = 0.2$.

2(a): If $l < 1$ or $T_l < 1 - \delta$ go to Step 2(c).

Set $l := l + 1$ and calculate

$$T_l = T(\boldsymbol{\theta}^j, \mu^l_j) = \frac{S_\infty(\boldsymbol{\theta}^j) - S_\infty(\boldsymbol{\theta}^j + \delta\boldsymbol{\theta}^j)}{S_\infty(\boldsymbol{\theta}^j) - \hat{d}^j_\infty(\mu^l_j)}$$

2(b): If $T_l < \delta$ and $over = true$ set $\mu^{l+1}_j = \frac{1}{2}(\tilde{\mu}_j + \mu^l_j)$

If $T_l < \delta$ and $over = false$ set $\mu^{l+1}_j = \beta\mu^l_j$

If $T_l < \delta$ set $\hat{\mu}_j = \mu^l_j$

If $l > 1$ and $T_l > 1 - \delta$ set $over := true$, $\mu^{l+1}_j = \frac{1}{2}(\hat{\mu}_j + \mu^l_j)$ and $\tilde{\mu}_j = \mu^l_j$

2(c): If $T_l < \delta$ or $[l > 1$ and $T_l > 1 - \delta]$ return to Step 2(b).

Otherwise if $l = 1$ set $\mu_{j+1} = \tau\mu^l_j$

If $l \neq 1$ set $\mu_{j+1} = \mu^l_j$

Step 3: set $\boldsymbol{\theta}^{j+1} = \boldsymbol{\theta}^j + \delta\boldsymbol{\theta}^j$ and return to Step 1. Repeat until certain convergence criteria are met.

3.5.2. Type II Algorithms

In this class we consider the sequential quadratic programming methods or single phase methods.

3.5.2.1 The Murray and Overton algorithm.

Murray and Overton (1980) developed a procedure for solving the nonlinear minimax problem. Overton (1982) adapted this approach to solve the nonlinear

L_∞-norm estimation problem. One could solve the L_∞-norm as a special case of the minimax problem by observing that

$$\min_{\boldsymbol{\theta}} \max_{1\le i\le n} |h_i(\boldsymbol{\theta})| \equiv \min_{\boldsymbol{\theta}} \max_{1\le i\le n} \{h_i(\boldsymbol{\theta}), -h_i(\boldsymbol{\theta})\}$$

For ease of notation we have chosen $h_i(\boldsymbol{\theta}) = y_i - f_i(\boldsymbol{x}_i, \boldsymbol{\theta})$. In the algorithm a descent direction is determined from an equality constrained quadratic program based on a projected Lagrangian function. Then a line search is performed using this direction to reduce the merit function, $S_\infty(\boldsymbol{\theta})$.

We shall partly adopt the Murray and Overton (1980) notation in this section. Without loss of generality it can be assumed that the constraints and their associated functions are set out with the first t constraints (and functions) being active.

- Active function set

$$A = \{1, \ldots, t\} = \{i | h_i(\boldsymbol{\theta}^*) = S_\infty(\boldsymbol{\theta}^*)\}$$

  Thus at any point $\boldsymbol{\theta} \neq \boldsymbol{\theta}^*$ we define A as the set of functions which we think will be active at the solution $\boldsymbol{\theta}^*$.

- Without loss of generality let $\hat{\boldsymbol{h}}(\boldsymbol{\theta}) = [h_1(\boldsymbol{\theta}), \ldots, h_t(\boldsymbol{\theta})]'$ be the vector of active functions.

- Active constraints

$$C_A(\boldsymbol{\theta}) = d_\infty \mathbf{1}_t + \hat{\Sigma}\hat{\boldsymbol{h}}(\boldsymbol{\theta})$$

$$\text{where} \quad \hat{\Sigma}_{t\times t} = diag\{\, sign[h_i(\boldsymbol{\theta})] : i = 1, \ldots, t\,\}$$

- The matrix of gradients of the active functions

$$\hat{V}_{k\times t} = (\,\nabla h_1(\boldsymbol{\theta}), \ldots, \nabla h_t(\boldsymbol{\theta})\,)$$

- The matrix of gradients of the active constraints

$$\nabla \hat{C}_A(\boldsymbol{\theta}) = \begin{pmatrix} -\hat{V}\hat{\Sigma} \\ \mathbf{1}_t' \end{pmatrix}$$

- The gradient of the objective function of (3.3) with respect to $(\boldsymbol{\theta}, d_\infty)$ is given by:

$$g_\infty(\boldsymbol{\theta}) = \begin{pmatrix} 0 \\ 0 \\ \vdots \\ 1 \end{pmatrix}$$

- Orthogonal decomposition of $\nabla\hat{C}_A(\boldsymbol{\theta})$ (Appendix 2F).

$$\nabla\hat{C}_A(\boldsymbol{\theta}) = [Y(\boldsymbol{\theta}) \;\; Z(\boldsymbol{\theta})] \begin{pmatrix} R(\boldsymbol{\theta}) \\ \cdots \\ O \end{pmatrix} = Q' \begin{pmatrix} R(\boldsymbol{\theta}) \\ \cdots \\ O \end{pmatrix}$$

where $Y(\boldsymbol{\theta})$ is $(k+1)\times t$, $Z(\boldsymbol{\theta})$ is $(k+1)\times(k+1-t)$ and $R(\boldsymbol{\theta})$ is $t \times t$. The matrices Y and Z are orthogonal i.e., $Y'Y = I_t$ and $Z'Z = I_{k+1-t}$ and $\nabla\hat{C}'_A Z = Y'Z = \boldsymbol{O}$. There are two subspaces that result:

> t-dimensional subspace defined by the first t columns of $\nabla\hat{C}_A$. Thus Y forms a basis for the range (column) space of $\nabla\hat{C}_A$.
>
> $(k+1-t)$-dimensional subspace (orthogonal complement). Its vectors are orthogonal to the rows of $\nabla\hat{C}_A$ or columns of $\nabla\hat{C}'_A$. Thus Z forms a basis of the null space of $\nabla\hat{C}'_A$.

- The Lagrangian function

$$L(d_\infty, \boldsymbol{\theta}, \boldsymbol{\lambda}) = d_\infty - \boldsymbol{\lambda}'\hat{C}_A(\boldsymbol{\theta})$$

- The gradient of $L(d_\infty, \boldsymbol{\theta}, \boldsymbol{\lambda})$ with respect to $(d_\infty, \boldsymbol{\theta})$ is given by:

$$g_\infty(\boldsymbol{\theta}) - \nabla\hat{C}_A(\boldsymbol{\theta})\boldsymbol{\lambda}$$

- The Hessian of the Lagrangian function with respect to $(d_\infty, \boldsymbol{\theta})$ is:

$$\nabla^2_{\boldsymbol{\theta}} L = \begin{pmatrix} \boldsymbol{W}_\infty(\boldsymbol{\theta}) & 0 \\ 0 & 0 \end{pmatrix}$$

$$\text{where} \qquad \boldsymbol{W}_\infty(\boldsymbol{\theta}) = \sum_{i=1}^{t} \lambda_i sign[(h_i(\boldsymbol{\theta})] \nabla^2 h_i(\boldsymbol{\theta})$$

- The projected Hessian of L onto the null space of $\nabla\hat{C}'_A$ is $Z'\boldsymbol{W}_\infty(\boldsymbol{\theta})Z$.

Several methods exist for calculating a feasible descent direction. Murray and Overton (1980) use a 2nd-derivative method which essentially solves an equality constrained QP problem [see Gill et al. (1981:159)]. A direct application of this method to the nonlinear L_∞-norm problem would be to solve for direction $\boldsymbol{d}$ in:

$$\text{Minimize } \frac{1}{2}\boldsymbol{d}'\boldsymbol{W}_\infty(\boldsymbol{\theta})\boldsymbol{d} + \boldsymbol{g}_\infty(\boldsymbol{\theta})'\boldsymbol{d} \tag{3.8}$$

$$\text{subject to} \qquad \nabla\hat{C}_A(\boldsymbol{\theta})\boldsymbol{d} + \hat{C}_A(\boldsymbol{\theta}) = 0$$

The QR factorization of $\nabla\hat{C}_A$ is then used to obtain a solution for $\boldsymbol{d}$.

In order to construct the Hessian $\boldsymbol{W}_\infty(\boldsymbol{\theta})$ reliable estimates of the Lagrange multiplier vector are needed. These are obtained by solving the linear least squares problem:

$$||\boldsymbol{g}_\infty(\boldsymbol{\theta}^j) - \nabla\hat{C}_A(\boldsymbol{\theta}^j)\boldsymbol{\lambda}||_2 \tag{3.9}$$

which is equivalent to the constrained least squares problem

$$\min ||\hat{V}\lambda||_2 \quad \text{subject to} \quad \sum_{i=1}^{n}\lambda_i = 1$$

The conceptual algorithm is stated very briefly:

The Algorithm

Step 0: Given an initial estimate $\boldsymbol{\theta}^1$ of the optimal $\boldsymbol{\theta}^*$, set $j = 1$.

Step 1: Calculate $\nabla\hat{C}_A(\boldsymbol{\theta}^j)$. Determine the Lagrange multiplier vector $\boldsymbol{\lambda}^j$ by solving:

$$||\boldsymbol{g}_\infty(\boldsymbol{\theta}^j) - \nabla\hat{C}_A(\boldsymbol{\theta}^j)\boldsymbol{\lambda}||_2$$

Compute $\boldsymbol{W}_\infty(\boldsymbol{\theta}^j)$.

Step 2: Calculate $\boldsymbol{d}^j$ by solving the QP problem:

$$\text{Minimize } \frac{1}{2}\boldsymbol{d}'\boldsymbol{W}_\infty(\boldsymbol{\theta})\boldsymbol{d} + \boldsymbol{g}_\infty(\boldsymbol{\theta})'\boldsymbol{d}$$

$$\text{subject to} \qquad \nabla\hat{C}_A(\boldsymbol{\theta})\boldsymbol{d} + \hat{C}_A(\boldsymbol{\theta}) = 0$$

evaluated at $\boldsymbol{\theta}^j$.

Step 3: Set $\boldsymbol{\theta}^{j+1} = \boldsymbol{\theta}^j + \gamma_j \boldsymbol{d}^j$ where γ_j is determined by line search. Return to Step 1. Repeat until certain convergence criteria are met.

Remarks:

(1) Murray and Overton (1980) provide a flowchart for their algorithm.

(2) Murray and Overton (1979) describe a special line search method that can be used in Step 3 of the algorithm.

3.5.2.2 The Han algorithm.

Han (1981) developed a method for solving the minimax problem (3.5). The method possesses the attractive features of the usual quasi-Newton methods of differentiable unconstrained optimization. It can therefore be considered as an extension of quasi-Newton methods to solve nondifferentiable problems.

In this section we adapt the Han algorithm to solve the L_∞-norm estimation problem. By examining the KKT conditions for (3.3) we see that at optimality $\sum_{i=1}^n (u_i + v_i) = 1$. Following, Han we define a type of Lagrangian function:

$$\tilde{L} = \sum_{i=1}^{n} (u_i - v_i)[y_i - f_i(\boldsymbol{x_i}, \boldsymbol{\theta})]$$

Suppose $\boldsymbol{\theta}^j$ is an estimate of the optimal point $\boldsymbol{\theta}^*$ at step j and that there exists a positive definite matrix B^j which approximates the Hessian

$$\nabla^2_{\boldsymbol{\theta}} \tilde{L} = \sum_{i=1}^{n} -(u_i - v_i) \nabla^2 f_i(\boldsymbol{x_i}, \boldsymbol{\theta}^j)$$

To determine the descent direction $\boldsymbol{d}$, the QP problem

$$\min \frac{1}{2} \boldsymbol{d}' B^j \boldsymbol{d} + d_\infty \tag{3.10}$$

$$\text{subject to} \quad \left.\begin{aligned} y_i - f_i(\boldsymbol{x_i}, \boldsymbol{\theta}^j) - \nabla f_i(\boldsymbol{x_i}, \boldsymbol{\theta}^j)' \boldsymbol{d} &\le d_\infty \\ -y_i + f_i(\boldsymbol{x_i}, \boldsymbol{\theta}^j) + \nabla f_i(\boldsymbol{x_i}, \boldsymbol{\theta}^j)' \boldsymbol{d} &\le d_\infty \end{aligned}\right\} \quad i = 1, \ldots, n$$

is solved with the solution denoted by $\boldsymbol{d}^j$. If n is large then solving this problem at each step may require excessive computation. Instead, the dual QP problem can be considered. The Lagrangian function of (3.10) is:

$$\begin{aligned} L(\boldsymbol{d}, d_\infty, \boldsymbol{\lambda}, \boldsymbol{\mu}) =& \frac{1}{2}\boldsymbol{d}' B^j \boldsymbol{d} + d_\infty + \sum_{i=1}^{n} \lambda_i [y_i - f_i(\boldsymbol{x}_i, \boldsymbol{\theta}^j) - \nabla f_i(\boldsymbol{x}_i, \boldsymbol{\theta}^j)'\boldsymbol{d} - d_\infty] \\ & - \sum_{i=1}^{n} \mu_i [y_i - f_i(\boldsymbol{x}_i, \boldsymbol{\theta}^j) - \nabla f_i(\boldsymbol{x}_i, \boldsymbol{\theta}^j)'\boldsymbol{d} + d_\infty] \\ =& \frac{1}{2}\boldsymbol{d}' B^j \boldsymbol{d} - (\boldsymbol{\lambda} - \boldsymbol{\mu})' J(\boldsymbol{\theta}^j)\boldsymbol{d} + [\boldsymbol{y} - \boldsymbol{f}(\boldsymbol{x}, \boldsymbol{\theta}^j)]'(\boldsymbol{\lambda} - \boldsymbol{\mu}) \\ & + d_\infty [1 - \mathbf{1}_n'(\boldsymbol{\lambda} + \boldsymbol{\mu})] \end{aligned}$$

The Wolfe (1961) dual formulation of (3.10) is:

$$\max L(\boldsymbol{d}, d_\infty, \boldsymbol{\lambda}, \boldsymbol{\mu})$$

$$\begin{aligned} \text{subject to} \quad \nabla_{\boldsymbol{d}} L(\boldsymbol{d}, d_\infty, \boldsymbol{\lambda}, \boldsymbol{\mu}) &= 0 \\ \frac{\partial L}{\partial d_\infty} &= 0 \\ \boldsymbol{\lambda},\ \boldsymbol{\mu} &\geq 0 \end{aligned}$$

This dual involves $\boldsymbol{d}, d, \boldsymbol{\lambda}$ and $\boldsymbol{\mu}$. Now

$$\nabla_{\boldsymbol{d}} L = B^j \boldsymbol{d} - J(\boldsymbol{\theta}^j)'(\boldsymbol{\lambda} - \boldsymbol{\mu}) = 0 \Rightarrow \boldsymbol{d} = (B^j)^{-1} J(\boldsymbol{\theta}^j)'(\boldsymbol{\lambda} - \boldsymbol{\mu})$$

and

$$\frac{\partial L}{\partial d_\infty} = 1 - \sum_{i=1}^{n} (\lambda_i + \mu_i) = 0$$

Substituting these two equations into $L(\boldsymbol{d}, d_\infty, \boldsymbol{\lambda}, \boldsymbol{\mu})$ and simplifying we obtain the dual problem involving the multipliers $\boldsymbol{\lambda}$ and $\boldsymbol{\mu}$ only:

$$\min \frac{1}{2}(\boldsymbol{\lambda} - \boldsymbol{\mu})' J(\boldsymbol{\theta}^j)(B^j)^{-1} J(\boldsymbol{\theta}^j)'(\boldsymbol{\lambda} - \boldsymbol{\mu}) - [\boldsymbol{y} - \boldsymbol{f}(\boldsymbol{x}, \boldsymbol{\theta}^j)]'(\boldsymbol{\lambda} - \boldsymbol{\mu}) \tag{3.11}$$

$$\begin{aligned} \text{subject to} \quad \sum_{i=1}^{n} (\lambda_i + \mu_i) &= 1 \\ \boldsymbol{\lambda},\ \boldsymbol{\mu} &\geq 0 \end{aligned}$$

Call the solution $(\boldsymbol{\lambda}^j, \boldsymbol{\mu}^j)$ and calculate $\boldsymbol{d}^j$ by solving

$$B^j \boldsymbol{d}^j = J(\boldsymbol{\theta}^j)'(\boldsymbol{\lambda}^j - \boldsymbol{\mu}^j)$$

Define

$$\boldsymbol{\delta}^j = \boldsymbol{\theta}^{j+1} - \boldsymbol{\theta}^j$$

$$\gamma^j = \nabla_{\boldsymbol{\theta}} \tilde{L}(\boldsymbol{\theta}^{j+1}) - \nabla_{\boldsymbol{\theta}} \tilde{L}(\boldsymbol{\theta}^j)$$

One could use the BFGS updates (see also Appendix 2C) to update B^j which approximates the Hessian:

$$B^{j+1} = B^j + \frac{\gamma^j(\gamma^j)'}{(\gamma^j)'\delta^j} - \frac{B^j\delta^j(\delta^j)'B^j}{(\delta^j)'B^j\delta^j}$$

To calculate the stepsize γ_j we adapt a line search strategy due to Han (1981). This rule provides a sufficiently large stepsize to ensure a sufficient decrease in $S_\infty(\boldsymbol{\theta})$ and ensures global convergence (Han proved convergence of the algorithm).

The Algorithm

Step 0: Given an initial estimate $\boldsymbol{\theta}^1$ of the optimal $\boldsymbol{\theta}^*$, set $B^1 = I_k$ and $j = 1$.

Step 1: Solve either QP problem (3.10) or (3.11) to obtain $\boldsymbol{d}^j$.

Step 2a: Choose γ_j as the first number in the set $\{1, \mu, \mu^2, \ldots\}$ where $0 < \mu < 1$ such that

$$S_\infty(\boldsymbol{\theta}^j + \gamma \boldsymbol{d}^j) \leq S_\infty(\boldsymbol{\theta}^j) - \omega\gamma \boldsymbol{d}^{j\prime} B^j \boldsymbol{d}^j$$

where $0 < \omega \ll \frac{1}{2}$.

Step 2b: Start with $\gamma_0^j = 1$, set $i = 1$. If γ_0^j does not satisfy the above inequality use quadratic interpolation to calculate a minimum of the quadratic $g(\gamma)$ which interpolates $S_\infty(\boldsymbol{\theta}^j + \gamma \boldsymbol{d}^j)$ at the points $\gamma = 0$ and $\gamma = \gamma_0^j$ such that $g'(0) = -\boldsymbol{d}^{j\prime} B^j \boldsymbol{d}^j$.

The minimum occurs at

$$\gamma_{min} = \frac{\boldsymbol{d}^{j\prime} B^j \boldsymbol{d}^j (\gamma_0^j)^2}{2[g(0) - g(\gamma_0^j) - \boldsymbol{d}^{j\prime} B^j \boldsymbol{d}^j \gamma_0^j]}$$

Continue in this fashion and choose $\gamma_{i+1}^j = \max[0.1\gamma_i^j, \gamma_{min}]$.

Once the above inequality is satisfied set $\boldsymbol{\theta}^{j+1} := \boldsymbol{\theta}^j + \gamma_j \boldsymbol{d}^j$.

Step 3: Update B^j:

$$B^{j+1} = B^j + \frac{\gamma^j(\gamma^j)'}{(\gamma^j)'\delta^j} - \frac{B^j\delta^j(\delta^j)'B^j}{(\delta^j)'B^j\delta^j}$$

set $j := j + 1$ and return to Step 1.

Example 3.6 Reconsider example 3.3. Choose $\theta^1 = 1$ and $B^1 = 1$. Thus

$$h(\theta^1) = \begin{pmatrix} -1 \\ -1.7183 \\ -3.3891 \end{pmatrix} \quad \text{and} \quad J(\theta^1) = \begin{pmatrix} 0 \\ -2.7183 \\ -14.7781 \end{pmatrix}$$

Step 1: Solve QP problem (3.10) i.e.,

$$\min \frac{1}{2}(d^1)^2 + d_\infty$$

$$\begin{aligned} \text{subject to} \quad -1 &\leq d_\infty \\ 1 &\leq d_\infty \\ -1.7183 - 2.7183d^1 &\leq d_\infty \\ 1.7183 + 2.7183d^1 &\leq d_\infty \\ -3.3891 - 14.7781d^1 &\leq d_\infty \\ 3.3891 + 14.7781d^1 &\leq d_\infty \end{aligned}$$

its solution is $d^1_\infty = 1$ with $d^1 = -0.2642$. Note that the direction is identical to the one obtained by linear programming when the Anderson-Osborne-Watson algorithm is applied.

Step 2: *Line search.* Let $\gamma^1_0 = 1, \omega = 0.01$.

$$S_\infty(\theta^1 + \gamma^1_0 d^1) = S_\infty(0.7358) = 1.0872$$

$$S_\infty(\theta^1) - \omega d^1 B^1 d^1 = 3.3891 - 0.01(-0.2642)^2 = 3.3884$$

We see that $S_\infty(\theta^1 + \gamma^1_0 d^1) < S_\infty(\theta^1) - \omega d^1 B^1 d^1$ and hence we select $\gamma_1 = 1$ and $\theta^2 = 0.7358$ etc.

3.5.3. Type III Algorithms

In this class we consider the hybrid or two phase methods.

3.5.3.1 The Watson algorithm.

Watson (1979) suggests a combination of a Levenberg-Marquardt type method and a Newton method for solving problem (3.2). The need for the Levenberg-Marquardt approach arises, since the Jacobian $J(\boldsymbol{\theta})$ is frequently not of full column rank.

In the first phase the following problem is solved:

$$\min_{\delta\boldsymbol{\theta}\in\Re^k} \|\tilde{\boldsymbol{r}}^j\|_\infty = \left\|\begin{pmatrix} \boldsymbol{y} - \boldsymbol{f}(\boldsymbol{x},\boldsymbol{\theta}^j) - J(\boldsymbol{\theta}^j)'\delta\boldsymbol{\theta} \\ \lambda_j\delta\boldsymbol{\theta}\end{pmatrix}\right\|_\infty = \left\|\begin{pmatrix} \boldsymbol{r}^j \\ \lambda_j\delta\boldsymbol{\theta}\end{pmatrix}\right\|_\infty \tag{3.12}$$

$$\|\boldsymbol{r}^j\|_\infty = \max_{1\le i\le n} |r_i^j|$$

is the Chebychev vector norm for the jth iteration.

Let $A = \{1,2,\ldots,r\}$, $r < k$ be the set of active functions at the optimum point $\boldsymbol{\theta}^*$. Once the set of active functions is identified the algorithm switches over to the second phase and $sign[y_i - f_i(\boldsymbol{x}_i,\boldsymbol{\theta}^*)]$ will be known and the k equations $y_i - f_i(\boldsymbol{x}_i,\boldsymbol{\theta}^*) = sign[y_i - f_i(\boldsymbol{x}_i,\boldsymbol{\theta}^*)]d_\infty$ will hold from the optimality conditions of Theorem 3.2. Essentially we then have a system of $k+r+1$ equations in $k+r+1$ unknowns, $\boldsymbol{\theta},\boldsymbol{\alpha}$ and d_∞. The system of equations is of the form:

$$\begin{aligned} \sum_{i=1}^r \alpha_i &= 1 \\ \sum_{i=1}^r \alpha_i sign[y_i - f_i(\boldsymbol{x}_i,\boldsymbol{\theta}^*)]\nabla f_i(\boldsymbol{x}_i,\boldsymbol{\theta}) &= 0 \\ \alpha_i &= 0, \quad i\notin A \\ y_i - f_i(\boldsymbol{x}_i,\boldsymbol{\theta}^*) &= sign[y_i - f_i(\boldsymbol{x}_i,\boldsymbol{\theta}^*)]d_\infty \end{aligned} \tag{3.13}$$

The optimal solution will be denoted by $(\boldsymbol{\theta}^*,\boldsymbol{\alpha}^*,d_\infty^*)$.

The Algorithm

Step 0: Choose $(\rho_1,\rho_2,\rho_3,\sigma_1,\sigma_2) = (0.01, 0.25, 0.25, 0.25, 2)$. Select the initial estimate $\boldsymbol{\theta}^1$ and an upper bound λ_1 on the solution length. Set $p_1 = 3$, $\epsilon_1 = 10^{-6}$, $\epsilon_2 = 10^{-6}$, $\epsilon_3 = 0.25$, $l = 1$ and $j = 1$.

Step 1: Solve the constrained linear L_∞-norm problem:

$$\min_{\delta\boldsymbol{\theta}} \max_{1\le i\le n} \{|y_i - f_i(\boldsymbol{x}_i,\boldsymbol{\theta}^j) - \nabla f_i(\boldsymbol{x}_i,\boldsymbol{\theta}^j)'\delta\boldsymbol{\theta}|\}$$

$$\text{subject to} \quad \max_{1\le i\le k} |\delta\theta_i| \le \lambda_j$$

Call the solution $\delta\boldsymbol{\theta}^j$ with objective function value S^j_∞.

Step 2: Determine λ_{j+1} using Madsen's procedure (§3.5.1.2).

Step 3: Set $\Delta = S_\infty(\boldsymbol{\theta}^j) - S^j_\infty$ and $E_l = max(10^{-2}, 10^{-2}\Delta)$

If $\Delta < \epsilon_1$, STOP;

If $\Delta > E_l$ set $j := j+1$ and go to Step 1.

Step 4: If $j < p_l$ then set $j := j+1$ and return to Step 1.

Step 5: Let $\sigma_i = \{i \mid |\mathbf{r}^j_i| = S^j_\infty \quad 1 \le i \le k+1\}$.

If $\sigma_i \cap \{1,2,\ldots,n\}$ (cardinality less than $k+1$) has not changed over p_l iterations, set $A = \sigma_i \cap \{1,2,\ldots,n\}$ and go to Step 6; otherwise set $j := j+1$ and return to Step 1.

Step 6: Compute the Newton direction for the system (3.13) call it $\boldsymbol{\Delta\theta}^j$ using the current estimates for the unknowns. Set $\gamma = 1$.

Step 7: If $\|\Delta\boldsymbol{\theta}^j\| < \epsilon_2$, STOP.

Step 8: If $\gamma < \epsilon_3$ set $p_l := p_l + j,\ j := j+1, l := l+1,\ E_l = \Delta/10$ and return to Step 2.

Step 9: Test for a reduction in the sum of squares of the residuals of (3.13) using γ in the direction $\boldsymbol{\Delta\theta}^j$. Should there be a reduction, go to Step 10, otherwise set $\gamma := \gamma/2$ and go to Step 8.

Step 10: Set $\boldsymbol{\theta}^{j+1} := \boldsymbol{\theta}^j + \gamma\Delta\boldsymbol{\theta}^j$; $j := j+1$. If $|y_i - f_i(\boldsymbol{x}_i, \boldsymbol{\theta}^j)| > S^j_\infty$ for all $i \notin A$ set $p_l := p_l + 1,\ l := l+1,\ E_l = \Delta/10$ and return to Step 1. Otherwise go to Step 6.

Example 3.7 Reconsider the Madsen example 3.4. Choose $\boldsymbol{\theta}^1 = (3,1)$ and $\lambda = 1.2$ and convergence parameters as before. Then

$$J(\boldsymbol{\theta}) = \begin{pmatrix} 2\theta_1+\theta_2 & 2\theta_2+\theta_1 \\ \cos\theta_1 & 0 \\ 0 & -\sin\theta_2 \end{pmatrix} \quad \text{or} \quad J(\boldsymbol{\theta}^1) = \begin{pmatrix} 7 & 5 \\ -0.989992 & 0 \\ 0 & -0.84147 \end{pmatrix}$$

Step 1: $(l = j = 1)$. Determine $\delta\boldsymbol{\theta}$ as a solution to the LP problem

$$\min \quad d_\infty$$

$$\text{subject to} \quad \begin{aligned} 7\delta\theta_1 + 5\delta\theta_2 - d_\infty &\le -13 \\ -7\delta\theta_1 - 5\delta\theta_2 - d_\infty &\le 13 \\ -0.989992\delta\theta_1 - d_\infty &\le -0.14112 \\ 0.989992\delta\theta_1 - d_\infty &\le 0.14112 \\ -0.841471\delta\theta_2 - d_\infty &\le -0.540302 \\ 0.841471\delta\theta_2 - d_\infty &\le 0.540302 \end{aligned}$$

$$\text{and} \quad \begin{aligned} -1.2 \le \delta\theta_1 \le 1.2 \\ -1.2 \le \delta\theta_2 \le 1.2 \end{aligned}$$

The solution is $\delta\boldsymbol{\theta}^1 = (-1.09791, -0.81731)$, $d_\infty^1 = 1.22808$.

Step 2: $\lambda_2 = 1.0979$ as in §3.5.1.2.

Step 3: $d_\infty^1 = S_\infty^1 = \max\{1.22808, 1.22804, 1.22804\} = 1.22808$

$S_\infty(\boldsymbol{\theta}^1) = \max\{13, 0.14112, 0.54032\} = 13$

$\Delta = 13 - 1.22808 = 11.7718$

$E_1 = \max\{0.01, 0.0117718\} = 0.0117718$

$\Delta > \epsilon_1$ and $\Delta > E_1$

Hence we return to Step 1. The algorithm proceeds in this way using the Madsen iteration until S_∞^j approximates $S_\infty(\boldsymbol{\theta}^j)$ sufficiently well so that the Newton iteration can be used. Assume that $A = \{1, 3\}$ and has remained the same over the past few iterations and that we are at iteration ℓ where $\boldsymbol{\theta}^\ell = (0.53, -0.91)$. Then we solve the set of nonlinear equations

$$\begin{aligned} \alpha_1 + \alpha_3 &= 1 \\ \alpha_1(2\theta_1 + \theta_2) + \alpha_3 \times 0 &= 0 \\ \alpha_1(\theta_1 + 2\theta_2) - \alpha_3 \sin\theta_2 &= 0 \\ \theta_1^2 + \theta_2^2 + \theta_1\theta_2 - d_\infty &= 0 \\ \cos\theta_2 - d_\infty &= 0 \end{aligned}$$

Note that $sign[h_i(\boldsymbol{\theta}^*)]$ will be determined by the last point $\boldsymbol{\theta}^\ell$. The optimal solution was found by program MINPACK [see Moré et al (1980)] to be

$$\alpha_1 = 0.3667 \quad \alpha_3 = 0.6333 \quad \theta_1 = 0.4533 \quad \theta_2 = -0.9066 \quad d_\infty = 0.616$$

When we solved this system of equations with $\boldsymbol{\theta}^\ell = (3,1)$ the alternative solution $\theta_1 = -0.4533$, $\theta_2 = 0.9066$ was found.

3.5.3.2 The Hald and Madsen algorithm.

Hald and Madsen (1981) considered problem (3.5). Their algorithm is similar to the Watson algorithm of §3.5.3.1 and the Hald and Madsen (1985) algorithm of §2.3.4.1. It consists of two phases. In the first phase a constrained linear minimax problem is solved. If a smooth valley is detected (phase 2) then a set of nonlinear equations and inequalities is solved using a quadratically convergent Newton iteration. To perform phase 2 successfully the set of active functions has to be chosen correctly; if an incorrect choice is made then the algorithm reverts back to phase 1.

Let $A^* = \{i \mid y_i - f_i(\boldsymbol{x}_i, \boldsymbol{\theta}^*) = h_{max}(\boldsymbol{\theta}^*)\}$ be the set of active functions with cardinality number s. The following equations will hold in a smooth valley *and* at the solution:

$$y_i - f_i(\boldsymbol{x}_i, \boldsymbol{\theta}) = y_{i_0} - f_{i_0}(\boldsymbol{x}_i, \boldsymbol{\theta}) \quad i, i_0 \in A^* \tag{3.14}$$

If the Haar condition holds at $\boldsymbol{\theta}^*$ ($\boldsymbol{\theta}^*$ will be unique) then the cardinality number $s > k$ and the Jacobian of (3.14) will have rank k. If $s \leq k$ then the Jacobian of (3.14) will be rank deficient and the first-order optimality conditions (Corollary 3.2) in conjunction with (3.14) will be solved.

The Algorithm

Step 0: Given an initial estimate $\boldsymbol{\theta}^1$ of the optimal $\boldsymbol{\theta}^*$, an upper bound Λ_1 on the direction length and parameters $\epsilon_1, \epsilon_2 > 0$, set $j = 1$, $H^j = I_{k\times k}$ and *_iphase_* $= 1$.

Step 1: Phase 1. Solve the constrained linear minimax problem:

$$\min_{\delta\boldsymbol{\theta}} \max_{1\leq i\leq n} \{|y_i - f_i(\boldsymbol{x}_i, \boldsymbol{\theta}^j) - \nabla f_i(\boldsymbol{x}_i, \boldsymbol{\theta}^j)'\delta\boldsymbol{\theta}|\}$$

$$\text{subject to} \quad \max_{1\leq i\leq k} |\delta\theta_i| \leq \Lambda_j$$

Let the solution be $\delta\boldsymbol{\theta}^j$

Step 2: If _*iphase*_ = 1 update the bound as follows:

$$\Lambda_j = \begin{cases} 0.5\max_{1\le i\le k}|\delta\theta_i^{j-1}| & \text{if active function set is incorrect} \\ \max_{1\le i\le k}|\delta\theta_i^{j-1}| & \text{if active function set is reasonably correct} \\ 2\max_{1\le i\le k}|\delta\theta_i^{j-1}| & \text{if active function set is correct} \end{cases}$$

If _*iphase*_ = 2 then set $\Lambda_{j+1} = \Lambda_l$ (l the latest phase 1 iteration).

If $h_{max}(\boldsymbol{\theta}^j + \boldsymbol{d}^j) > h_{max}(\boldsymbol{\theta}^j)$ set $\boldsymbol{\theta}^{j+1} = \boldsymbol{\theta}^j$ otherwise set $\boldsymbol{\theta}^{j+1} = \boldsymbol{\theta}^j + \delta\boldsymbol{\theta}^j$.

Step 3: Define $A_j = \{i \mid h_{max}(\boldsymbol{\theta}^j) - y_i - f_i(\boldsymbol{x}_i, \boldsymbol{\theta}^j) \le \epsilon_1\}$.

Calculate the first-order multiplier estimate $\boldsymbol{\alpha}^j$ by solving the linear least squares problem

$$\min_{\boldsymbol{\alpha}} \|\sum_{i\in A_j} \alpha_i \nabla f_i(\boldsymbol{x}_i, \boldsymbol{\theta}^j)\|_2$$

Step 4: If $\boldsymbol{\theta}^{j+1} \ne \boldsymbol{\theta}^j$, calculate

$$\boldsymbol{\delta}^j = \boldsymbol{\theta}^{j+1} - \boldsymbol{\theta}^j$$

$$\gamma^j = \sum_{i\in A_j} \alpha_i^j [\nabla f_i(\boldsymbol{x}_i, \boldsymbol{\theta}^{j+1}) - \nabla f_i(\boldsymbol{x}_i, \boldsymbol{\theta}^j)]$$

update H^j as follows:

$$H^{j+1} = \left(I_k - \frac{\boldsymbol{\delta}^j(\gamma^j)'}{(\gamma^j)'\boldsymbol{\delta}^j}\right) H^j \left(I_k - \frac{\gamma^j(\boldsymbol{\delta}^j)'}{(\gamma^j)'\boldsymbol{\delta}^j}\right) + \frac{\boldsymbol{\delta}^j(\boldsymbol{\delta}^j)'}{(\gamma^j)'\boldsymbol{\delta}^j}$$

Step 5: Phase 2. Suppose the last 3 different iterates that were calculated in Step 2 are $\boldsymbol{\theta}^l$, $\boldsymbol{\theta}^{l-j_1}$ and $\boldsymbol{\theta}^{l-j_2}$. Suppose the active set has also remained constant:

$$A_l = A_{l-j_1} = A_{l-j_2}$$

and the multiplier estimates are such that

$$\sum_{\ell\in A_j} \alpha_\ell = 1$$

$$\max_{1\le i\le k} |\delta\theta_i^j| = \Lambda_j$$

$$A_j = A_{j-k_1} = A_{j-k_2}$$

$$\|\sum_{\ell \in A_j} \alpha_\ell \nabla f_\ell(\boldsymbol{x}_\ell, \boldsymbol{\theta}^j)\|_2 \leq \epsilon_2$$

then go to Step 6, otherwise return to Step 1 with *_iphase_* = 1.

Step 6: Perform one iteration of the BFGS method with H^l, $\boldsymbol{\theta}^l$ and $\boldsymbol{\alpha}^{l-j_1}$ as starting values to solve the system of equations

$$\sum_{i \in A_j} \alpha_i \nabla f_i(\boldsymbol{x}_i, \boldsymbol{\theta}) = 0$$

$$\sum_{i \in A_j} \alpha_i = 1$$

$$y_i - f_i(\boldsymbol{x}_i, \boldsymbol{\theta}) = y_{i_0} - f_{i_0}(\boldsymbol{x}_i, \boldsymbol{\theta}) \quad i, i_0 \in A_j, i \neq i_0$$

Call the next iterate $(\boldsymbol{\theta}^{l+1}, \boldsymbol{\alpha}^{l+1})$.

Step 7: Denote the residual vector of the system of equations in Step 6 by $r(\boldsymbol{\theta}, \boldsymbol{\alpha})$. If the following three conditions hold

$\|r(\boldsymbol{\theta}^{l+1}, \boldsymbol{\alpha}^{l+1})\| \leq 0.999\|r(\boldsymbol{\theta}^l, \boldsymbol{\alpha}^{l-j_1}\|$

$h_{max}(\boldsymbol{\theta}^{j+1}) = \max\{y_i - f_i(\boldsymbol{x}_i, \boldsymbol{\theta}^{j+1}), i \in A_j\}$

$\alpha_i^{l+1} \geq 0 \qquad i \in A_j$

then return to Step 6. If not then return to Step 1 with *_iphase_* = 2.

Remarks

(1) A_j determined in Step 3 is an approximation of the optimal set A^*.

(2) In Step 2 no line search is performed.

Convergence

Assuming the regularity condition holds:

(a) Hald and Madsen show that the algorithm will either terminate as a quadratically convergent phase 1 iteration (Step 1) or as a superlinearly convergent phase 2 iteration (Step 6) depending on whether the Haar condition is satisfied at the solution or not.

(b) If $s = k+1$ then a stationary point $\boldsymbol{\theta}^*$ is unique if and only if the Haar condition holds at $\boldsymbol{\theta}^*$.

3.5.4. Type IV algorithms

3.5.4.1 The Charalambous acceleration algorithm.

Charalambous (1979) utilizes the augmented Lagrangian function approach of NLP in which the following modified objective function is minimized in a sequential fashion:

$$U(\boldsymbol{\theta}, \boldsymbol{\alpha}, p, \xi) = \begin{cases} M(\boldsymbol{\theta}, \xi)\left[\sum_{i \in S} \alpha_i \left(\frac{h_i(\boldsymbol{\theta}) - \xi}{M(\boldsymbol{\theta}, \xi)}\right)^{q(p)}\right]^{1/q(p)} & \text{if } M(\boldsymbol{\theta}, \xi) \neq 0 \\ 0 & \text{otherwise;} \end{cases}$$

with $\alpha_i \geq 0$ and $\sum_{i=1}^{n} \alpha_i = 1$. where

$$M(\boldsymbol{\theta}, \xi) = \max_{1 \leq i \leq n} h_i(\boldsymbol{\theta}) - \xi$$

$$q = p\, sign[M(\boldsymbol{\theta}, \xi)] \qquad 1 < p < \infty$$

and

$$S = \begin{cases} \{i \mid h_i(\boldsymbol{\theta}) - \xi > 0, \quad i = 1, \ldots, n\} & \text{if } M(\boldsymbol{\theta}, \xi) > 0; \\ N & \text{if } M(\boldsymbol{\theta}, \xi) < 0; \end{cases}$$

where $N = \{1, \ldots, n\}$, the set of all the model function indices. p and ξ are fixed when U is minimized. The author observes that if $\boldsymbol{\alpha}^j$ is chosen close to the optimal $\boldsymbol{\alpha}^*$ then $\boldsymbol{\theta}^j$ will be close to $\boldsymbol{\theta}^*$ even for values of p not too large, thereby overcoming the inherent ill-conditioning in the penalty function approach of Charalambous and Bandler (1976). The algorithm proceeds as follows:

Algorithm

Step 0: Choose $p > 1, \quad \xi^1 = 0, \quad \alpha_i^1 = 1, \quad i = 1, \ldots, n$ and set $j = 1$.

Step 1: Minimize $U(\boldsymbol{\theta}, \boldsymbol{\alpha}^1, p, \xi^j)$ with respect to $\boldsymbol{\theta}$ call the solution $\hat{\boldsymbol{\theta}}^j$.

Step 2: Set $\alpha^{j+1} = \boldsymbol{v}^{j+1} / \sum_{i=1}^{n} v_i^{j+1}$ where

$$v_i^{j+1} = \begin{cases} \alpha_i^j \left(\frac{h_i(\boldsymbol{\theta}^j) - \xi}{M(\boldsymbol{\theta}^j, \xi)}\right)^{q-1} & \text{if } i \in S(\boldsymbol{\theta}^j, \xi); \\ 0 & \text{otherwise} \end{cases}$$

Step 3: Set $p = \beta p, \quad (\beta \geq 1)$; $j = j + 1$ and return to Step 1. Continue until certain convergence criteria are met.

Convergence properties:

(1)

$$\text{If} \quad \frac{q-1}{h_{max}(\boldsymbol{\theta}^*) - \xi} \geq c \geq 0$$

for c sufficiently large then θ^* is a strong local minimum of $U(\boldsymbol{\theta}, \boldsymbol{\alpha}^*, p, \xi)$.

(2) The Hessian $\nabla_{\boldsymbol{\theta}} U(\boldsymbol{\theta}^*, \boldsymbol{\alpha}^*, p, \xi)$ is positive definite for c sufficiently large.

(3) The algorithm is locally convergent.

The FORTRAN program CHARALAMBOUS is listed in Appendix 3A.

Remark:

Charalambous and Bandler (1976) proposed two algorithms for solving the minimax problem (3.5). In each algorithm a sequence of least p^{th} optimization problems is constructed with a constant and finite value of $p > 1$. This approach is novel since the exponent p assumes the role of a penalty parameter, thus the optimal solution is reached as $p \to \infty$. The following objective function is minimized in a sequential fashion.

$$U(\boldsymbol{\theta}, p, \xi) = \begin{cases} M(\boldsymbol{\theta}, \xi)\left[\sum_{i \in S}\left(\frac{h_i(\boldsymbol{\theta}) - \xi}{M(\boldsymbol{\theta}, \xi)}\right)^{q(p)}\right]^{1/q(p)} & \text{if } M(\boldsymbol{\theta}, \xi) \neq 0 \\ 0 & \text{otherwise;} \end{cases}$$

and is similar to $U(\boldsymbol{\theta}, \boldsymbol{\alpha}, p, \xi)$. The above objective function is a generalization of the usual p^{th} objective function but becomes increasingly more ill-conditioned as p becomes larger. Our numerical experience indicates that the Charalambous and Bandler (1976) and Charalambous (1979) algorithms are about equally efficient.

3.5.4.2 The Zang algorithm.

Zang (1980) considers the minimax problem (3.5) and proposes that $h_{max}(\boldsymbol{\theta})$ be approximated by a continuously differentiable function, denoted by $h_{max}(\boldsymbol{\theta}, \beta)$, with the parameter β a real number which controls the accuracy of the approximation. The function $h_{max}(\boldsymbol{\theta}, \beta)$ approximates $h_{max}(\boldsymbol{\theta})$ at points of derivative discontinuity while $h_{max}(\boldsymbol{\theta}, \beta) = h_{max}(\boldsymbol{\theta})$ otherwise. Again assuming for notational convenience that $h_i(\boldsymbol{\theta}) = y_i - f_i(\boldsymbol{x_i}, \boldsymbol{\theta})$.

The approximation is:

$$h_{max}(\boldsymbol{\theta}, \beta) = h_1(\boldsymbol{\theta}) + \max[\beta, \{h_2(\boldsymbol{\theta}) - h_1(\boldsymbol{\theta})\} + \cdots + \max[\beta, \{h_n(\boldsymbol{\theta}) - h_{n-1}(\boldsymbol{\theta})\}] \cdots]$$

where, for $i \geq 2$

$$\max[\beta, h_i(\boldsymbol{\theta}) - h_{i-1}(\boldsymbol{\theta})] = \begin{cases} 0 & \text{if } h_i(\boldsymbol{\theta}) - h_{i-1}(\boldsymbol{\theta}) \leq -\beta \\ b_j[\beta, h_i(\boldsymbol{\theta}) - h_{i-1}(\boldsymbol{\theta})] & \text{if } -\beta \leq h_i(\boldsymbol{\theta}) - h_{i-1}(\boldsymbol{\theta}) \leq \beta \\ h_i(\boldsymbol{\theta}) - h_{i-1}(\boldsymbol{\theta}) & \text{if } \beta \leq h_i(\boldsymbol{\theta}) - h_{i-1}(\boldsymbol{\theta}) \end{cases}$$

with $b_j(\beta, h_i(\boldsymbol{\theta}) - h_{i-1}(\boldsymbol{\theta}))$ a polynomial of degree $2j$. Zang gives expressions for b_j which can be used to give first, second, third and fourth-order smooth approximations to $h_{max}(\boldsymbol{\theta})$. For example, a first-order smooth approximation is given by

$$b_1(\beta, t) = \frac{t^2}{4\beta} + \frac{t}{2} + \frac{\beta}{4}$$

Hence instead of (3.5) the following approximate problem is solved

$$\min_{\boldsymbol{\theta}} h_{max}(\boldsymbol{\theta}, \beta) \tag{3.15}$$

$$\text{where} \quad \lim_{\beta \to 0} h_{max}(\boldsymbol{\theta}, \beta) = h_{max}(\boldsymbol{\theta})$$

The Algorithm

Step 0: Choose values $\epsilon > 0$, $\alpha > 1$. Given an initial estimate $\beta^1 > 0$, set $j = 1$.

Step 1: Solve $\min h_{max}(\boldsymbol{\theta}, \beta^j)$. Let the solution be $\boldsymbol{\theta}^j$

Step 2: If $h_{max}(\boldsymbol{\theta}^j, \beta^j) = h_{max}(\boldsymbol{\theta}^j)$ or $\beta^j < \epsilon$ then $\boldsymbol{\theta}^j$ is the solution of (3.5) therefore STOP.

Step 3: Let $\beta^{j+1} = \beta^j / \alpha$ and return to Step 1.

Convergence properties.

(1) If $\boldsymbol{\theta}^*$ solves problem (3.5) and $\hat{\boldsymbol{\theta}}$ solves (3.15) , then

$$0 \leq h_{max}(\hat{\boldsymbol{\theta}}, \beta) - h_{max}(\boldsymbol{\theta}^*) \leq \min\{\beta, (n-1)b(\beta, 0)\}$$

(2) Suppose $\{\beta^j\} \to 0$ monotonically, $\hat{\boldsymbol{\theta}}^j$ solves (3.15) and $\hat{\boldsymbol{\theta}}$ is an accumulation point of $\{\boldsymbol{\theta}^j\}$ then $h_{max}(\hat{\boldsymbol{\theta}}) = h_{max}(\boldsymbol{\theta}^*)$.

(3) Let $\boldsymbol{\theta}^*$ be a local minimum point of (3.5) where $h_{max}(\boldsymbol{\theta}^*) = h_l(\boldsymbol{\theta}^*)$, for only one value $1 \leq l \leq n$. Then there exists a $\bar{\beta} > 0$ such that $\boldsymbol{\theta}^*$ is a local minimum point of (3.15) for every β satisfying $0 \leq \beta \leq \bar{\beta}$.

Remarks

(1) Any efficient unconstrained minimization procedure can be used in Step 1.

(2) For $n > 2$ the value of $h_{max}(\boldsymbol{\theta}, \beta)$ can depend on the order of the model functions $\{h_1 \prec h_2 \prec \cdots \prec h_n\}$. This dependence can be removed by choosing β sufficiently small. However, should β be chosen too small then the unconstrained problem can become ill-conditioned.

Appendix 3A: FORTRAN Programs

Program MINMAXABS

```
C
C         THIS PROGRAM IS BASED ON THE ALGORITHM BY OSBORNE AND
C         WATSON (1969),
C
C         IT USES SUBROUTINE CHEB DUE TO BARRODALE AND PHILLIPS (1975)
C
C         SUBROUTINE LINAND PERORMS THE ANDERSON AND OSBORNE LINE
C         SEARCH RULE: ANDERSON D.H. AND OSBORNE, M.R. (1977A). DISCRETE
C         NONLINEAR APPROXIMATION PROBLEMS IN POLYHEDRAL NORMS.
C         NUMER. MATH. 28, PP 143-156.
C
C         OPTIMIZATION PROBLEM DATA
C         N = NUMBER OF PARAMETERS
C         M = NUMBER OF OF PROBLEM FUNCTIONS (OR OBSERVATIONS)
C         NITER = MAXIMUM NUMBER OF ITERATIONS REQUIRED
C         THETA(.)  = PARAMETER VECTOR
C
C         THIS EXAMPLE IS DUE TO :
C
C         MADSEN, K. (1975).  AN ALGORITHM FOR MINIMAX SOLUTION TO
C         OVERDETERMINED SYSTEMS OF NON-LINEAR EQUATIONS. J. INST.
C         MATH. APPLIC. 16, PP 321-328.
C
          IMPLICIT REAL*8 (A-H,O-Z)
          DIMENSION F(3),THETA(2),DTHETA(2),ALP(2),T(40)
          DIMENSION A(40,40),Y(90),B(40),YHAT(90)
          TOL=1D-15
          M=3
          N=2
          MDIM=40
          NDIM=40
          NITER=100
          THETA(1)=3.0
          THETA(2)=1.0
          KOUNT=0
          CALL FUNMAX(THETA,FOLD)
          WRITE(6,98)(THETA(I),I=1,N),FOLD
98        FORMAT(' INITIAL THETA PARAMETERS: ',2F8.3,' FNORM',F10.5)
          SJ=1D10
15        CONTINUE
          KOUNT=KOUNT+1
          CALL FUNC(THETA,A,F)
          DO 999 I=1,M
          Y(I)=0D0
999       B(I)=Y(I)-F(I)
          RELERR=0D0
          CALL CHEB(M,N,MDIM,NDIM,A,B,TOL,RELERR,T,RANK,RESMAX,ITER, OCODE)
```

```
      WRITE(6,80)KOUNT,ITER
80    FORMAT(' ITERATION NUMBER = ',I4,' NUMBER OF SIMPLEX ITERATIONS ',I4)
      WRITE(6,776) (T(I),I=1,N)
776   FORMAT(1X,' DELTA THETA =',2F12.6)
      SJ=RESMAX
      IF(KOUNT.EQ.NITER) GO TO 200
      F1=FOLD
      CALL LINAND(N,T,THETA,SJ,FNEW,F1)
      WRITE(6,150)(THETA(I),I=1,N),SJ,FNEW
150   FORMAT(' THETA ' ,2F12.5/' DHATJ',F13.7,' SBARJ+1 ',F13.7)
      IF(DABS((FNEW-FOLD)/FOLD).LT.1D-10)GO TO 200
      IF(DABS(FNEW-SJ).LT.1D-6)GO TO 200
      FOLD=FNEW
      GO TO 15
200   STOP
      END
      SUBROUTINE FUNC(THETA,A,F)
      IMPLICIT REAL*8 (A-H,O-Z)
      DIMENSION A(40,40),F(1),THETA(1)
C
C     A(I,J) IS THE DERIVATIVE OF THE J-TH MODEL FUNCTION F(J) W.R.T
C     THE I-TH PARAMETER THETA(I)
C
      F(1)=THETA(1)**2 +THETA(1)*THETA(2) +THETA(2)**2
      F(2)=DSIN(THETA(1))
      F(3)=DCOS(THETA(2))
      A(1,1)=2DO*THETA(1)+THETA(2)
      A(1,2)=DCOS(THETA(1))
      A(1,3)=ODO
      A(2,1)=THETA(1)+2DO*THETA(2)
      A(2,2)=ODO
      A(2,3)=-DSIN(THETA(2))
      RETURN
      END
      SUBROUTINE FUNMAX(THETA,FNEW)
      IMPLICIT REAL*8 (A-H,O-Z)
      DIMENSION F(100),THETA(1)
      COMMON/DATA/M
      F(1)=THETA(1)**2 +THETA(1)*THETA(2) +THETA(2)**2
      F(2)=DSIN(THETA(1))
      F(3)=DCOS(THETA(2))
      FNEW=DMAX1(DABS(F(1)),DABS(F(2)),DABS(F(3)))
      RETURN
      END
C
C     SUBROUTINE LINAND(NPAR,T,THETA,SJ,FNEW,F1)
C     THIS IS AN INACCURATE LINE SEARCH DUE TO ANDERSON AND OSBORNE
C     (1977)
C
      SUBROUTINE LINAND(NPAR,T,THETA,SJ,FNEW,F1)
      IMPLICIT REAL*8 (A-H,O-Z)
      DIMENSION THETA(1),T(1),DTHETA(5)
```

```
          DELTA = 1D-4
          GAM=1D0
1         CONTINUE
          DO 10 I=1,NPAR
10        DTHETA(I)=THETA(I) + GAM*T(I)
          CALL FUNMAX(DTHETA,FNEW)
          PSI=(F1 - FNEW)/(GAM*(F1-SJ))
C         WRITE(6,900)PSI
          IF(PSI.GE.DELTA) GO TO 30
          GAM = GAM*0.1D0
          GO TO 1
30        DO 35 I=1,NPAR
35        THETA(I)=DTHETA(I)
          CALL FUNMAX(THETA,FNEW)
C         WRITE(6,900)GAM,F1,FNEW
900       FORMAT(3F14.4)
          RETURN
          END
```

```
INITIAL THETA PARAMETERS: 3.000 1.000 FNORM 13.00000
ITERATION NUMBER = 1 NUMBER OF SIMPLEX ITERATIONS = 3
DELTA THETA = -1.097914 -0.817312
THETA 1.90209 0.18269
DHATJ 1.2280463 SBARJ+1 3.9987966
ITERATION NUMBER = 2 NUMBER OF SIMPLEX ITERATIONS = 3
DELTA THETA = -0.420958 -0.545962
THETA 1.48113 -0.36327
DHATJ 1.0825458 SBARJ+1 1.7876531
ITERATION NUMBER = 3 NUMBER OF SIMPLEX ITERATIONS = 3
DELTA THETA = -1.012158 -0.082720
THETA 0.46897 -0.44599
DHATJ 0.9053455 SBARJ+1 0.9021828
ITERATION NUMBER = 4 NUMBER OF SIMPLEX ITERATIONS = 3
DELTA THETA = -0.880845 -1.317728
THETA 0.38089 -0.57777
DHATJ 0.3337755 SBARJ+1 0.8376850
ITERATION NUMBER = 5 NUMBER OF SIMPLEX ITERATIONS = 3
DELTA THETA = -0.965796 -0.572814
THETA 0.28431 -0.63505
DHATJ 0.5248403 SBARJ+1 0.8050437
ITERATION NUMBER = 10 NUMBER OF SIMPLEX ITERATIONS = 3
DELTA THETA = -1.299755 -0.239126
THETA 0.52292 -0.93393
DHATJ 0.4249372 SBARJ+1 0.6573021
```

```
ITERATION NUMBER = 20 NUMBER OF SIMPLEX ITERATIONS = 3
DELTA THETA = -1.198173 0.009223
THETA 0.46455 -0.92896
DHATJ 0.6059808 SBARJ+1 0.6472266
ITERATION NUMBER = 50 NUMBER OF SIMPLEX ITERATIONS = 3
DELTA THETA= -1.191021 -0.014010
THETA 0.46483 -0.91433
DHATJ 0.5993372 SBARJ+1 0.6270551
```

Program CHARALAMBOUS

```
C
C          THIS PROGRAM USES FLETCHER'S ALGORITHM (1970)
C          HARWELL SUBROUTINE VA10AD ( NUMERICAL DERIVATIVES)
C          TO SOLVE THE GENERAL MINIMAX PROBLEM
C          BY MEANS OF THE AUGMENTED LAGRANGIAN FUNCTION APPROACH OF
C          CHARALAMBOUS (1979)
C
           IMPLICIT REAL*8(A-H,O-Z)
           DIMENSION X(100),G(100),H(210),W(100),XM(100)
           COMMON/DATA/P,Q,ZETA,M
           COMMON/LAGRAN/U(10),V(10)
C
C          OPTIMIZATION PROBLEM DATA
C          N = NUMBER OF PARAMETERS
C          M = NUMBER OF OF PROBLEM FUNCTIONS (OR OBSERVATIONS)
C          X(.)  = PARAMETER VECTOR
C          P AND Q ARE THE EXPONENT PARAMETERS USED IN THE ALGORITHM
C          F IS THE OBJECTIVE FUNCTION VALUE U(THETA,P,ZETA)
C          DFN, MAXFN, EPS, IPRINT, MODE ARE CONTROL
C          PARAMETERS OF SUBROUTINE VA10AD
C
C          THIS EXAMPLE IS DUE TO :
C
C          MADSEN, K. (1975).  AN ALGORITHM FOR MINIMAX SOLUTION TO
C          OVERDETERMINED SYSTEMS OF NON-LINEAR EQUATIONS. J. INST. MATH.
C          APPLIC. 16, PP 321-328.
C
           M=3
           N=2
           X(1)=3.00
           X(2)=1.0
           U(1)=1.0
           U(2)=1.0
           U(3)=1.0
           DO 20 I=1,N
20         XM(I)=1D0
           HH=1D-6
           EPS=1.D-8
           DFN=0.
           MAXFN=100
```

```
      MODE=1
      CALL FUNMAX(X,HMAX)
      ZETA = ODO
      ALP = 1D10
      P = 2DO
30    CONTINUE
      IPRINT=MAXFN + 1
      CALL VA10AD(N,X,F,G,H,W,DFN,XM,HH,EPS,1,MAXFN,IPRINT,IEXIT)
C
C     UPDATING THE MULTIPLIERS U(I)
C
      VSUM = ODO
      DO 40 I =1,M
40    VSUM = VSUM + V(I)
      DO 50 I =1,M
50    U(I) =V(I)/VSUM
C
C     UPDATING P
C
      P = 2DO*P
      WRITE(6,1003) (X(I),I=1,N)
      WRITE(6,1004) (U(I),I=1,M)
1003  FORMAT(1X,'THETA =',5F20.8)
1004  FORMAT(1X,'U(I) =',5F20.8)
      WRITE(6,1002)F,ZETA,Q
1002  FORMAT(1X,'U(THETA,P,ZETA)=',F16.8,' ZETA =',F12.4,' Q =',F12.4)
      GO TO 500
26    ALP=F
      GO TO 30
500   STOP
      END
      SUBROUTINE FUNCT(N,X,F)
      IMPLICIT REAL*8 (A-H,O-Z)
      DIMENSION X(1),FF(3)
      COMMON/DATA/P,Q,ZETA,M
      COMMON/LAGRAN/U(10),v(10)
      CALL FUNMAX(X,HMAX)
      HMMAX = HMAX - ZETA
      F=ODO
      FF(1)=X(1)**2 +X(1)*X(2) +X(2)**2
      FF(2)=DSIN(X(1))
      FF(3)=DCOS(X(2))
      Q = P*DSIGN(1DO,HMMAX)
      IF(HMMAX.EQ.ODO) RETURN
      IF (HMMAX.GT.ODO) GO TO 5
      IF (HMMAX.LT.ODO) GO TO 6
5     DO 2 I=1,M
      V(I) = 0.0
      IF((FF(I)-ZETA).LT.ODO) GO TO 2
      V(I) = U(I)*((FF(I)-ZETA)/HMMAX)**(Q-1)
      F = F + U(I)*((FF(I)-ZETA)/HMMAX)**Q
2     CONTINUE
```

```
      F = F**(1DO/Q)
      GO TO 7
6     DO 3 I=1,M
      V(I) = U(I)*((FF(I)-ZETA)/HMMAX)**(Q-1)
      F = F + U(I)*((FF(I)-ZETA)/HMMAX)**Q
3     CONTINUE
      F = F**(1DO/Q)
7     F = F*HMMAX
      RETURN
8     F = ODO
      RETURN
      END
      SUBROUTINE FUNMAX(X,FNEW)
      IMPLICIT REAL*8 (A-H,O-Z)
      DIMENSION F(100),X(1)
      F(1)=X(1)**2 +X(1)*X(2) +X(2)**2
      F(2)=DSIN(X(1))
      F(3)=DCOS(X(2))
      FNEW = DMAX1(F(1),F(2),F(3))
      RETURN
      END
```

```
THETA = 0.15543724 -0.69456378
U(I) = 0.30158079 0.11712539 0.58129381
U(THETA,P,ZETA) = 0.87931738 ZETA = 0.0000 Q = 2.0000
THETA = 0.41048706 -0.91278605
U(I) = 0.34618412 0.03466345 0.61915243
U(THETA,P,ZETA) = 0.60147589 ZETA = 0.0000 Q = 4.0000
THETA = 0.45009053 -0.90788675
U(I) = 0.36480273 0.00312199 0.63207528
U(THETA,P,ZETA) = 0.61388271 ZETA = 0.0000 Q = 8.0000
THETA = 0.45330413 -0.90665378
U(I) = 0.36668203 0.00001856 0.63329940
U(THETA,P,ZETA) = 0.61631244 ZETA = 0.0000 Q = 16.0000
THETA = 0.45329640 -0.90659281
U(I) = 0.36669726 0.00000000 0.63330274
U(THETA,P,ZETA) = 0.61643208 ZETA = 0.0000 Q = 32.0000
THETA = 0.45329624 -0.90659247
U(I) = 0.36669729 0.00000000 0.63330271
U(THETA,P,ZETA) = 0.61643244 ZETA = 0.0000 Q = 64.0000
THETA = 0.45329624 -0.90659247
U(I) = 0.36669677 0.00000000 0.63330323
U(THETA,P,ZETA) = 0.61643244 ZETA = 0.0000 Q = 128.0000
```

Chapter 3: Additional notes

§3.1

1 The L_∞-norm problem is considered in Rice (1969), Watson (1980) & Powell (1981). More recently the latter norm was used in forecasting problems arising in the field of metereology, see for example Pilibossian & Der Megreditchian (1983).

2 Formulation (2.4) in Chapter 2 should not be used to define an equivalent formulation involving n equality constraints and $k+2$ variables. Inspection of such a "*formulation*" shows that the equality constraints are inconsistent and hence the problem is infeasible.

§3.2

1 Various authors considered the necessary and sufficiency conditions for L_∞-norm problems; for example Powell & Yuan (1984). Womersley & Fletcher (1986) also use a certain regularity assumption. They consider the more general *composite nonsmooth* optimization problem of which the L_∞, minimax and L_1-norm problems are special cases.

2 Theorem 3.2: There are two possibilities for d_∞^*. Either $d_\infty^* = 0$ or $d_\infty^* > 0$. In the first case the model interpolates the data exactly and hence from KKT condition (a) $h_i(\boldsymbol{\theta}^*) = 0$ for all i. This case is not of interest.

§3.3

1 Sometimes the terms "minimax" and "L_∞-norm" are incorrectly used synonomously.

2 The minimax problem has received more attention in the literature than the nonlinear L_∞-norm problem, the argument being that the latter is a special case of the former. Note that the latter problem will always have a solution whilst the former may not have a solution.

3 Minimax problems are considered in Zuhovickii et al. (1963), Zangwill (1967), Dem'yanov (1968), Ishizake & Watanabe (1968), Shor & Shabashova (1972), Dem'yanov & Malozemov (1974), Charalambous & Moharram (1978), Wierzibicki (1978) and Drezner (1982).

4 The following authors give certain optimality conditions for the minimax problem: Charalambous & Bandler (1976), Charalambous & Conn (1978), Han (1981) and Ben-Tal & Zowe (1982).

5 Nonlinearly constrained minimax problems have also received attention; the interested reader is referred to the papers by Dutta and Vidyasagar (1977),

Coleman (1978), Madsen & Schjær-Jacobsen (1978), Fletcher & Watson (1980), Bazaraa & Goode (1982) and Shanno & Whitaker (1985).

§3.5

In the literature the focus has mainly been on algorithms which solve the minimax problem (3.3). These algorithms can then be used to solve the L_∞-norm problem by substituting $\{h_i(\boldsymbol{\theta}), -h_i(\boldsymbol{\theta})\}$ for $h_i(\boldsymbol{\theta})$ in the relevant minimax algorithm. It has been argued by Murray & Overton (1980) that it is better to design the algorithm to solve the Chebychev problem directly rather than by making the above substitution.

§3.5.1

1 Charalambous & Conn (1978) solve problem (3.6) directly by means of gradient projection. The algorithm is similar to the method of Zangwill (1967) and involves two search directions:

The *horizontal direction* which attempts to reduce the maximum deviation whilst at the same time keeping those deviations which are numerically close to the maximum deviation approximately equal.

The *vertical direction* which attempts to reduce the error of the remaining inactive functions by means of linearization.

2 Zangwill (1967) used the gradient projection technique of Rosen (1960,1961) to solve minimax problems.

§3.5.1.2

Madsen (1975) observed that the Gauss-Newton type algorithms will only converge as long as the resulting linear system is bounded away from singularity (matrix $J'J$ is nonsingular in the least squares context). Should this not be the case, convergence of the iterates $\boldsymbol{\theta}^j$ to a nonstationary point of (3.2) can occur. One way of overcoming the singularity problem is the Levenberg-Marquardt approach.

§3.5.1.3

1 The Levenberg-Marquardt modification of the Anderson-Osborne (1977a) algorithm is similar to the algorithm described in §2.3.2.2 in Chapter 2 for the L_1-norm case.

2 Step 1 can also be written as

$$\min_{\delta\boldsymbol{\theta}} \max_{1 \le i \le n} \{|h_i(\boldsymbol{\theta}^j) + \nabla h_i(\boldsymbol{\theta}^j)'\delta\boldsymbol{\theta}|, \lambda_j|\delta\boldsymbol{\theta}|\}$$

3 Note that the Madsen and the Anderson-Osborne-Levenberg-Marquardt algorithms differ only with respect to the restriction on the length of $\delta\boldsymbol{\theta}$. The

former states

$$\delta\boldsymbol{\theta} \le \lambda_j \mathbf{1}_k$$
$$-\delta\boldsymbol{\theta} \le \lambda_j \mathbf{1}_k$$

whilst the latter stipulates

$$\lambda_j \delta\boldsymbol{\theta} \le d_\infty \mathbf{1}_k$$
$$-\lambda_j \delta\boldsymbol{\theta} \le d_\infty \mathbf{1}_k$$

§3.5.2.2

1 Subroutine QPROG in the IMSL library may be used to solve both QP problems. If the quadratic parts of the objective functions become semi-definite one may run into numerical problems with this routine.

2 The dual problem (3.12) has $2n$ nonnegative variables and a single constraint whilst the primal problem (3.11) has $k+1$ variables and $2n$ constraints. For problems with a large number of functions, solving the dual would be advantageous.

3 Han (1981) mentions that B^j should be updated at each iteration but provides no further detail. Presumably any quasi-Newton updating rule, e.g., the BFGS rule would suffice.

4 Certain conditions are, however, placed on B^j viz that it is uniformly positive definite and bounded i.e., for any positive numbers ρ_1 and ρ_2

$$\rho_1 ||\boldsymbol{x}||^2 \le \boldsymbol{x}' B^j \boldsymbol{x} \le \rho_2 ||\boldsymbol{x}||^2 \quad \text{for any} \quad \boldsymbol{x} \in \Re^n$$

This would then make the convergence analysis independent of the updating rule used for B^j. Furthermore B^j will be a good approximation of the Hessian $\nabla^2_{\boldsymbol{\theta}} L$ at convergence.

§3.5.3.1

1 To ensure that problem (3.13) is of full rank and hence that a solution exists, we must have $\lambda_j \ge 0$ (If J is of full rank then $\lambda = 0$). At each iteration λ_j can be determined by the Anderson and Osborne (1977b) procedure. Another approach, due to Madsen (1975), is available to determine the optimal $\delta\boldsymbol{\theta}$ and λ_j (see §3.4.2).

2 Variants of this algorithm may be found in Watson (1976,1980).

§3.5.3.2

1 In this second stage the algorithm exhibits superlinear convergence.

2 If an incorrect choice of the active set is made this could entail several switches back and forth between stages 1 and 2. In Hald and Madsen's experience only a few such switches take place.

3 The authors are rather vague about the updating of λ_j in Step 1. Their terms "unsuccessful", "not unsuccessful" and "successful" were replaced by "A^* is correct" etc. Instead of the rule in Step 1, Madsen's rule for updating λ_j may be used.

4 The Newton iteration is approximate since 2nd-derivative information is obtained by finite differences.

5 The conditions in Step 2 ensure that unnecessary iterations in Step 1 are avoided before switching to the Newton (or quasi-Newton) step.

6 The condition $A_j = A_{j-k_1} = A_{j-k_2}$ tests the stability of the active set and the last condition holds when $\boldsymbol{\theta}^j$ is close to $\boldsymbol{\theta}^*$, with $A^* = A_j$.

7 It is not required for $h_{max}(\boldsymbol{\theta})$ to decrease in the Newton iteration. In fact it may often increase.

8 Hettich (1976), Han (1978) and Watson (1979) noted that 2nd-derivative information is necessary to enhance convergence. They observed that when the minimum is well determined and no smooth valley traverses the minimum then 1st-derivative information will be sufficient. In the presence of such a smooth valley then 2nd derivatives have to be used. This argument is similar to the one in least squares which will result in the matrix $J'J$ being singular. The stipulation that the the Haar condition holds ensures that no smooth valley traverses the solution.

§3.5.4.1

1 The augmented Lagrangian approach to NLP was independently studied by Hestenes (1969), Powell (1969), Haarhoff and Buys (1970), Buys (1972), Rockafellar (1973) and Bertsekas (1976). If the minimax problem is written in its NLP form then the algorithm is similar to the augmented Lagrangian method applied to the resulting NLP problem.

2 Charalambous (1979) makes the following assumptions:

(a) The functions $h_i(\boldsymbol{\theta})$ are twice continuously differentiable.

(b) Regularity assumption: The gradients of the active constraint set at the optimum solution $\boldsymbol{\theta}^*$

$$\begin{pmatrix} -1 \\ \nabla h_j(\boldsymbol{\theta}^*) \end{pmatrix} \qquad j \in A = \{i \mid h_i(\boldsymbol{\theta}) = h_{max}(\boldsymbol{\theta}^*)\}$$

are linearly independent.

(c) The second-order sufficiency conditions for a local minimum of problem (3.3) hold.

(d) Strict complementarity holds (the multipliers $\alpha_i > 0$ for $i \in A$).

3 Charalambous does not mention any other value of ξ except $\xi = 0$ nor does he mention whether ξ should be updated or not.

§3.5.4.2

Zang (1980) reports his numerical experience on two examples. A starting value of $\beta = 0.1$ was used, $\epsilon = 10^{-5}$ and 3 values $\alpha = 10, 100, 1000$ were chosen to investigate how β should be decreased. Both 1st- and 2nd-degree smoothing functions $b_1(\beta, t)$ and $b_2(\beta, t)$ were evaluated. A value of $\alpha = 100$ with $b_1(\beta, t)$ was found to be preferable.

Bibliography Chapter 3

Anderson, D.H. and Osborne, M.R. (1977a). Discrete, nonlinear approximation problems in polyhedral norms. *Numer. Math.* **28**, pp 143–156.

Anderson, D.H. and Osborne, M.R. (1977b). Discrete, nonlinear approximation problems in polyhedral norms. A Levenberg-like algorithm. *Numer. Math.* **28**, pp 157–170.

Barrodale, I. and Phillips, C. (1975). Solution of an overdetermined system of linear equations in the Chebychev norm. *ACM Trans. on Math. Software* **1**, pp 264–270.

Bazaraa, M.S. and Goode, J. (1982). An algorithm for constrained minimax problems. *Eur. J. Oper. Res.* **11**, pp 158–166.

Ben-Tal, A. and Zowe, J. (1982). Necessary and sufficient conditions for a class of nonsmooth minimization problems. *Math. Programming* **24** pp 70–88.

Bertsekas, D.P. (1976). Multiplier methods: A survey. *Automatica* **12**, pp 133–145.

Buys, J.D. (1972). Dual algorithms for constrained optimization problems. *PhD Thesis. Leiden University, Leiden, Netherlands.*

Charalambous, C. (1979). Acceleration of the least pth algorithm for minimax optimization with engineering applications. *Math. Programming* **17**, pp 270–297.

Charalambous, C. and Bandler, J.W. (1976). Non-linear minimax optimization as a sequence of least pth optimization with finite values of p. *Int. J. Systems Sci.* **7**, pp 377–391.

Charalambous, C. and Conn, A.R. (1978). An efficient method to solve the minimax problem directly. *SIAM J. Numer. Anal.* **15**, pp 162–187.

Charalambous, C. and Moharram, O. (1978). A new approach to minimax optimization. In: *G.J. Savage and P.H. Roe, eds. Large Eng. Systems* **2** pp 169–172.

Chatelon, J.A., Hearn, D.W. and Lowe, T.J. (1978). A subgradient algorithm for certain minimax and minisum problems. *Math. Programming* **15** pp 130–145.

Coleman, T.F. (1978). A note on 'New algorithms for constrained minimax optimization'. *Math. Programming* **15**, pp 239–242.

Cromme, L. (1978). Strong uniqueness: A far reaching criterion for the convergence analysis of iterative procedures. *Numer. Math.* **29**, pp 179–194.

Dem'yanov V.F. (1968). Algorithms for some minimax problems. *J. Comp. Syst. Sci.* **2**, pp 342–380.

Dem'yanov V.F. and Malozemov, V.N. (1974). Introduction to minimax. *John Wiley, New York.*

Drezner, Z. (1982). On minimax optimization problems. *Math. Programming* **22**, pp 227–230.

Dutta, S.R.K. and Vidyasagar, M. (1977). New algorithms for constrained minimax optimization. *Math. Programming* **13**, pp 140–155.

Fletcher, R. and Watson, G.A. (1980). First and second order conditions for a class of nondifferentiable optimization problems. *Math. Programming* **18**, pp 291–307.

Gill, P.E. , Murray, W. and Wright, M. (1981). Practical optimization. *Academic Press, London.*

Haarhoff, P.C. and Buys, J.D. (1970). A new method for the optimization of a nonlinear function subject to nonlinear constraints. *Comput. J.* **13**, pp 178–184.

Hald, J. and Madsen, K. (1978). A 2-stage algorithm for minimax optimization. In: *Lecture notes in Control and Information Sciences* **15**, *Springer Verlag* pp 225–239.

Hald, J. and Madsen, K. (1981). Combined LP and Quasi-Newton methods for minmax optimization. *Math. Programming* **20**, pp 49–62.

Hald, J. and Madsen, K. (1985). Combined LP and Quasi-Newton methods for nonlinear l_1 optimization. *SIAM J. Numer. Anal.* **22**, pp 68–80.

Han, S.P. (1978). Superlinear convergence of a minimax method. *Cornell University, Computer Science TR 78-336.*

Han, S.P. (1981). Variable metric methods for minimizing a class of nondifferentiable functions. *Math. Programming* **20**, pp 1–13.

Hestenes, M.R. (1969). Multiplier and gradient methods. *J. Optim. Theory Appl.* **4**, pp 303–320.

Hettich, R. (1976). A Newton method for nonlinear Chebyshev approximation, In: R. Schaback and K. Scherer eds., *Approximation theory, Lecture Notes in Mathematics, 556, Springer Verlag, Berlin*, pp 222–236.

Ishizake, Y. and Watanabe, H. (1968). An iterative Chebychev approximation method for network design. *IEEE Trans. Circuit Theory CT* **15**, pp 326–336.

Joe, B. and Bartels, R. (1983). An exact penalty method for constrained, discrete, linear ℓ_∞ data fitting. *SIAM J. Sci. Stat. Comput.* **4**, pp 69–84.

Madsen, K. (1975). An algorithm for minimax solution of overdetermined systems of non-linear equations. *J. Inst. Maths. Applic.* **16**, pp 321–328.

Madsen, K. and Schjær-Jacobsen, H. (1978). Linearly constrained minimax optimization. *Math. Programming* **14**, pp 208–223.

McLean, R.A. and Watson, G.A. (1980). Numerical methods for nonlinear discrete L_1 approximation problems. In: L. Collatz, G. Meinardus and H. Werner, eds. *Numerical methods of approximation theory, Birkhäuser Verlag, Basel.*

Moré, J.J., Garbow, B. and Hillstrom, K.E. (1980). *User guide for MINPACK-1. Argonne National Laboratory, Argonne, Illinois.*

Murray, W. and Overton, M (1979). Steplength algorithms for minimizing a class of nondifferentiable functions. *Computing* **23**, pp 309–331.

Murray, W. and Overton, M (1980). A projected Lagrangian algorithm for nonlinear minimax optimization. *SIAM J. Sci. Stat. Comput.* **1**, pp 345–370.

Osborne, M.R. and Watson, G.A. (1969). An algorithm for minimax approximation in the nonlinear case. *Comput. J.* **12**, pp 63–68.

Overton, M. (1982). Algorithms for nonlinear l_1 and l_∞ fitting. In: M.J.D. Powell ed., *Nonlinear optimization, Academic Press, London.*

Pilibossian, P. and Der Megreditchian, G. (1983). Research of a regression function by means of minimax criterion (French). *Publications of Institute of Statistics, University of Paris IV, Paris.*

Powell, M.J.D. (1969). A method for nonlinear constraints in minimization problems, In: R. Fletcher, ed., Optimization, *Academic Press, London*, pp 283–298.

Powell, M.J.D. (1981). Approximation theory and methods. *Cambridge University Press, Cambridge, UK.*

Powell, M.J.D. and Yuan, Y. (1984). Conditions for superlinear convergence in ℓ_1 and ℓ_∞ solutions of overdetermined non-linear equations. *IMA J. Numer. Anal.* **4**, pp 241–251.

Rice, J.R. (1969). The approximation of functions, vol 2: Nonlinear and multivariate theory. *Addison-Wesley, Reading, Massachusetts.*

Rockafellar, R.T. (1973). The multiplier method of Hestenes and Powell applied to convex programming. *J. Optim. Theor. Appl.* **12**, pp 555–562.

Rosen, J.B. (1960). The gradient projection method for nonlinear programming, Part I: Linear constraints. *J. Soc. Ind. Appl. Math.* **8**, pp 181–217.

Rosen, J.B. (1961). The gradient projection method for nonlinear programming, Part II: Nonlinear constraints. *J. Soc. Ind. Appl. Math.* **9**, pp 514–532.

Shanno, D.F. and Whitaker, L. (1985). A note on solving nonlinear minimax problems via a differentiable penalty function. *J. Optim. Theory Appl.* **46**, pp 95–103.

Shor, N.Z. and Shabashova, L.P. (1972). Solution of minimax problems by the method of generalised gradient descent with dilatation of the space. *Cybernetics* **1**, pp 88–94.

Shrager, R.I. and Hill, E. (1980). Nonlinear curve–fitting in the L_1 and L_∞ norms. *Math. Comput.* **34**, pp 529–541.

Watson G.A. (1976). A method for calculating best non-linear Chebychev approximations. *J. Inst. Math. Applic.* **18**, pp 351–360.

Watson G.A. (1979). The minimax solution of an overdetermined system of nonlinear equations. *J. Inst. Math. Applic.* **23**, pp 167–180.

Watson G.A. (1980). Approximation theory and numerical methods. *John Wiley, New York.*

Watson G.A. (1986). Methods for best approximation and regression problems. *Numerical analysis Report NA/96, University of Dundee, Scotland.*

Wierzibicki, A.P. (1978). Lagrangian functions and nondifferentiable optimization. *Rept. WP-78-63, Intl. Inst. Appl. Sys. Anal., Laxenburg, Austria.*

Wolfe, P. (1961). A duality theorem for nonlinear programming. *Q. Appl. Math.* **19**, pp 239–244.

Womersley, R.S. and Fletcher, R. (1986). An algorithm for composite nonsmooth optimization problems. *J. Optim. Theory Appl.* **48** pp 493–523.

Zang, I. (1980). A smoothing-out technique for min-max optimization. *Math. Programmming* **19**, pp 61–77.

Zangwill, W.I. (1967). An algorithm for the Chebyshev problem - with an application to concave programming. *Management Science,* **14**, pp 58–78.

Zangwill, W.I. (1969). Nonlinear Programming: A unified approach. *Prentice–Hall, New Jersey.*

Zuhovickii, S.I., Polyak, R.A. and Primak, M.E. (1963). An algorithm for the solution of the problem of convex Chebychev approximations. *Soviet Math.* **4**, pp 901–904.

4

The Nonlinear L_p-Norm Estimation Problem

In Chapters 2 and 3 the two extreme cases in nonlinear L_p-norm estimation, the nondifferentiable L_1- and L_∞-norm estimation problems were considered. In this chapter the differentiable nonlinear L_p-norm $(1 < p < \infty)$ estimation problem will be examined. The least squares approach $(p = 2)$ is a special case. Necessary as well as sufficient conditions for $\boldsymbol{\theta}^*$ to be optimal are given. These conditions are stated in terms of the first- and second-order partial derivatives of the *nonlinear L_p-norm objective function*, $S_p(\boldsymbol{\theta})$. Algorithms for solving the nonlinear L_p-norm estimation problem will also be discussed.

4.1. The nonlinear L_p-norm estimation problem $(1 < p < \infty)$

The L_p-norm estimation problem for the nonlinear model:

$$y_i = f_i(\boldsymbol{x_i}, \boldsymbol{\theta}) + e_i, \qquad i = 1, \ldots, n \quad (n \geq k) \tag{4.1}$$

is defined as:

Find parameters $\boldsymbol{\theta}$ which minimize

$$S_p(\boldsymbol{\theta}) = \sum_{i=1}^{n} |y_i - f_i(\boldsymbol{x_i}, \boldsymbol{\theta})|^p \quad \text{where} \quad 1 < p < \infty \tag{4.2}$$

An equivalent formulation is the nonlinear programming (NLP) problem:

$$\text{minimize} \sum_{i=1}^{n} u_i \tag{4.3}$$

$$\text{subject to} \quad \left.\begin{aligned} y_i - f_i(\boldsymbol{x_i}, \boldsymbol{\theta}) - u_i^{1/p} &\leq 0 \\ -y_i + f_i(\boldsymbol{x_i}, \boldsymbol{\theta}) - u_i^{1/p} &\leq 0 \\ u_i &\geq 0 \end{aligned}\right\} \quad i = 1, \ldots, n$$

Note that the constraints are equivalent to stating that $|y_i - f_i(\boldsymbol{x_i}, \boldsymbol{\theta})|^p \leq u_i$ for all i, and the equivalence of (4.3) to (4.2) follows. The original problem in k unknowns has now been defined as an NLP problem in $2n$ nonlinear inequality constraints, n nonnegative variables u_i and k unconstrained variables. In this Chapter we shall discuss algorithms that take the structure of (4.2) directly into account.

4.2. Optimality conditions for the nonlinear L_p-norm problem

We introduce the matrix notation due to Gonin (1986). Let $\nabla^2 f_i$ be the Hessian matrix of f_i with respect to $\boldsymbol{\theta}$ and define

- p-Jacobian matrix $J_p = J_p(\boldsymbol{\theta})$ with $(i,j)^{th}$ element

$$|y_i - f_i|^{\frac{1}{2}p-1}\frac{\partial f_i}{\partial \theta_j} \qquad i = 1,\ldots,n \quad j = 1,\ldots,k$$

- p-residual vector

$$(\boldsymbol{y} - \boldsymbol{f})_p = [|y_i - f_i|^{\frac{1}{2}p-1}(y_i - f_i)] \qquad i = 1,\ldots,n$$

- $k \times k$ matrix

$$B_p(\boldsymbol{\theta}) = \sum_{i=1}^{n} |y_i - f_i|^{p-2}(f_i - y_i)\nabla^2 f_i$$

Theorem 4.1 The first- and second-order partial derivatives of $S_p(\boldsymbol{\theta})$ are given by:

$$\nabla S_p(\boldsymbol{\theta}) = -pJ_p{}'(\boldsymbol{y} - \boldsymbol{f})_p \tag{4.4}$$

$$\nabla^2 S_p(\boldsymbol{\theta}) = p[(p-1)J_p{}'J_p + B_p(\boldsymbol{\theta})] \tag{4.5}$$

Proof Define $F_i = |y_i - f_i(\boldsymbol{x_i},\boldsymbol{\theta})|^p$ then $F_i^{1/p} = |y_i - f_i(\boldsymbol{x_i},\boldsymbol{\theta})|$ for $i = 1,\ldots,n$. By means of implicit differentiation we find that:

$$\frac{1}{p}F_i^{1/p-1}\frac{\partial F_i}{\partial \theta_\ell} = -sign(y_i - f_i)\frac{\partial f_i}{\partial \theta_\ell}$$

for some index ℓ when $y_i \neq f_i$. Thus

$$\begin{aligned}\frac{\partial F_i}{\partial \theta_\ell} &= -p|y_i - f_i|^{p-1}sign(y_i - f_i)\frac{\partial f_i}{\partial \theta_\ell}\\ &= -p|y_i - f_i|^{p-2}(y_i - f_i)\frac{\partial f_i}{\partial \theta_\ell}\end{aligned}$$

By summation over $i = 1, \dots, n$, we find

$$\frac{\partial S_p}{\partial \theta_\ell} = -p \sum_{i=1}^{n} |y_i - f_i|^{p-2}(y_i - f_i)\frac{\partial f_i}{\partial \theta_\ell} \quad \text{for} \quad \ell = 1, \dots, k$$

By inspection we can see that the gradient

$$\nabla S_p(\boldsymbol{\theta}) = -pJ_p'(\boldsymbol{y} - \boldsymbol{f})_p$$

Similarly

$$\begin{aligned}\frac{\partial^2 F_i}{\partial \theta_\ell \partial \theta_s} &= -p|y_i - f_i|^{p-2}(y_i - f_i)\frac{\partial^2 f_i}{\partial \theta_\ell \partial \theta_s} + p|y_i - f_i|^{p-2}\frac{\partial f_i}{\partial \theta_\ell}\cdot\frac{\partial f_i}{\partial \theta_s} + \\ &\quad p(p-2)|y_i - f_i|^{p-4}(y_i - f_i)^2\frac{\partial f_i}{\partial \theta_\ell}\cdot\frac{\partial f_i}{\partial \theta_s} \\ &= p(p-1)|y_i - f_i|^{p-2}\frac{\partial f_i}{\partial \theta_\ell}\cdot\frac{\partial f_i}{\partial \theta_s} - p|y_i - f_i|^{p-2}(y_i - f_i)\frac{\partial^2 f_i}{\partial \theta_\ell \partial \theta_s}\end{aligned}$$

By summation over $i = 1, \dots, n$, we find the desired expression for

$$\frac{\partial^2 S_p}{\partial \theta_\ell \partial \theta_s} = p\sum_{i=1}^{n} |y_i - f_i|^{p-2}\left\{(p-1)\frac{\partial f_i}{\partial \theta_\ell}\cdot\frac{\partial f_i}{\partial \theta_s} + (f_i - y_i)\frac{\partial^2 f_i}{\partial \theta_\ell \partial \theta_s}\right\}$$

By inspection we can see that the Hessian matrix

$$\nabla^2 S_p(\boldsymbol{\theta}) = p[(p-1)J_p{}'J_p + B_p(\boldsymbol{\theta})]$$

and the theorem is proved. ∎

Corollary 4.1 In L_2-norm estimation (least squares) the 1st- and 2nd-order partial derivatives of $S_2(\boldsymbol{\theta})$ are given by:

$$\nabla S_2(\boldsymbol{\theta}) = -2J'(\boldsymbol{y} - \boldsymbol{f})$$

$$\nabla^2 S_2(\boldsymbol{\theta}) = 2[J'J + B_2(\boldsymbol{\theta})]$$

where J is the usual Jacobian matrix and matrix

$$B_2(\boldsymbol{\theta}) = \sum_{i=1}^{n}(f_i - y_i)\nabla^2 f_i$$

The following necessary and sufficient conditions for optimality can now be stated. These conditions are simply the usual optimality conditions encountered in unconstrained optimization; see for example McCormick (1983:Chapter 2).

Theorem 4.2 Suppose $S_p(\boldsymbol{\theta})$ is once continuously differentiable in an open set $\Theta \subseteq \Re^k$ and that $\boldsymbol{\theta}^*$ is a local minimum point of problem (4.2). Then a necessary condition for $\boldsymbol{\theta}^*$ to be an optimal point is that:

$$\nabla S_p(\boldsymbol{\theta}^*) = -pJ_p{}'(\boldsymbol{y} - \boldsymbol{f})_p = 0$$

Theorem 4.3 Suppose $S_p(\boldsymbol{\theta})$ is twice continuously differentiable in an open set $\Theta \subseteq \Re^k$. If $\boldsymbol{\theta}^*$ is such that

$$\nabla S_p(\boldsymbol{\theta}^*) = -pJ_p{}'(\boldsymbol{y} - \boldsymbol{f})_p = 0$$

and there is a neigbourhood N of $\boldsymbol{\theta}^*$ such that for all $\boldsymbol{\theta} \in N$

$$\boldsymbol{d}'\nabla^2 S_p(\boldsymbol{\theta})\boldsymbol{d} = \boldsymbol{d}'[p\{(p-1)J_p{}'J_p + B_p(\boldsymbol{\theta})\}]\boldsymbol{d} > 0$$

$\forall \boldsymbol{d} \in \Re^k$ and $\boldsymbol{d} \neq 0$, then $\boldsymbol{\theta}^*$ is an isolated local minimum point of problem (4.2).

Example 4.1 Given the problem:

$$\min_{\boldsymbol{\theta}} \sum_{i=1}^{3} |h_i(\boldsymbol{\theta})|^p$$

$$\text{where} \qquad h_1(\boldsymbol{\theta}) = \theta_1^2 + \theta_2^2 \qquad h_2(\boldsymbol{\theta}) = \sin\theta_1 \qquad h_3(\boldsymbol{\theta}) = \cos\theta_2$$

The p-Jacobian matrix is:

$$J_p = -\begin{pmatrix} 2\theta_1|\theta_1^2+\theta_2^2|^{\frac{1}{2}p-1} & 2\theta_2|\theta_1^2+\theta_2^2|^{\frac{1}{2}p-1} \\ \cos\theta_1|\sin\theta_1|^{\frac{1}{2}p-1} & 0 \\ 0 & -\sin\theta_2|\cos\theta_2|^{\frac{1}{2}p-1} \end{pmatrix}$$

The p-residual vector is:

$$(\boldsymbol{h})_p = -\begin{pmatrix} |\theta_1^2+\theta_2^2|^{\frac{1}{2}p-1}(\theta_1^2+\theta_2^2) \\ \sin\theta_1|\sin\theta_1|^{\frac{1}{2}p-1} \\ \cos\theta_2|\cos\theta_2|^{\frac{1}{2}p-1} \end{pmatrix}$$

Then

$$\nabla S_p(\boldsymbol{\theta}^*) = -p \begin{pmatrix} 2\theta_1|\theta_1^2+\theta_2^2||p-2(\theta_1^2+\theta_2^2)+|\sin\theta_1|^{p-2}\sin\theta_1\cos\theta_1 \\ 2\theta_2|\theta_1^2+\theta_2^2||p-2(\theta_1^2+\theta_2^2)-|\cos\theta_2|^{p-2}\sin\theta_2\cos\theta_2 \end{pmatrix} = 0$$

Since $\boldsymbol{\theta}^* = (0,0)'$ clearly satisfies these equations $\boldsymbol{\theta}^*$ is therefore an optimal point.

4.3. Algorithms for nonlinear L_p-norm estimation problems

Least squares is the appropriate method of solution when the error variates are normally distributed with expected value and variance, $\mathcal{E}(e_i) = 0$ and $\text{Var}(e_i) = \sigma^2$ respectively. The error variates are frequently not normally distributed in which case an alternative such as L_p-norm estimation should be considered.

The solution to the L_p-norm problem can either be obtained directly by numerical minimization or by transformation of the original problem into the NLP problem (4.3). Although there are efficient numerical methods for solving problems in nonlinear constraints, see for example McCormick (1983), this approach remains a cumbersome way of solving the original estimation problem in view of the number of nonlinear constraints and additional number of constrained variables. Another approach would be to treat it as an unconstrained minimization problem and to solve it using a quasi-Newton method. However, algorithms which take the special structure of problem (4.2) directly into account converge substantially faster than the general unconstrained procedures.

Algorithms for solving various types of nonlinear least squares problems are cited in the Additional notes (§4.3) at the end of this Chapter.

4.3.1. Watson's algorithm

In an attempt to overcome the slow convergence of the polyhedral-norm methods for solving the L_1- and L_∞-norm problems, Watson (1979, 1980) proposed a procedure for solving the dual of the general nonlinear approximation problem:

Find the parameters $\boldsymbol{\theta} \in \Re^k$ which

$$\text{minimize} \quad \|\boldsymbol{h}(\boldsymbol{\theta})\|$$

with $\boldsymbol{y} - \boldsymbol{f}(\boldsymbol{x},\boldsymbol{\theta}) = \boldsymbol{h} : \Re^k \to \mathcal{M}$, where $\mathcal{M}$ is a normed real linear vector space and

$\|\cdot\|$ a general norm. The nonlinear L_p-norm estimation problem (4.2) is a special case and has the dual problem:

Find the solution $\boldsymbol{\theta}$ to the system of nonlinear equations

$$\boldsymbol{h}(\boldsymbol{\theta})'UJ = 0$$

where $U_{n\times n} = diag(u_1, \ldots, u_n)$

$$\text{with} \qquad u_i = \begin{cases} |h_i(\boldsymbol{\theta})|^{p-2}\|\boldsymbol{h}(\boldsymbol{\theta})\|^{1-p} & \text{if } h_i(\boldsymbol{\theta}) \neq 0 \\ 0 & \text{otherwise} \end{cases}$$

Remarks:

(1) Though not stated by Watson, this approach is the same as solving the system of equations $\nabla S_p(\boldsymbol{\theta}) = 0$. Since it follows from Theorem 4.1 that

$$0 = \|\boldsymbol{h}(\boldsymbol{\theta})\|^{p-1}\boldsymbol{h}(\boldsymbol{\theta})'UJ = J_p\boldsymbol{h}(\boldsymbol{\theta}) = pJ_p\boldsymbol{h}(\boldsymbol{\theta}) = \nabla S_p(\boldsymbol{\theta})$$

In terms of polyhedral norms the dual problem can be shown to be the solution to the KKT conditions. The first-order gradient method of Gonin (1986) can also be regarded as a dual method in the context of Watson [see §4.3.2].

(2) For the L_1- and L_∞-norms the dual problem is simply the first-order necessary conditions of optimality as discussed in Chapters 2 and 3.

Example 4.2 Reconsider example 2.3. The L_p-norm fit of the curve $y = e^{\theta x}$ to the three points $(0,0)$, $(1,1)$, $(2,4)$ will now be calculated by means of Watson's approach.

We find

$$\boldsymbol{h}(\theta) = \begin{pmatrix} -1 \\ 1 - e^{\theta} \\ 4 - e^{2\theta} \end{pmatrix}$$

$$U = [1 + |1 - e^{\theta}|^{1-p} + |4 - e^{2\theta}|^{1-p}] \begin{pmatrix} 1 & 0 & 0 \\ 0 & |1 - e^{\theta}|^{p-2} & 0 \\ 0 & 0 & |4 - e^{2\theta}|^{p-2} \end{pmatrix}$$

$$J = \begin{pmatrix} 0 \\ -e^{\theta} \\ -2e^{2\theta} \end{pmatrix}$$

thus we must solve

$$\begin{aligned} \boldsymbol{h}(\theta)'UJ &= [1 + |1 - e^{\theta}|^{1-p} + |4 - e^{2\theta}|^{1-p}] \\ &\quad \times \{-e^{\theta}(1 - e^{\theta})|1 - e^{\theta}|^{p-2} - 2e^{2\theta}(4 - e^{2\theta})|4 - e^{2\theta}|^{p-2}\} \\ &= 0 \end{aligned}$$

for θ. For $p = 2$ and 3 the solutions are $\theta^* = 0.66193$ and 0.632775 respectively. Note that

$$S_p'(\theta) = -p[e^{\theta}(1 - e^{\theta})|1 - e^{\theta}|^{p-2} + 2e^{2\theta}(4 - e^{2\theta})|4 - e^{2\theta}|^{p-2}]$$

Thus solving the dual is equivalent to solving the first-order optimality conditions. The graphs of $S_1(\theta)$, $S_2(\theta)$, $S_3(\theta)$ and $S_\infty(\theta)$ are given in Fig. 4.1 below.

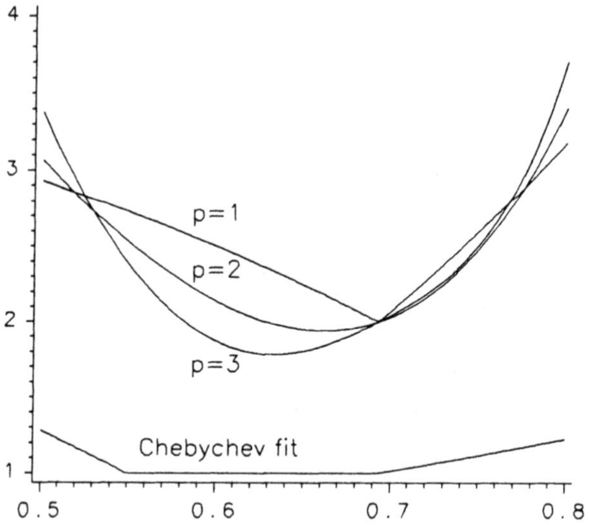

Figure 4.1 $S_p(\theta)$ vs p for values $p = 1, 2, 3, \infty$.

4.3.2. A first-order gradient algorithm

This algorithm is due to Gonin (1986) and is efficient in solving fairly well-behaved, small residual nonlinear L_p-norm estimation problems when $1 < p < \infty$. It uses the structure of the L_p-norm estimation problem and is an extension of the

Gauss-Newton method for solving nonlinear least squares problems. In the algorithm only the first-order partial derivatives of $S_p(\boldsymbol{\theta})$ are used.

To guarantee that the objective function $S_p(\boldsymbol{\theta})$ decreases at each succeeding iteration a descent direction must be used. A necessary condition for this is that the Hessian matrix $\nabla^2 S_p(\boldsymbol{\theta})$ should be positive definite at each iteration (Appendix 2B). A continuously differentiable function will be strictly convex if and only if its corresponding Hessian matrix is positive definite. It is also reasonable to assume that $S_p(\boldsymbol{\theta})$ behaves like a quadratic function in the vicinity of a local minimum. It may happen that the Hessian will not be positive definite at each iteration, especially at points $\boldsymbol{\theta}$ distant from the optimum. In this event a factorization procedure due to Gill and Murray (1974) will be used. This procedure transforms a given (Hessian) matrix into one which is positive definite (see Appendix 4A). Specifically, given a nonpositive definite matrix A, a new positive definite matrix $\tilde{A} = A + E = L\tilde{D}L'$ is formed where E is a diagonal matrix, L is a unit lower triangular matrix and $\tilde{D}$ is a diagonal matrix. This decomposition is known as the modified Cholesky factorization. The following notation will be used:

$$\boldsymbol{g}^j = \nabla S_p(\boldsymbol{\theta}^j) = pJ_p{}'(\boldsymbol{f} - \boldsymbol{y})_p = pJ_p(\boldsymbol{\theta}^j)'(\boldsymbol{f} - \boldsymbol{y})_p \tag{4.6}$$

$$G^j = p(p-1)J_p{}'J_p = p(p-1)J_p(\boldsymbol{\theta}^j)'J_p(\boldsymbol{\theta}^j) \tag{4.7}$$

The norm $||\cdot||$ is the usual Euclidean (L_2) norm.

In the Gauss-Newton method for nonlinear least squares a sequence of iterates $\boldsymbol{\theta}^j$ is constructed so that $\boldsymbol{\theta}^{j+1} = \boldsymbol{\theta}^j + \gamma_j \boldsymbol{d}^j$ where γ_j is the steplength and $\boldsymbol{d}^j$ a direction of search satisfying the equation:

$$J'J\boldsymbol{d}^j = J'(\boldsymbol{y} - \boldsymbol{f}) \tag{4.8}$$

Note that only first-derivative information is (given by J) taken into account since for small residual problems the norm $||B_2(\boldsymbol{\theta})||$ is small compared to the norm $||J'J||$. If we examine the Hessian, $\nabla^2 S_p(\boldsymbol{\theta})$, we see that for $p > 2$ we may ignore the second derivative term $B_p(\boldsymbol{\theta})$ in expression (4.5) when $\boldsymbol{f} \approx \boldsymbol{y}$. If $p < 2$ we may experience slow convergence of the algorithm especially when $\boldsymbol{f} \approx \boldsymbol{y}$. Indeed $\frac{\partial^2 F_i}{\partial \theta_\ell \partial \theta_s}$ will have a singularity at a point $\boldsymbol{\theta}$ where $f_i = y_i$.

When it is feasible to ignore the second derivative term $B_p(\boldsymbol{\theta})$, the Hessian can be approximated by $p(p-1)J_p{}'J_p$. The first-order gradient iteration, for solving nonlinear L_p-norm estimation problems is therefore:

$$(p-1)J_p{}'J_p\boldsymbol{d}^j = -J_p{}'(\boldsymbol{f}-\boldsymbol{y})_p \tag{4.9}$$

It is clear that the Gauss-Newton method is imbedded in this method, since iteration (4.9) is expressed in terms of p, the p-Jacobian and the p-residual vector, as opposed to the Jacobian and residual vector used in (4.8). The approximation $J_p{}'J_p$ of the Hessian $\nabla^2 S_p(\boldsymbol{\theta})$ need not be positive definite at each iteration in which case the modified Cholesky factorization procedure is used.

The Algorithm

Step 0: Let ϵ_1, ϵ_2 and ϵ_3 $(=10^{-10})$ be prescribed tolerances. Select the initial estimate $\boldsymbol{\theta}^1$ of $\boldsymbol{\theta}^*$ and set $j := 1$.

Step 1: Calculate

(a) the p-Jacobian $J_p(\boldsymbol{\theta}^j)$

(b) the p-residual vector $(\boldsymbol{y}-\boldsymbol{f})_p$

(c) $\boldsymbol{g}^j$ and G^j using (4.6) and (4.7).

Step 2: Compute the modified Cholesky factorization of G^j:

$L^j\tilde{D}^j(L^j)' = G^j + E^j$.

Step 3:

(a) If $\|\boldsymbol{g}^j\| \le \epsilon_1$ and $\|E^j\| = 0$ then $\boldsymbol{\theta}^j$ is optimal, STOP.

(b) If $\|\boldsymbol{g}^j\| > \epsilon_1$ determine the search direction by solving the set of linear equations for $\boldsymbol{d}^j$:

$G^j\boldsymbol{d}^j = L^j\tilde{D}^j(L^j)'\boldsymbol{d}^j = -\boldsymbol{g}^j$

Go to Step 4.

(c) If $\|\boldsymbol{g}^j\| \le \epsilon_1$ and $\|E^j\| \ne 0$ determine $\boldsymbol{d}^j$ as follows:

Solve for $\boldsymbol{u}$ by backward substitution in

$$(L^j)'\boldsymbol{u} = \boldsymbol{e}_q \quad (\boldsymbol{e}_q \text{ is a } q^{th} \text{ unit vector})$$

where $\tilde{d}_q^j - E_{qq}^j = \min\{\tilde{d}_i^j - E_{ii}^j \mid 1 \le i \le k\}$, with $\tilde{D}^j = diag(\tilde{d}_i^j)$.

$$\text{Set} \quad \boldsymbol{d}^j = \begin{cases} -sign[\boldsymbol{u}'\boldsymbol{g}^j]\boldsymbol{u} & \text{if } \|\boldsymbol{g}^j\| > 0 \\ \boldsymbol{u} & \text{if } \|\boldsymbol{g}^j\| = 0 \end{cases}$$

Step 4: Compute γ_j by means of Fletcher's (1970) line search procedure [Appendix 4C] and set $\boldsymbol{\theta}^{j+1} = \boldsymbol{\theta}^j + \gamma_j \boldsymbol{d}^j$.

Step 5: Continue until the following convergence criteria are met:

(a) $|g_i^j| < \epsilon_2$ for $i = 1, \ldots, k$

(b) $\frac{|S_p(\boldsymbol{\theta}^{j+1}) - S_p(\boldsymbol{\theta}^j)|}{S_p(\boldsymbol{\theta}^j)} < \epsilon_3$

in, for example, 4 consecutive iterations. Otherwise set $j := j + 1$ and return to Step 1.

Remarks

(1) The generalised gradient $\boldsymbol{g}^*$ [due to Dennis (1977)] defined by

$$g_i^* = \frac{g_i^j}{\sqrt{(p-1)(J_p'J_p)_{ii}S_p(\boldsymbol{\theta}^j)}}$$

may be used as an alternative in Step 5(a).

(2) Direction $\boldsymbol{d}^j$ in Step 3(c) is known as a direction of negative curvature.

(3) Note that in this algorithm a diagonal matrix E^j is added to G^j if the latter is insufficiently positive definite. In the Levenberg-Marquardt approach a diagonal matrix λI is added to G^j. At convergence $\lambda \to 0$. See for example the trust region approach of Moré (1977).

Convergence of the algorithm will now be derived via a number of theorems which are simply extensions of those stated by Dennis and Schnabel (1983:222–223) and Gill and Murray (1974). We shall initially consider the convergence properties of the first-order gradient iteration (4.9).

Theorem 4.4 Suppose $S_p(\boldsymbol{\theta})$ is twice continuously differentiable in an open convex set $\Theta \subset \Re^k$, that $J_p(\boldsymbol{\theta})$ is Lipschitz continuous in Θ with $\|J_p(\boldsymbol{\theta})\| \le \alpha$ for all $\boldsymbol{\theta} \in \Theta$ and that there exist a $\boldsymbol{\theta}^* \in \Theta$ and $\mu, \eta \ge 0$ such that $J_p(\boldsymbol{\theta}^*)'(\boldsymbol{y} - \boldsymbol{f})_p^* = 0$, μ being the smallest eigenvalue of $J_p{}'J_p$, and

$$||[J_p(\boldsymbol{\theta}) - J_p(\boldsymbol{\theta}^*)]'(\boldsymbol{y} - \boldsymbol{f})_p^*|| \leq \eta||\boldsymbol{\theta} - \boldsymbol{\theta}^*||$$

for all $\boldsymbol{\theta} \in \Theta$. If $\eta < \mu$ then there exists an $\epsilon > 0$ for any constant $1 < c < \frac{\mu}{\eta}$ such that the sequence $\{\boldsymbol{\theta}^j\}$ generated by (4.9) is well defined, converges to $\boldsymbol{\theta}^*$, whilst obeying

$$||\boldsymbol{\theta}^{j+1} - \boldsymbol{\theta}^*|| \leq \frac{c\eta}{\mu}||\boldsymbol{\theta}^j - \boldsymbol{\theta}^*|| + \frac{c\alpha\gamma}{2\mu}||\boldsymbol{\theta}^j - \boldsymbol{\theta}^*||^2$$

$$\text{and} \quad ||\boldsymbol{\theta}^{j+1} - \boldsymbol{\theta}^*|| \leq \frac{c\eta + \mu}{2\mu}||\boldsymbol{\theta}^j - \boldsymbol{\theta}^*|| < ||\boldsymbol{\theta}^j - \boldsymbol{\theta}^*||$$

for all $\boldsymbol{\theta}^1$ belonging to the neigbourhood $N(\boldsymbol{\theta}^*, \epsilon)$.

The proof is as in Dennis and Schnabel (1983:222–223).

Corollary 4.2 Let the assumptions of the previous theorem hold. If $(\boldsymbol{y} - \boldsymbol{f})_p^* = 0$ then there exists an $\epsilon > 0$ such that for all $\boldsymbol{\theta}^1 \in N(\boldsymbol{\theta}^*, \epsilon)$ the sequence $\{\boldsymbol{\theta}^j\}$ generated by (4.9) is well defined and converges quadratically to $\boldsymbol{\theta}^*$.

Proof If $(\boldsymbol{y} - \boldsymbol{f})_p^* = 0$, choose $\eta = 0$ in $||[J_p(\boldsymbol{\theta}) - J_p(\boldsymbol{\theta}^*)]'(\boldsymbol{y} - \boldsymbol{f})_p^*|| \leq \eta||\boldsymbol{\theta} - \boldsymbol{\theta}^*||$. Since $||\boldsymbol{\theta}^{j+1} - \boldsymbol{\theta}^*|| \leq \frac{1}{2}||\boldsymbol{\theta}^j - \boldsymbol{\theta}^*|| < ||\boldsymbol{\theta}^j - \boldsymbol{\theta}^*||$, the sequence $\{\boldsymbol{\theta}^j\}$ converges to $\boldsymbol{\theta}^*$ and since

$$||\boldsymbol{\theta}^{j+1} - \boldsymbol{\theta}^*|| \leq \frac{c\alpha\gamma}{2\mu}||\boldsymbol{\theta}^j - \boldsymbol{\theta}^*||^2$$

the convergence rate is quadratic. ∎

Define

$$\Omega = \{\boldsymbol{\theta} | \boldsymbol{\theta} \in \Theta, S_p(\boldsymbol{\theta}) < S_p(\boldsymbol{\theta}^1)\}$$

denote its closure by $\bar{\Omega}$, the closed convex hull of $\bar{\Omega}$ by $\text{Co}[\bar{\Omega}]$ and let the set of stationary points of $S_p(\boldsymbol{\theta})$ be

$$S = \{\boldsymbol{\theta}^* | \boldsymbol{\theta}^* \in \Omega, \nabla S_p(\boldsymbol{\theta}^*) = 0\}$$

The following two theorems relate to the convergence of the algorithm as stated.

Theorem 4.5 Suppose $S_p(\boldsymbol{\theta})$ is twice continuously differentiable in an open convex set $\Theta \subset \Re^k$, S is finite and $\boldsymbol{\theta}^1$ is chosen such that

(a) $\bar{\Omega}$ is compact (closed *and* bounded).

(b) $\text{Co}[\bar{\Omega}] \in \Theta$.

(c) The norm of the Hessian, $||\nabla^2 S_p(\boldsymbol{\theta})|| \leq \rho$ for all $\boldsymbol{\theta} \in \bar{\Omega}$.

If $\epsilon_1 = 0$ in Step 3 of the algorithm then the sequence $\{\boldsymbol{\theta}^j\}$ is well defined and converges to $\boldsymbol{\theta}^* \in S$.

The proof is as in Gill and Murray (1974).

Theorem 4.6 Let ϵ_1 be defined as in Step 3 of the algorithm. and the assumptions of Theorem 4.5 hold. Then

$$\lim_{\epsilon_1 \to 0} \lim_{k \to \infty} \boldsymbol{\theta}^j(\epsilon_1) = \boldsymbol{\theta}^*$$

where $\boldsymbol{\theta}^*$ is an isolated local minimum point of problem (4.2) provided that $\nabla^2 S_p(\boldsymbol{\theta})$ is not positive semi-definite at any stationary point of S.

The proof is as in Gill and Murray (1974).

Remark: If the sequence $\{\boldsymbol{\theta}^j\}$ ultimately lies in a region where $\nabla^2 S_p(\boldsymbol{\theta})$ is uniformly positive definite and converges to an isolated local minimum point of (4.2) then the ultimate convergence rate will be superlinear.

4.3.2.1 Examples.

Example 4.3 This is the well-known sum of squared terms example due to Rosenbrock (1960). The problem is:

$$\min_{\boldsymbol{\theta}} \sum_{i=1}^{2} |h_i(\boldsymbol{\theta})|^p$$

$$\text{where} \quad \begin{aligned} h_1(\boldsymbol{\theta}) &= 10(\theta_2 - \theta_1^2) \\ h_2(\boldsymbol{\theta}) &= 1 - \theta_1 \end{aligned}$$

The usual starting value $\boldsymbol{\theta}^1 = (-1.2, 1)$ was used. When $p = 2$ the optimal solution is $\boldsymbol{\theta}^* = (1, 1)$. The intermediate calculations of one iteration of the first-order gradient algorithm will now be illustrated.

At $\boldsymbol{\theta}^1 = (-1.2, 1)'$, $S_2(\boldsymbol{\theta}^1) = 24.2$.

Step 1:

$$J_2(\boldsymbol{\theta}^1) = J = \begin{pmatrix} 24 & 10 \\ -1 & 0 \end{pmatrix} \qquad (\boldsymbol{y} - \boldsymbol{f})_2 = (\boldsymbol{y} - \boldsymbol{f}) = \begin{pmatrix} 4.4 \\ -2.2 \end{pmatrix}$$

$$\boldsymbol{g}^1 = \begin{pmatrix} -215.6 \\ -88 \end{pmatrix} \qquad G^1 = \begin{pmatrix} 1154 & 480 \\ 480 & 200 \end{pmatrix}$$

$$\|\boldsymbol{g}^1\| = 232.8674 \qquad \|E^1\| = 0.0$$

Step 3b:

$$\boldsymbol{d}^1 = \begin{pmatrix} 2.20 \\ -4.84 \end{pmatrix}$$

Step 4:

$$\gamma_1 = 0.07181, \qquad \boldsymbol{\theta}^2 = \begin{pmatrix} -1.042014 \\ 0.652432 \end{pmatrix} \quad \text{and} \quad S_2(\boldsymbol{\theta}^2) = 22.95007$$

The algorithm performed 31 function and gradient evaluations to reach the optimal solution. Betts (1976), Gill and Murray (1978) and Dennis et al. (1981) reported 27, 31 and 26 evaluations respectively. The BFGS method [Fletcher (1970)] and Jacobson and Oksman (1972) algorithms are established methods for general optimization problems. These two methods performed 47 and 69 function evaluations respectively. The best reported result by Jones (1970) is 17. His method is specifically designed for least squares problems. The solutions for other values of p are given in Table 4.1.

Table 4.1: Evaluations for different values of p.

	p					
$p \rightarrow$	1.5	1.75	2.0	2.5	2.75	3.0
θ_1	1.0000	1.0000	1.0000	1.0000	1.0000	1.0000
θ_2	1.0000	1.0000	1.0000	1.0000	1.0000	0.9999
$S_p(\boldsymbol{\theta})$	$0.163^{(-9)}$	$0.861^{(-8)}$	$0.464^{(-28)}$	$0.127^{(-9)}$	$0.132^{(-10)}$	$0.173^{(-10)}$
evaluations	29	34	31	38	41	40

Example 4.4 This example is due to Powell (1962) and has a singular Hessian at the optimal solution. The algorithm performed 33 function and gradient evaluations. Betts (1976), Gill and Murray (1978) and Dennis et al. (1981) reported 16,

12 and 20 evaluations respectively. The starting point $\boldsymbol{\theta}^1 = (3, -1, 0, 1)$ was used. The solutions for other values of p are given in Table 4.2.

Table 4.2: Evaluations for different values of p.

	p					
$p \rightarrow$	1.5	1.75	2.0	2.5	2.75	3.0
θ_1	0.0000	0.0000	0.0000	0.0000	0.0001	0.0002
θ_2	0.0000	0.0000	0.0000	0.0000	0.0000	0.0000
θ_3	0.0000	0.0000	0.0000	0.0000	0.0000	0.0000
θ_4	0.0000	0.0000	0.0000	0.0000	0.0000	0.0000
$S_p(\boldsymbol{\theta})$	$0.152^{(-6)}$	$0.373^{(-15)}$	$0.260^{(-17)}$	$0.279^{(-21)}$	$0.245^{(-21)}$	$0.630^{(-21)}$
evaluations	13	14	33	28	31	33

4.3.3. The mixture method for large residual and ill-conditioned problems

This algorithm is due to Gonin and Du Toit (1987) and is an extension and modification of the nonlinear least squares algorithm of Gill and Murray (1978). It uses a mixture of 1st-order derivative (Gauss-Newton – GN) and 2nd-order derivative (Newton – N) search directions and is specifically designed for problems in which $y_i \neq f_i$ for some i and $\|B_p(\boldsymbol{\theta})\| \gg 0$. Thus 2nd-order derivative information, $B_p(\boldsymbol{\theta})$ must be taken into account especially when $\|B_p(\boldsymbol{\theta})\| \gg \|J_p{}'J_p\|$. Problems in which this occurs are known as large residual problems. Gill and Murray (1978) observed that a large value of $S_2(\boldsymbol{\theta})$ may be due to the scaling of $S_2(\boldsymbol{\theta})$ and they therefore define a large residual problem as one in which $S_2(\boldsymbol{\theta})/\|J'J\|$ is large. The poor convergence behaviour of the conventional GN methods may therefore be explained by the fact that in large residual problems the effect of $B_2(\boldsymbol{\theta})$ is ignored.

In the mixture method the singular values of J_p are examined. A first-order gradient (GN) direction $\bar{\boldsymbol{d}}_1$ corresponding to the dominant singular values is then computed. This is valid since the dominant singular values correspond to that part of J_p (and indirectly the Hessian) which is well-conditioned. The direction $\bar{\boldsymbol{d}}_1$ is therefore based on 1st-order derivative information only. For the remaining nondominant singular values 2nd-order derivative information is used and a direction

$\bar{\boldsymbol{d}}_2$ based on 1st- and 2nd-order derivative information is calculated. In geometrical terms $\bar{\boldsymbol{d}}_1$ may be regarded as the 1st-order gradient (GN) direction in the subspace spanned by the dominant singular vectors of J_p and $\bar{\boldsymbol{d}}_2$ the Newton (N) direction in the subspace spanned by the nondominant singular vectors of J_p. The sum of the directions, $\bar{\boldsymbol{d}}_1 + \bar{\boldsymbol{d}}_2 = \boldsymbol{d}$ is then used as a descent direction in the algorithm. This choice of $\bar{\boldsymbol{d}}_1$ and $\bar{\boldsymbol{d}}_2$ constitutes the *mixture* algorithm.

The algorithm judiciously selects the r most dominant singular values of J_p, thereby enhancing the convergence considerably. It was argued that for well conditioned small residual problems r will equal k at every iteration. However, for ill-conditioned and large residual problems $r < k$ when $\boldsymbol{\theta}$ is far away from the optimal solution and r will be close to k in the vicinity of the optimal solution $\boldsymbol{\theta}^*$. The Newton search direction $\boldsymbol{d}^j$ can be calculated from:

$$\nabla^2 S_p(\boldsymbol{\theta}^j)\boldsymbol{d}^j = -\nabla S_p(\boldsymbol{\theta}^j) \tag{4.10}$$

In large residual problems 2nd-order derivative information is incorporated and the Newton direction, $\boldsymbol{d}^j$ is given by:

$$[(p-1)J_p{}'J_p + B_p(\boldsymbol{\theta})]\boldsymbol{d}^j = J_p{}'(\boldsymbol{y}-\boldsymbol{f})_p = -J_p{}'(\boldsymbol{f}-\boldsymbol{y})_p \tag{4.11}$$

A numerically stable method for calculating $\boldsymbol{d}^j$ from the system of equations (4.11) involves the singular-value decomposition process [see Appendix 4B]. The singular-value decomposition of matrix J_p enables us to write matrix J_p as the product of three matrices:

$$J_p = UDV' = U\begin{pmatrix} D(s) \\ \dots \\ 0 \end{pmatrix} V' \tag{4.12}$$

where $U_{n\times n}$ and $V'_{k\times k}$ are orthogonal matrices and $D(s)_{k\times k}$ is a diagonal matrix. Let $B_p = B_p(\boldsymbol{\theta})$ and substitute (4.12) into (4.11) thus yielding:

$$[(p-1)VD'U'UDV' + B_p]\boldsymbol{d}^j = -VD'U'(\boldsymbol{f}-\boldsymbol{y})_p$$

which becomes

$$[(p-1)VD^2(s)V' + B_p]\boldsymbol{d}^j = -V[D(s):0]U'(\boldsymbol{f}-\boldsymbol{y})_p$$

Premultiplication by V' gives

$$[(p-1)D^2(s)V' + V'B_p]\boldsymbol{d}^j = -[D(s):0]U'(\boldsymbol{f}-\boldsymbol{y})_p \tag{4.13}$$

Define $\boldsymbol{d}^j = V\boldsymbol{z}^j$, then (4.13) may be rewritten as:

$$[(p-1)D^2(s) + V'B_pV]\boldsymbol{z}^j = -[D(s):0]U'(\boldsymbol{f}-\boldsymbol{y})_p \tag{4.14}$$

The following distinct cases have to be considered at this stage:

(a) $(p-1)J_p'J_p + B_p(\boldsymbol{\theta})$ is positive definite and $||B_p(\boldsymbol{\theta})|| \ll ||(p-1)J_p'J_p||$.

(b) $J_p'J_p$ is ill-conditioned.

(c) $(p-1)J_p'J_p + B_p(\boldsymbol{\theta})$ is indefinite.

(d) $(p-1)J_p'J_p + B_p(\boldsymbol{\theta})$ is negative definite.

Case (a) In this event the matrix

$$(p-1)D^2(s) + V'B_pV \tag{4.15}$$

is positive definite and the Cholesky factorization $L\tilde{D}L' = (p-1)D^2(s) + V'B_pV$ can therefore be used to calculate $\boldsymbol{z}^j$ from (4.14). The notation $\tilde{D}$ for the diagonal matrix is used to avoid confusion with matrix D in (4.12). If the matrix in (4.15) is insufficiently positive definite, the modified Cholesky method of Gill and Murray (1974) may be used. In this case the gradient method of §4.3.2 results.

Case (b) In the event that $J_p'J_p$ is ill-conditioned; this ill-conditioning will be reflected in the matrix in (4.15) especially when $||B_p(\boldsymbol{\theta})|| \ll ||(p-1)J_p'J_p||$. Such ill-conditioning often occurs in data-fitting problems with small residuals. A sum of a GN and N direction will be calculated, where the GN direction is based on the first r singular values and the N direction on the remaining $k-r$ singular values.

Definition 4.1 The number of dominant singular values, r, is called the grade of J_p.

The reader may well ask: How is the grade r determined ? The term " dominant " may well be subjective. The singular values $\{s_1, \ldots, s_k\}$ of J_p are calculated using the singular value decomposition routine SVDRS due to Lawson and Hanson (1974).

Note that for $i = 1, \ldots, k-1$, $s_{i+1} \leq s_i$. Let *toler* (cf. Step 3 of the algorithm) be a quantity which is a function of k, the number of parameters.

$$\text{If} \quad \frac{s_k}{s_1} > toler \quad \text{set grade} \quad r = k$$

$$\text{If} \quad \frac{s_k}{s_1} \leq toler \quad \text{and} \quad \frac{s_k + s_{k-1}}{s_1} > toler \quad \text{set} \quad r = k-1$$

and so on. The above can be reformulated as:

$$r = \begin{cases} k, & \text{if } s_k/s_1 > toler; \\ k - j^*, & \text{if } \sum_{j=1}^{j^*} s_{k-j+1}/s_1 < toler. \end{cases}$$

where j^*, $1 \leq j^* \leq k$ is the smallest integer for which $\sum_{j=1}^{j^*} s_{k-j+1}/s_1 < toler$.

This procedure provides a judicious choice of r and works well in practice. Now partition matrix $D(s)$ into two submatrices:

$$D_1 : r \times r = diag(s_1, \ldots, s_r)$$

$$D_2 : (k-r) \times (k-r) = diag(s_{r+1}, \ldots, s_k)$$

Similarly partition V into

$$V = \begin{pmatrix} V_1 : k \times r & V_2 : k \times (k-r) \end{pmatrix}$$

$$z = \begin{pmatrix} z_1 : r \times 1 \\ z_2 : (k-r) \times 1 \end{pmatrix}$$

and

$$U'(f - y)_p = \begin{pmatrix} f_1 : r \times 1 \\ f_2 : (k-r) \times 1 \\ f_3 : (n-k) \times 1 \end{pmatrix}$$

Direction d^j will be the sum of two components:

$$d^j = V_1 z_1 + V_2 z_2 = \bar{d}_1 + \bar{d}_2$$

Substitution of these partitions into (4.14) yields the system of linear equations:

$$(p-1) D_1^2 z_1 + V_1' B_p (V_1 z_1 + V_2 z_2) = -D_1 f_1 \tag{4.16}$$

$$[(p-1) D_2^2 + V_2' B_p V_2] z_2 + V_2' B_p V_1 z_1 = -D_2 f_2 \tag{4.17}$$

Recall that the partitioning proceeded according to the number of dominant singular values r and that $\|B_p(\boldsymbol{\theta})\| \ll \|J_p{}'J_p\|$, or to be more precise $\|B_p(\boldsymbol{\theta})\|$ is small in comparison with $\|(p-1)J_p{}'J_p\|$. Hence the contribution of the term

$$V_1{}'B_p(V_1\boldsymbol{z}_1 + V_2\boldsymbol{z}_2) = V_1{}'B_p(\bar{\boldsymbol{d}}_1 + \bar{\boldsymbol{d}}_2) = V_1{}'B_p\boldsymbol{d}^j$$

in expression (4.16) will be small and may be neglected. Solve for $\boldsymbol{z}_1$ from the approximate system of equations:

$$(p-1)D_1^2\boldsymbol{z}_1 \approx -D_1\boldsymbol{f}_1 \quad \text{i.e.,} \quad \boldsymbol{z}_1 \approx -\frac{D_1^{-1}\boldsymbol{f}_1}{p-1}$$

and calculate

$$\bar{\boldsymbol{d}}_1 = V_1\boldsymbol{z}_1 = -\frac{V_1D_1^{-1}\boldsymbol{f}_1}{p-1} \tag{4.18}$$

Substitution of (4.18) into (4.17) yields

$$[(p-1)D_2^2 + V_2{}'B_pV_2]\boldsymbol{z}_2 = -D_2\boldsymbol{f}_2 + \frac{V_2{}'B_pV_1D_1^{-1}\boldsymbol{f}_1}{p-1} \tag{4.19}$$

One can therefore solve for $\boldsymbol{z}_2$ from (4.19) once $\bar{\boldsymbol{d}}_1$ is known and calculate $\bar{\boldsymbol{d}}_2 = V_2\boldsymbol{z}_2$. Our direction is therefore $\boldsymbol{d}^j = \bar{\boldsymbol{d}}_1 + \bar{\boldsymbol{d}}_2$ which is the sum of a GN and N direction respectively.

Case (c) In this case $\boldsymbol{d}^j$ will no longer be a descent direction. The modified Cholesky method will provide a direction of negative curvature [Gill and Murray (1974)]. Recall that information on the indefiniteness of matrix $(p-1)J_p{}'J_p+B_p$ or $(p-1)D_2^2+V_2{}'B_pV_2$ will be available in its modified Cholesky factorization. The computation of such a direction of negative curvature was described in step 3(c) of the algorithm in §4.3.2.

Case (d) In this event the modified Cholesky factorization of $(p-1)J_p{}'J_p + B_p$ will simply reverse the direction to one of descent. If $(p-1)J_p{}'J_p + B_p$ is negative definite then the diagonal elements of $\tilde{D}$ in $L\tilde{D}L'$ will be negative and in the modified Cholesky method $\tilde{d}_j$ will be replaced by $|\tilde{d}_j|$. This is equivalent to replacing matrix $(p-1)J_p{}'J_p + B_p$ by $(1-p)J_p{}'J_p - B_p$ thus yielding a descent search direction.

Another important point has to be considered. Suppose $(p-1)J_p{}'J_p + B_p$ has grade r and that it is negative definite (i.e., k negative eigenvalues). The modified

Cholesky method is only applied to the portion $(p-1)D_2^2 + V_2{}'B_pV_2$, hence only $k-r$ eigenvalues will be made positive. As a safeguard Gill and Murray (1978) suggest the use of the projected gradient $(\boldsymbol{g}^j)'\boldsymbol{d}^j$. If the quantity

$$-\frac{(\boldsymbol{g}^j)'\boldsymbol{d}^j}{\|\boldsymbol{g}^j\|\,\|\boldsymbol{d}^j\|} < \eta < \|D_1^{-2}\|$$

with η some small positive constant, then the direction $\boldsymbol{d}^j$ is recomputed by setting $r=0$ and proceeding as before. In this event B_p need not be recomputed and $\bar{\boldsymbol{d}}_1$ will be set equal to zero.

The Algorithm

Step 0: Let tol, ϵ and η be prescribed tolerances ($tol = 10^{-9}$, $\epsilon = 10^{-9}$, $\eta = 0.0001$). Select the initial estimate $\boldsymbol{\theta}^1$ of $\boldsymbol{\theta}^*$ and set $j := 1$.

Step 1: Calculate

(a) the p-Jacobian $J_p(\boldsymbol{\theta}^j)$

(b) the p-residual vector $(\boldsymbol{y}-\boldsymbol{f})_p$

(c) the gradient of $S_p(\boldsymbol{\theta}^j)$: $\boldsymbol{g}^j = -pJ_p(\boldsymbol{\theta}^j)'(\boldsymbol{y}-\boldsymbol{f})_p$

Step 2: Compute the singular-value decomposition of $J_p(\boldsymbol{\theta}^j)$ viz:

$$U\begin{pmatrix} D(s) \\ \cdots \\ 0 \end{pmatrix} V'$$

Step 3: Compute the 2nd-derivative matrix $B_p(\boldsymbol{\theta}^j)$ numerically. Calculate the grade r as follows: Define

$\rho_1 = \|diag(B_p)\|/\|(p-1)J_p{}'J_p\|$

$\rho_2 = S_p(\boldsymbol{\theta}^j)/\|(p-1)J_p{}'J_p\|$

$$toler = \begin{cases} (1/k^2)\min(\rho_1,\rho_2), & \text{if } \min(\rho_1,\rho_2) < 1; \\ (1/k^2), & \text{if } \min(\rho_1,\rho_2) \geq 1; \end{cases}$$

If $\frac{s_k}{s_1} > toler$ then $r = k$ (full GN direction)

Otherwise set $r = k - j^*$ where j^* is the smallest integer satisfying $\sum_{j=1}^{j^*} \frac{s_{k-j+1}}{s_1} < toler$.

Step 4:

(a) If grade $r = 0$ then $\bar{\boldsymbol{d}}_1 = 0$ (full N direction). Go to Step 5.

For $r > 0$, calculate the 1st-order gradient direction

$\bar{\boldsymbol{d}}_1 = -\frac{V_1 D_1^{-1} \boldsymbol{f}_1}{p-1}$

(b) If grade $0 < r < k$ go to Step 5.

(c) If grade $r = k$ set $\boldsymbol{d}^j = \bar{\boldsymbol{d}}^1$, $\bar{\boldsymbol{d}}^2 = 0$ and take a full first-order gradient step. Go to Step 8.

Step 5: Compute the modified Cholesky factorization of

$$(p-1)D_2^2 + V_2{}'B_p V_2 + E^j = L^j \tilde{D}^j (L^j)'$$

Step 6:

(a) If $\|\boldsymbol{g}^j\| \leq \epsilon$ and $\|E^j\| = 0$ then $\boldsymbol{\theta}^j$ is optimal and STOP.

(b) If $\|\boldsymbol{g}^j\| > \epsilon$ determine the search direction by solving the set of linear equations for $\boldsymbol{z}_2$:

$L^j \tilde{D}^j (L^j)' \boldsymbol{z}_2 = -D_2 \boldsymbol{f}_2 - V_2{}' B_p \bar{\boldsymbol{d}}_1$

Go to Step 7.

(c) If $\|\boldsymbol{g}^j\| \leq \epsilon$ and $\|E^j\| \neq 0$ determine $\boldsymbol{z}_2$ by means of the following search procedure:

Let q be the subscript for which

$\tilde{d}_q^j - E_{qq}^j = \min\{\tilde{d}_i^j - E_{ii}^j \mid r+1 \leq i \leq k\}$

Solve $(L^j)' \boldsymbol{u} = \boldsymbol{e}_q$ by backward substitution for $\boldsymbol{u}$ ($\boldsymbol{e}_q$ is a qth unit vector).

$$\text{Set} \quad \boldsymbol{z}_2 = \begin{cases} -sign[\boldsymbol{u}'\boldsymbol{g}^j]\boldsymbol{u} & \text{if } \|\boldsymbol{g}^j\| > 0 \\ \boldsymbol{u} & \text{if } \|\boldsymbol{g}^j\| = 0 \end{cases}$$

Step 7: Compute the direction $\bar{\boldsymbol{d}}_2 = V_2 \boldsymbol{z}_2$ and set $\boldsymbol{d}^j = \bar{\boldsymbol{d}}_1 + \bar{\boldsymbol{d}}_2$.

Step 8: If the negative of the normalised projected gradient

$-\frac{(\boldsymbol{g}^j)'\boldsymbol{d}^j}{\|\boldsymbol{g}^j\|\,\|\boldsymbol{d}^j\|} < \eta < \|D_1^{-2}\|$

return to Step 5 with grade $r = 0$. Otherwise proceed to Step 9.

Step 9: Compute γ_j using Fletcher's (1970) line search procedure and set

$\boldsymbol{\theta}^{j+1} = \boldsymbol{\theta}^j + \gamma_j \boldsymbol{d}^j$.

Step 10: Continue until the convergence criteria are met:

(a) $|g_i^j| < tol$ for $i = 1, \ldots, k$

(b) $\frac{|S_p(\boldsymbol{\theta}^{j+1}) - S_p(\boldsymbol{\theta}^j)|}{S_p(\boldsymbol{\theta}^j)} < tol$

in 4 consecutive iterations. Otherwise set $j := j + 1$ and return to Step 1.

Remark As an alternative to test 10(a) the generalised gradient $\boldsymbol{g}^*$ due to Dennis (1977) may be used.

4.3.3.1 Examples.

Gonin and Du Toit (1987) considered 10 test examples and reported their numerical experience. We shall consider 3 of those examples.

Example 4.5 The intermediate steps of one iteration of the mixture method applied to the Rosenbrock problem will now be illustrated. The value $p = 2$ is again used. The value of $toler = 1/k = 0.5$ was used throughout.

$$\boldsymbol{\theta}^1 = (-1.2, 1)' \text{ and } S_p(\boldsymbol{\theta}^1) = 95.832.$$

Step 1:

$$J_p(\boldsymbol{\theta}^1) = \begin{pmatrix} 24 & 10 \\ -1 & 0 \end{pmatrix} \qquad (\boldsymbol{y} - \boldsymbol{f})_p = \begin{pmatrix} 4.4 \\ -2.2 \end{pmatrix}$$

$$\boldsymbol{g}^1 = \begin{pmatrix} -215.6 \\ -88 \end{pmatrix}$$

Step 3:

$$B_p = \begin{pmatrix} -88 & 0 \\ 0 & 0 \end{pmatrix}$$

$$s_1 = 26.2 \qquad s_2 = 0.3844 \qquad toler = 0.5 \qquad \frac{s_2}{s_1} = 0.0147 < 0.5$$

$$\frac{s_1 + s_2}{s_1} = 1.015 > 0.5$$

Thus $j^* = 1$ and $r = 1$.

Step 4:

$$\bar{\boldsymbol{d}}_1 = \begin{pmatrix} 0.1588 \\ 0.0661 \end{pmatrix}$$

Step 5: $||E^1|| = 0$

Step 6b: $||\boldsymbol{g}^1|| = 232.87 > 10^{-9}$

$$\boldsymbol{z}_2 = \begin{pmatrix} 0.3504 \\ 0.3844 \end{pmatrix} \qquad \bar{\boldsymbol{d}}_2 = \begin{pmatrix} -0.1346 \\ 0.3235 \end{pmatrix} \qquad \boldsymbol{d}^1 = \begin{pmatrix} 0.0242 \\ 0.3896 \end{pmatrix}$$

Step 8: $(\boldsymbol{g}^1)'\boldsymbol{d}^1 = -39.51$ and

$$-\frac{(\boldsymbol{g}^1)'\boldsymbol{d}^1}{\|\boldsymbol{g}^1\|\,\|\boldsymbol{d}^1\|} = 0.4346 > \eta = 0.0001$$

Step 9: $\gamma_1 = 1$

$$\boldsymbol{\theta}^2 = \begin{pmatrix} -1.2 \\ 1 \end{pmatrix} + \begin{pmatrix} 0.0242 \\ 0.3896 \end{pmatrix} = \begin{pmatrix} -1.1758 \\ 1.3896 \end{pmatrix}$$

and so on.

Example 4.6 This example is due to Osborne (1972). The usual starting value $\boldsymbol{\theta}^1 = (0.5, 1.5, -1.0, 0.01, 0.02)$ was used. Betts (1976), Gill and Murray (1978) and Dennis et al. (1981) reported 10, 11 and 27 evaluations respectively. The mixture algorithm took 9 function and gradient evaluations. Solutions for other values of p are given in Table 4.3.

Table 4.3: Evaluations for different values of p.

			p			
$p \rightarrow$	1.5	1.75	2.0	2.5	2.75	3.0
θ_1	0.376210	0.375789	0.375410	0.374549	0.373899	0.373441
θ_2	2.036996	1.981291	1.935829	1.845625	1.786680	1.747961
θ_3	-1.567076	-1.510663	-1.464699	-1.373441	-1.313823	-1.274608
θ_4	0.013062	0.012957	0.012867	0.012674	0.012535	0.012439
θ_5	0.021745	0.021946	0.022123	0.022508	0.022789	0.022989
$S_p(\boldsymbol{\theta})$	$.12055^{(-2)}$	$.25313^{(-3)}$	$.54650^{(-4)}$	$.27164^{(-5)}$	$.61803^{(-6)}$	$.14137^{(-6)}$
evaluations	21	15	9	10	14	21

Example 4.7 The following curve-fitting problem was considered by Gonin and Du Toit (1987). The data describe the physiological relationship between oxygen saturation (y) and oxygen tension (x). This problem and the data are considered in more detail in Chapter 6 (§6.2). The following model was fitted

$$y = C_5 C_1^{C_2^{(1-C_3^x)^{C_4}}}$$

$$\text{where} \qquad C_i = \frac{1}{1+e^{\theta_i}} \qquad i = 1,2,3 \qquad C_4 = 1 + e^{\theta_4} \qquad C_5 \;\; \theta_5$$

The starting value $\boldsymbol{\theta}^1 = (4.5, 6.0, -4.2, -1.0, 100)$ was used. The solutions for other values of p are given Table 4.4.

Table 4.4: Evaluations for different values of p.

	p					
$p \rightarrow$	1.5 †	1.75†	2.0†	2.5 †	2.75 †	3.0
θ_1	4.365812	4.371020	4.370218	4.361674	4.356185	4.350555
θ_2	7.303459	7.255965	7.207015	7.113816	7.071245	7.031252
θ_3	-4.157157	-4.148084	-4.138477	-4.119685	-4.110944	-4.102682
θ_4	-1.288798	-1.281238	-1.270184	-1.245082	-1.232678	-1.220805
θ_5	99.560517	99.583881	99.607959	99.653185	99.674093	99.694334
$S_p(\boldsymbol{\theta})$	2.787901	1.903481	1.314361	0.644671	0.456706	0.325514
evaluations	19	13	9	13	13	13

†$toler = 1/k^4$.

CONCLUSION:

Current research in nonlinear least squares procedures concentrates on hybrid methods (mixture of Gauss-Newton and quasi-Newton methods). These ideas have also been used in nonlinear L_1- and L_∞-norm problems as we have seen. The extension to the general nonlinear L_p-norm problem will follow naturally. One would expect these methods to perform well.

This concludes the discussion of algorithms for solving nonlinear L_p-norm estimation problems. The remaining chapters will deal with the statistical properties of the estimates, the choice of p and the application of L_p-norm estimation in practice.

Appendix 4A: Cholesky decomposition of symmetric matrices

Consider the system of equations

$$A\boldsymbol{x} = \boldsymbol{b} \tag{1}$$

where $A_{m\times m}$ is a symmetric positive definite matrix and $\boldsymbol{b}$ an m-vector of known constants. A numerically stable (i.e., minimizing the effect of rounding errors) method for calculating $\boldsymbol{x}$ is to factorize A by the method of Cholesky into the form

$$A = LDL' \tag{2}$$

where L is a unit lower-triangular matrix and D a diagonal matrix. The vector $\boldsymbol{x}$ is then computed using forward and backward substitution. Let A_{ij}, ℓ_{ij} denote the ij^{th} element of A and L respectively and d_j the jj^{th} element of D (d_j not to be confused with the direction $\boldsymbol{d}^j$).

The jth step of Cholesky's method is then given by

$$d_j = a_{jj} - \sum_{k=1}^{j-1} \ell_{jk}^2 d_k \tag{3}$$

$$\ell_{ij} = C_{ij}/d_j \qquad i = j+1, \ldots, m \tag{4}$$

with the auxiliary quantities C_{ij} given by

$$C_{ij} = a_{ij} - \sum_{k=1}^{j-1} \ell_{jk}\ell_{ik}d_k, \qquad i = j+1, \ldots, m \tag{5}$$

Note that $d_1 = a_{11}$ and $C_{i1} = a_{i1}$, $i = 2, \ldots, m$. Set $T = LD$ then (2) may be written as $TL'\boldsymbol{x} = \boldsymbol{b}$. If we set $\boldsymbol{y} = L'\boldsymbol{x}$ then the m unknowns $\boldsymbol{y}$ in the system of linear equations $T\boldsymbol{y} = \boldsymbol{b}$ can be computed by forward substitution:

$$y_i = \frac{b_i - \sum_{k=1}^{i-1} t_{ik} y_k}{t_{ii}} \qquad i = 1, \ldots, m \tag{6}$$

Finally, the m unknowns $\boldsymbol{x}$ in the system $L'\boldsymbol{x} = \boldsymbol{y}$ can be computed by backward substitution:

$$\begin{aligned} x_i &= \frac{y_i - \sum_{k=i+1}^{m} \ell_{ki} x_k}{\ell_{ii}} \qquad i = m, m-1, \ldots, 1 \\ &= y_i - \sum_{k=i+1}^{m} \ell_{ki} x_k \end{aligned} \tag{7}$$

since L is a unit lower triangular matrix.

The modified Cholesky factorization method for symmetric matrices

The modified Cholesky factorization method constructs a positive definite matrix from a given square matrix which may be indefinite, negative definite or insufficiently positive definite and was derived by Gill and Murray (1974).

Matrix A is insufficiently positive definite:

Equations (4) and (5) show that ℓ_{ij} may be too large whenever d_j is too small. This will happen when A is insufficiently positive definite. Since A is positive definite it follows that $A = (LD^{\frac{1}{2}})(LD^{\frac{1}{2}})'$. The diagonal and off-diagonal elements of $LD^{\frac{1}{2}}$ are therefore $\sqrt{d_j}$ and $\ell_{ij}\sqrt{d_j}$ respectively. From equation (3)

$$\sqrt{d_j} = \sqrt{a_{jj} - \sum_{k=1}^{j-1} \ell_{jk}^2 d_k}$$

Now, since d_j is positive it follows that each $\ell_{jk}\sqrt{d_k} < \sqrt{a_{jj}}$. This means that no element of $LD^{\frac{1}{2}}$ can exceed $\sqrt{a_{jj}}$. Thus, a large element or elements in $LD^{\frac{1}{2}}$ can only occur if A has a large diagonal element or elements. A Cholesky decomposition will now be described with the following properties:

(a) All elements of $LD^{\frac{1}{2}}$ are bounded above by a value β.

(b) All elements of D are bounded below by a value δ.

We now consider how β and δ are determined.

$$\begin{aligned}
\text{Let } \varsigma &= \text{maximum in absolute value of the off-diagonal elements of } A \\
&= \max\{|a_{ij}| : i = j+1, \ldots, m\} \\
\eta &= \text{maximum in absolute value of the diagonal elements of } A \\
&= \max\{|a_{jj}| : j = 1, \ldots, m\} \\
\rho &= \max\{|C_{ij}| : i = j+1, \ldots, m\} \\
EPS &= \text{relative machine precision} = 2^{-56} \text{ in IBM double precision} \\
&= (\text{the smallest positive machine representable floating point number}).
\end{aligned}$$

Gill and Murray (1974) show that the choice

$$\beta^2 = \max\{\varsigma/m, \eta, EPS\}, \quad \delta = \max\{EPS\|A\|, EPS\} \quad d_j^* = \max\{\delta, |d_j|, \rho^2/\beta^2\}$$

automatically ensures that $|\ell_{ij}\sqrt{d_j^*}| \leq \beta$. This simply means that the terms of $LD^{\frac{1}{2}}$ are bounded above by β. We see that the choice of β^2 and d_j^* prevents the off-diagonal elements of A from becoming too large and the diagonal elements of A from becoming too small. We shall see that if the matrix A is sufficiently positive definite (i.e., when $d_j > \delta$) no modification to d_j is made, thus $d_j^* = d_j = |d_j| > 0$. Hence the stability of the factorization is assured. We therefore select:

$$d_j^* = \begin{cases} \delta, & \text{if } \delta \geq \max\{|d_j|, \rho^2/\beta^2\} \\ |d_j| & \text{if } |d_j| \geq \max\{\delta, \rho^2/\beta^2\} \\ \rho^2/\beta^2 & \text{if } \rho^2/\beta^2 \geq \max\{\delta, |d_j|\} \end{cases} \tag{8}$$

Thus the elements d_j of D are either retained or modified as above. The factors obtained by the modified procedure are identical to those obtained by applying Cholesky's method to the matrix $A^* = A + E$ where E is a diagonal matrix with a typical element:

$$E_{ii} = \begin{cases} \delta - d_j, & \text{if } \delta \geq \max\{|d_j|, \rho^2/\beta^2\} \\ |d_j| - d_j, & \text{if } |d_j| \geq \max\{\delta, \rho^2/\beta^2\} \\ \rho^2/\beta^2 - d_j, & \text{if } \rho^2/\beta^2 \geq \max\{\delta, |d_j|\} \end{cases} \tag{9}$$

If A is sufficiently positive definite then $E = 0$. This can be seen if d_j is the largest of the quantities ρ^2/β^2 and δ, then $d_j = |d_j|$ whence $E_{ii} = 0$.

Matrix A is negative definite:

In this case $d_j < 0$ and is replaced by $d_j^* = |d_j|$.

Matrix A is indefinite:

Let us substitute $G^j = \nabla^2 S_p(\boldsymbol{\theta}^j)$ for A. Let $\boldsymbol{g}^j = \nabla S_p(\boldsymbol{\theta}^j)$ then the direction

$$\boldsymbol{d}^j = -(\tilde{G}^j)^{-1}\boldsymbol{g}^j = (G^j + E^j)^{-1}\boldsymbol{g}^j$$

will be zero when the gradient $\boldsymbol{g}^j = 0$. Furthermore, if G^j is indefinite then $\boldsymbol{\theta}^j$ will be a saddlepoint and not be a local minimum. An alternative direction or so-called direction of negative curvature has to be used, whenever the norm $\|\boldsymbol{g}^j\| > 0$, to prevent convergence to saddlepoints. The information regarding the indefiniteness of G^j is already available in the modified Cholesky factorization of G^j.

Definition 1 A direction $\boldsymbol{d}$ is a *direction of negative curvature* with respect to the indefinite matrix G^j if $\boldsymbol{d}'G^j\boldsymbol{d} < 0$.

The following postulates are stated without proof [see Gill and Murray (1974)].

(1) Let s be any integer such that $d_s^* \leq \min\{d_i^* \mid 1 \leq i \leq m\}$. If G^j is indefinite then $d_s^* \leq 0$.

(2) If G^j is indefinite and $d_s^* = 0$, then for $j = 1, \ldots, q-1, q+1, \ldots, s-1$ we can choose $\ell_{sj} = 0$ and $\ell_{sq} d_q^* = \beta^2$ for some q such that $1 \leq q \leq m$.

(3) Suppose $\boldsymbol{\theta}^j$ is such that $||\boldsymbol{g}^j|| = 0$ and G^j is indefinite. Let $\boldsymbol{u}$ be a solution to the equation $L'\boldsymbol{u} = \boldsymbol{e}_s$ ($\boldsymbol{e}_s$ is the sth unit co-ordinate vector) with d_s^* defined as before, then $\boldsymbol{u}$ is a direction of negative curvature.

Remark:

The following is not explicitly stated in Gill and Murray (1974). In postulate (3) it is assumed that $||\boldsymbol{g}^j|| = 0$. If, however, $||\boldsymbol{g}^j|| > 0$ and $(\boldsymbol{u})'\boldsymbol{g}^j > 0$ ($S_p(\boldsymbol{\theta}^j)$ increases in the direction $\boldsymbol{u}$) then $-\boldsymbol{u}$ will be a descent direction. The following decision rule is therefore used:

$$\boldsymbol{u} = \begin{cases} -sign[(\boldsymbol{u})'\boldsymbol{g}^j]\boldsymbol{u} & \text{if } 0 < ||\boldsymbol{g}^j|| < \epsilon \\ \boldsymbol{u} & \text{if } ||\boldsymbol{g}^j|| = 0 \end{cases} \tag{10}$$

where $\epsilon > 0$ is a small positive number.

Appendix 4B: The singular value decomposition (SVD) of a matrix

We shall digress to discuss the *eigenvalue-eigenvector decomposition* of a symmetric matrix A since the two decompositions are interrelated.

Theorem 1 Let $A_{n\times n}$ be a real symmetric matrix. Then A has an eigenvalue-eigenvector decomposition of A of the form $A = VDV'$. Where the columns of V are the eigenvectors of A and D is a diagonal matrix with diagonal elements the eigenvalues of A; ordered as follows

$$\lambda_1 \geq \lambda_2 \geq \ldots \geq \lambda_n$$

Proof By definition of the eigenvectors, V is an orthogonal matrix i.e., $VV' = V'V = I$. Let $V = [\boldsymbol{v}_1, \cdots, \boldsymbol{v}_n]$ then $A\boldsymbol{v}_i = \lambda_i \boldsymbol{v}_i$ for $i = 1, \ldots, n$ or in matrix notation $AV = VD$. Hence $A = AVV' = VDV'$ since V is orthogonal. ∎

The *singular value decomposition* of matrix $A_{n\times k}$ hinges on the following theorem:

Theorem 2 Suppose the rank$(A) = r < k$, then there exists an orthogonal matrix $U_{n\times n}$ partitioned as

$$U = \begin{pmatrix} U_1 : n \times r & U_2 : n \times (n-r) \end{pmatrix}$$

and an orthogonal matrix $V_{k\times k}$ partitioned as

$$V = \begin{pmatrix} V_1 : k \times r & V_2 : k \times (k-r) \end{pmatrix}$$

such that $A = UDV'$ where

$$D_{n\times k} = \begin{pmatrix} diag(s_1, \ldots, s_r) & 0 \\ 0 & 0 \end{pmatrix} = \begin{pmatrix} D(s) & 0 \\ 0 & 0 \end{pmatrix}$$

with $s_1 \geq s_2 \geq \ldots \geq s_r > 0$.

Proof The matrix $A'A$ will either be positive or positive semidefinite of rank r. It will therefore possess an eigenvalue-eigenvector decomposition of the form:

$$A'A = V'D^2V' = \begin{pmatrix} D^2(s) & 0 \\ 0 & 0 \end{pmatrix} V'$$

Thus $V'A'AV = D^2$ where $s_1 \geq s_2 \geq \ldots \geq s_r > 0$.

Denote the first r columns of V by V_1 and the remaining $(k-r)$ columns by V_2. Then $I = VV' = V_1V_1' + V_2V_2'$ or $V_1V_1' = I - V_2V_2'$. Since $V'A'AV = D^2$ it follows that $V_1'A'AV_1 = D^2(s)$ and $V_2A'AV_2 = 0$. Thus it follows that $AV_2 = 0$. We see that

$$\begin{aligned} A &= A - (AV_2)V_2' \\ &= A(I - V_2V_2') \\ &= AV_1V_1' \\ &= AV_1[D(s)^{-1}D(s)]V_1' \\ &= [(AV_1(D(s)^{-1}]D(s)V_1' \\ &= U_1D(s)V_1' \end{aligned}$$

By the QR factorization (Appendix 2D) it is possible to construct a matrix U_2 such that $U = \begin{pmatrix} U_1 & U_2 \end{pmatrix}$ is orthogonal. (Note that U_1 is columnwise orthogonal i.e., $U_1'U_1 = D(s)^{-1}V_1'A'AV_1D(s)^{-1} = I$). Observe that

$$\begin{aligned} UDV' &= \begin{pmatrix} U_1 & U_2 \end{pmatrix} \begin{pmatrix} D(s) & 0 \\ 0 & 0 \end{pmatrix} \begin{pmatrix} V_1' \\ V_2' \end{pmatrix} \\ &= U_1D(s)V_1' \end{aligned}$$

and the proof is complete. ∎

Remarks:

(1) Matrices U and V consist of the orthonormalized eigenvectors of AA' and $A'A$ respectively and the singular values s_i are the nonnegative square roots of the eigenvalues of $A'A$.

(2) Subroutine SVDRS of Lawson and Hanson (1974) [see also Kennedy and Gentle (1980:278–286)] calculates the singular value decomposition of matrix A of rank $r \leq k$ in two stages:

(I) It constructs a sequence of Householder transformations (Appendix 2D)

$$P_i,\ i = 1, \ldots, k \quad \bar{P}_i,\ i = 1, \ldots k-1$$

so that

$$P_kP_{k-1}\ldots P_1A\bar{P}_1\bar{P}_2\ldots\bar{P}_{k-1} \equiv P'A\bar{P} = Q$$

where Q is an $n \times k$ bidiagonal matrix:

$$\begin{pmatrix} a_1 & b_1 & 0 & \cdots & 0 \\ 0 & a_2 & b_2 & & \vdots \\ \vdots & & a_3 & \ddots & 0 \\ \vdots & & & \ddots & b_{k-1} \\ 0 & \cdots & \cdots & 0 & a_k \\ & & & & \\ & & \mathit{0} & & \end{pmatrix}$$

with the same singular values as A.

(II) The singular values of Q are then computed with the aid of a special QR algorithm for computing singular values.

Subroutine SVDRS uses the following subroutines: H12: Constructs a Householder transformation and applies the transformation to a given vector. QRBD: Computes the singular value decomposition of the bidiagonal matrix Q. G1, G2: Construction and application of rotation matrices. DIFF: Termination criterion.

Appendix 4C: Fletcher's line search algorithm

The line search algorithm due to Fletcher (1970) uses gradient information. The following condition is required to ensure that the magnitude of the gradient is sufficiently increased away from $\boldsymbol{\theta}^j$, i.e.,

$$|\boldsymbol{g}(\boldsymbol{\theta}^j + \gamma_j \boldsymbol{d}^j)'\boldsymbol{d}^j| > -\rho(\boldsymbol{g}^j)'\boldsymbol{d}^j \quad \text{for } 0 < \rho < 1$$

A value of $\rho = 0.9$ gives a weak line search and $\rho = 0.1$ gives an accurate line search.

The Algorithm

Step 0: Let ΔS_p be a user supplied estimate of the likely reduction in $S_p(\boldsymbol{\theta})$. Set $S_1 = S_p(\boldsymbol{\theta}^j)$ and $\boldsymbol{g}^j = \boldsymbol{g}(\boldsymbol{\theta}^j)$.

Step 1: Calculate $\gamma = \min\{1, -2\Delta S_p/(\boldsymbol{g}^j)'\boldsymbol{d}^j\}$.

Step 2: Compute $\boldsymbol{\theta}_N = \boldsymbol{\theta}^j + \gamma \boldsymbol{d}^j$, $S_2 = S_p(\boldsymbol{\theta}_N)$ and $\boldsymbol{g}_2'\boldsymbol{d}^j = \boldsymbol{g}(\boldsymbol{\theta}_N)'\boldsymbol{d}^j$.

Step 3: If $S_2 \geq S_1$ go to Step 7.

Step 4: If $|\boldsymbol{g}_2'\boldsymbol{d}^j/(\boldsymbol{g}^j)'\boldsymbol{d}^j| < \rho$, STOP. The optimal γ has been found.

Step 5: If $\boldsymbol{g}_2'\boldsymbol{d}^j > 0$ go to Step 7.

Step 6: If $(\boldsymbol{g}^j)'\boldsymbol{d}^j < \boldsymbol{g}_2'\boldsymbol{d}^j$ set

$$\gamma := \gamma \min\left\{10, \frac{\boldsymbol{g}_2'\boldsymbol{d}^j}{(\boldsymbol{g}^j)'\boldsymbol{d}^j - \boldsymbol{g}_2'\boldsymbol{d}^j}\right\}$$

if $(\boldsymbol{g}^j)'\boldsymbol{d}^j > \boldsymbol{g}_2'\boldsymbol{d}^j$ set $\gamma := 10\gamma$ and return to Step 2.

Step 7: Cubic interpolation. Calculate

$$z = \frac{3}{\gamma}(S_1 - S_2) + \boldsymbol{g}_2'\boldsymbol{d}^j + (\boldsymbol{g}^j)'\boldsymbol{d}^j$$

$$w = \sqrt{z^2 - \boldsymbol{g}_2'\boldsymbol{d}^j(\boldsymbol{g}^j)'\boldsymbol{d}^j}$$

$$\gamma := \gamma\left(1 - \frac{w - z + \boldsymbol{g}_2'\boldsymbol{d}^j}{\boldsymbol{g}_2'\boldsymbol{d}^j - (\boldsymbol{g}_j)'\boldsymbol{d}^j + 2w}\right)$$

Return to Step 2.

Chapter 4: Additional notes

§4.1

1 The uniqueness of nonlinear L_p approximations for $1 < p < 2$ is considered by Aklaghi and Wolfe (1982).

2 More theoretical discussions of nonlinear L_p approximations may be found in the books by Rice (1969), Watson (1980) and Powell (1981).

§4.3

1 The literature on nonlinear least squares is fairly extensive. A discussion of numerical procedures may be found in Dennis and Schnabel (1983) and Kennedy and Gentle (1980). A survey of large residual problems is given by Nazareth (1980). A numerical comparison of various nonlinear least squares algorithms was undertaken by Hiebert (1979), Dennis et al. (1981) and Moré et al. (1981) and Lindström (1983).

2 Al-Baali and Fletcher (1985) consider Newton-like methods which minimize a variational measure which estimates the error in an inverse Hessian approximation. A hybrid method which combines the Gauss-Newton and BFGS quasi-Newton approach is also considered. Fletcher and Xu (1986) show by counter example that this variational method will not always be superlinearly convergent as conjectured. They propose two GN/BFGS hybrid methods which are superlinearly convergent under certain mild conditions.

3 Toint (1987) considers the large scale nonlinear least squares problem (up to 100 parameters and 1000 observations). Extensions to classical algorithms are proposed. These extensions involve the concept of partially separable structures which have proved their usefulness in large scale optimization. An adaptable algorithm is considered and preliminary numerical experience indicates that the actual numerical solution of a large class of problems involving several hundred parameters is possible at a reasonable cost. Dennis and Steihaug (1986) suggest a generalised Gauss-Seidel approach for solving sparse nonlinear (and linear) least squares problems.

4 The fitting of sums of exponential functions in the least squares and L_1 case respectively are considered by Varah (1985) and Watson (1985).

5 Bates and Watts (1987) present a generalised Gauss-Newton algorithm (which includes a Levenberg-Marquardt option) for multi-response parameter estimation problems.

6 The total least squares problem (errors in all variables) was considered by Golub and Van Loan (1980) and more recently by Schwetlick and Tiller (1985).

7 The nonlinearly constrained least squares problem was examined by Holt and Fletcher (1979) and Lindström (1983). Wright and Holt (1985) propose 2 algorithms for the nonlinear least squares problem with linear inequality constraints.

8 Barrodale et al. (1970) considered the L_p-norm estimation problem (for $p = 1, 2$ and ∞) for a class of approximating functions nonlinear in only one of several parameters (the others obviously enter linearly). Their method proceeds as follows:

For a given value (not necessarily optimal) of the single nonlinear parameter the best L_p-approximation can be calculated using one of the available linear approximation algorithms (linear programming or least squares). The resulting error of approximation will then be a function of the single nonlinear parameter only. This univariate function can then be minimized by line search methods (Appendix 2B).

§4.3.1

Watson observed that some methods for solving the general norm problem involve the sequential solution of a number of linear subproblems whose solutions converge to the solution of the general nonlinear approximation problem (e.g., the Osborne and Watson (1971) and Anderson and Osborne (1977a,1977b) algorithms). The convergence would also be norm-dependent. So for convex monotonic norms the convergence rate would generally be linear whilst for polyhedral norms the rate would be quadratic (strong uniqueness).

§4.3.3

1 The procedure for calculating the grade r is due to Gonin and Du Toit (1987).

2 The ratio $\frac{s_k}{s_1}$ used in Step 3 of the mixture algorithm may be regarded as the reciprocal of the "*condition number*" of J_p. Troskie and Conradie (1986) derived the distribution of the condition number of a matrix.

Bibliography Chapter 4

Akhlaghi, M. and Wolfe, J.M. (1982). Nonlinear L_p approximation for $1 < p < 2$. *J. Approx. Theory.* **34**, pp 12–36.

Al-Baali, M. and Fletcher, R. (1985). Variational methods for non-linear least squares. *J. Opl. Res. Soc.* **36**, pp 405–421.

Anderson, D.H. and Osborne, M.R. (1977a). Discrete, nonlinear approximation problems in polyhedral norms. *Numer. Math.* **28**, pp 143–156.

Anderson, D.H. and Osborne, M.R. (1977b). Discrete, nonlinear approximation problems in polyhedral norms. A Levenberg-like algorithm. *Numer. Math.* **28**, pp 157–170.

Barrodale, I., Roberts, F.D.K. and Hunt, C.R. (1970). Computing best ℓ_p approximations by functions nonlinear in one parameter. *Computer J.* **13**, pp 382–386.

Bates, D.M. and Watts, D.G. (1987). A generalised Gauss-Newton procedure for multi-response parameter estimation. *SIAM J. Sci. Stat. Comput.* **8**, pp 49–55.

Betts, J.T. (1976). Solving the nonlinear least squares problem: Application of a general method. *J. Optim. Theory Appl.***18**, pp 469–483.

Dennis, J.E. (1977). Non-linear least squares and equations. In: D. Jacobs, ed., *The state of the art in numerical analysis, Academic Press, London.*

Dennis, J.E., Gay, D.M. and Welsch, R.E. (1981). An adaptive nonlinear least-squares algorithm. *Trans. Math. Software* **7**, pp 348–368.

Dennis, J.E. and Schnabel, R.B. (1983). Numerical methods for unconstrained optimization and nonlinear equations. *Prentice-Hall, Englewood Cliffs, New Jersey.*

Dennis, J.E. and Steihaug, T. (1986). On the successive projections approach to least-squares problems. *SIAM. J. Numer. Anal.* **23**, pp 717–733.

Fletcher, R. (1970). A new approach to variable metric algorithms. *Comput. J.* **13**, pp 317–322.

Fletcher, R. and Xu, C. (1986). Hybrid methods for nonlinear least squares. *Numerical analysis Report NA/92, University of Dundee, Scotland.*

Gill, P.E. and Murray, W. (1974). Newton–type methods for unconstrained and linearly constrained optimization. *Math. Programming* **7**, pp 311–350.

Gill, P.E. and Murray, W. (1978). Algorithms for the solution of the nonlinear least-squares problem. *SIAM. J. Numer. Anal.* **15**, pp 977–992.

Golub, G.H. and Van Loan, C.F. (1980). An analysis of the total least squares problem. *SIAM J. Numer. Anal.* **17**, pp 883–893.

Gonin, R. (1983). A contribution to the solving of nonlinear estimation problems. *PhD Thesis, University of Cape Town, Cape Town.*

Gonin, R. (1986). Numerical algorithms for solving nonlinear L_p-norm estimation problems: Part I - A first-order gradient algorithm for well-conditioned small residual problems. *Commun. Statist.-Simula. Computa.* **15**, pp 801–813.

Gonin, R. and Du Toit, S.H.C. (1987). Numerical algorithms for solving nonlinear L_p-norm estimation problems: Part II - A mixture method for large residual and ill-conditioned problems. *Commun. Statist.-Theor. Meth.* **16**, pp 969–986.

Hiebert, K.L. (1979). A comparison of nonlinear least squares software. *Sandia Laboratories Report SAND 79–0483, Albuquerque, New Mexico.*

Holt, J.N. and Fletcher, R. (1979). An algorithm for constrained non-linear least squares. *J. Inst. Maths. Applic.* **23**, pp 449–463.

Jacobson, D.H. and Oksman W. (1972). An algorithm that minimizes homogeneous functions of N variables in $N+2$ iterations and rapidly minimizes general functions, *J. Math. Anal. Appl.* **38**, pp 535–552.

Jones, A.P. (1970). SPIRAL–A new algorithm for nonlinear parameter estimation using least squares. *Comput. J.* **13**, pp 301–308.

Kennedy, W.J. and Gentle, J.E. (1980). Statistical Computing. *Marcel Dekker, New York.*

Lawson, C.L. and Hanson, R.J. (1974). Solving least squares problems, *Prentice–Hall, New Jersey.*

Lindström, P. (1983). Algorithms for nonlinear least squares – particularly problems with constraints. *PhD Thesis. University of Umeå, Umeå, Sweden.*

McCormick, G.P. (1983). Nonlinear programming: Theory, algorithms and applications. *John Wiley, New York.*

Moré, J.J. (1977). The Levenberg-Marquardt algorithm: implementation and theory, In: G.A. Watson, ed. *Lecture Notes in Mathematics 630*, Springer Verlag, Berlin, pp 105–116.

Moré, J.J., Garbow, B.S. and Hillstrom, K.E. (1981). Testing unconstrained optimization software. *ACM Trans. Math. Softw.* **7**, pp 17–41.

Nazareth, L. (1980). Some recent approaches to solving large residual nonlinear least squares problems. *SIAM Rev.* **22**, pp 1–11.

Osborne, M.R. (1972). Some aspects of non-linear least squares calculations. In: F. Lootsma, ed. *Numerical methods for nonlinear optimization, Academic Press, New York*, pp 171–189.

Osborne, M.R. and Watson, G.A. (1971). On an algorithm for discrete nonlinear L_1 approximation. *Comput. J.* **14**, pp 184–188.

Powell, M.J.D. (1962). An iterative method for finding stationary values of a function of several variables. *Comput. J.* **5**, pp 147–151.

Powell, M.J.D. (1981). Approximation theory and methods. *Cambridge University Press, Cambridge.*

Rice, J.R. (1969). The approximation of functions: vol. 2, Nonlinear and multivariate theory. *Addison-Wesley, Reading, Massachusetts.*

Rosenbrock, H.H. (1960). An automatic method for finding the greatest or the least value of a function. *Comput. J.* **3**, pp 175–184.

Schwetlick, H. and Tiller, V. (1985). Numerical methods for estimating parameters in nonlinear models with errors in the variables. *Technometrics* **27**, pp 17–24.

Toint, PH.L. (1987). On large scale nonlinear least squares calculations. *SIAM J. Sci. Stat. Comput.* **8**, pp 416–435.

Troskie, C.G. and Conradie, W.J. (1986). The distribution of the ratios of characteristic roots (condition numbers) and their application in principal component or ridge regression. *Lin. Alg. Applic.* **82**, pp 255–279.

Varah, J.M. (1985). On fitting exponentials by nonlinear least squares. *SIAM J. Sci. Stat. Comput.***6**, pp 30–44.

Watson G.A. (1979). Dual methods for nonlinear best approximation problems. *J. Approx. Theory* **26**, pp 142–150.

Watson G.A. (1980). Approximation theory and numerical methods. *John Wiley, New York.*

Watson G.A. (1985). Data fitting by positive sums of exponentials *Numerical analysis Report NA/90, University of Dundee, Scotland.*

Watson G.A. (1986). Methods for best approximation and regression problems. *Numerical analysis Report NA/96, University of Dundee, Scotland.*

Wright, S.J., Holt, J.N. (1985). Algorithms for nonlinear least squares with linear inequality constraints. *SIAM J. Sci. Stat. Comput.* **6**, pp 1033–1048.

5

Statistical Aspects of L_p-Norm Estimators

In Chapters 2 to 4 we concentrated on numerical procedures to estimate the unknown parameters $\boldsymbol{\theta}$ of the nonlinear model (2.1). Relatively little appears to have been published with regard to the statistical aspects of nonlinear L_p-norm estimation.

In this Chapter we shall be concerned with statistical inference, the statistical properties of the estimator $\hat{\boldsymbol{\theta}}$ of the true parameter vector $\boldsymbol{\theta}$ and finally critique (regression diagnostics) of the model $f(\boldsymbol{x}, \boldsymbol{\theta})$. In §5.1, 5.2 and 5.3 nonlinear least squares, L_1-norm and L_p-norm estimations will be considered respectively.

In §5.1.1 three conventional approaches for constructing confidence intervals for: $\hat{\theta}_i$; and joint confidence regions for $\hat{\boldsymbol{\theta}}$ will be discussed. The jackknife procedure as a robust alternative in setting up confidence regions and intervals for small samples will be described. Furthermore, confidence intervals for the conditional mean response and an individual value of the response variable y_i are proposed. A Wald type test statistic for hypothesis testing is considered. The bias in $\hat{\boldsymbol{\theta}}$ as well as a technique for reducing bias are also discussed.

In §5.2 and 5.3, following on from a conjecture concerning the asymptotic distribution of $\sqrt{n}(\hat{\boldsymbol{\theta}} - \boldsymbol{\theta})$, in the L_1-norm and L_p-norm cases, confidence intervals for: $\hat{\theta}_i$; the conditional mean response and an individual value of the response variable y_i are proposed. An expression for the Bias($\hat{\boldsymbol{\theta}}$) follows from this conjecture. Furthermore, a Wald type test statistic for hypothesis testing is also proposed.

The choice of the appropriate value of p to be used in fitting a nonlinear model is considered (§5.4). Three formulae which relate the exponent p to the sample kurtosis of the errors in a nonlinear model are given. One formula is motivated from theoretical considerations while the other two are established through a simulation study which involves the generation of random deviates from *symmetric* distributions (Appendix 5A). The estimates of p obtained by the above mentioned formulae are shown to be asymptotically normally distributed (§5.5). An adaptive procedure is described for the selection of p when the error distribution is *unknown* – (§5.6). Finally in §5.7 some general guidelines for the critique of the model, i.e., the regression diagnostics for identifying influential observations and outliers are given.

The application of the concepts examined in this Chapter will be illustrated by means of practical examples in Chapter 6.

5.1. Nonlinear least squares

Jennrich (1969) showed that for the least squares estimators, $\sqrt{n}(\hat{\boldsymbol{\theta}} - \boldsymbol{\theta})$ is asymptotically normally distributed with mean zero and variance $\sigma^2 Q^{-1}$ where

$$Q = \lim_{n \to \infty} \frac{J(\boldsymbol{\theta})' J(\boldsymbol{\theta})}{n} \tag{5.1}$$

If σ^2 is unknown then $s^2 = S_2(\hat{\boldsymbol{\theta}})/(n-k)$ as an estimate of $\mathrm{Var}(e_i) = \sigma^2$ is used. Jennrich (1969) derived conditions for consistency of the estimators $\hat{\boldsymbol{\theta}}$. His results may be used to construct confidence intervals for $\hat{\boldsymbol{\theta}}$.

5.1.1. Statistical inference

5.1.1.1 Confidence intervals and joint confidence regions.

There are currently four approaches available for constructing confidence regions and intervals viz the linearization, likelihood, lack-of-fit for large samples and the jackknife and bootstrap approach for small samples.

In the *linearization* approach these intervals and regions are based on first-order derivative information and are therefore approximate. Computationally this procedure is inexpensive. If the contours of $S_2(\boldsymbol{\theta})$ are approximately elliptical (exact ellipses in the linear case), then these approximations will be adequate.

These intervals hold asymptotically, specifically, a $100(1-\alpha)\%$ confidence interval for θ_j is given by

$$\hat{\theta}_j \pm z_{\alpha/2} \sqrt{\sigma^2 (\hat{J}'\hat{J})^{jj}} \tag{5.2}$$

where $(\hat{J}'\hat{J})^{jj}$ denotes the jth diagonal element of $[J(\hat{\boldsymbol{\theta}})'J(\hat{\boldsymbol{\theta}})]^{-1}$, $z_{\alpha/2}$ is the appropriate percentile of the standard normal distribution and $\hat{\boldsymbol{\theta}}$ is the estimate of $\boldsymbol{\theta}$. If σ^2 is unknown we use s^2 as denoted above.

A 100(1 − α)% joint confidence region is given by

$$\{\boldsymbol{\theta} \mid (\boldsymbol{\theta} - \hat{\boldsymbol{\theta}})'\hat{J}'\hat{J}(\boldsymbol{\theta} - \hat{\boldsymbol{\theta}}) \leq s^2 k F_{k,n-k,1-\alpha}\} \tag{5.3}$$

Beale (1960) used the *likelihood* approach and suggested the following 100(1 − α)% joint confidence region for $\boldsymbol{\theta}$.

$$\{\boldsymbol{\theta} \mid S_2(\boldsymbol{\theta}) \leq S_2(\hat{\boldsymbol{\theta}})[1 + \frac{k}{n-k}F_{k,n-k,1-\alpha}]\} \tag{5.4}$$

This result is based on the observation that

$$\frac{(n-k)\{S_2(\boldsymbol{\theta}) - S_2(\hat{\boldsymbol{\theta}})\}}{kS_2(\hat{\boldsymbol{\theta}})}$$

follows an F distribution with k and $n - k$ degrees of freedom when n is sufficiently large. This approach is computationally expensive since $S_2(\boldsymbol{\theta})$ has to be calculated at a significant number of points to produce a contour.

Since these intervals and regions are based on first-order derivative information, they may be quite inaccurate especially when the contours of $S_2(\boldsymbol{\theta})$ are markedly nonelliptical. Clarke (1980) derived an expression for the variance-covariance matrix of $\hat{\boldsymbol{\theta}}$ which takes into account third- and fourth-order derivative information. This expression facilitates the calculation of more accurate confidence intervals. Its derivation is, however, extremely involved. Clarke (1987) proposes a two-step procedure for calculating approximate confidence limits for some specified parameter function in nonlinear regression. The first stage consists of an iterative search for the k Ross-transform points. In the second stage the parameter function is minimized and maximized on the surface of the ellipsoid in the transform space.

The *lack-of-fit* approach is exact and is based on the fact that the following quadratic forms

$$\frac{\hat{e}'\hat{J}'(\hat{J}'\hat{J})^{-1}\hat{J}'\hat{e}}{s^2}$$

$$\frac{\hat{e}'[I - \hat{J}'(\hat{J}'\hat{J})^{-1}\hat{J}']\hat{e}}{s^2}$$

are independently, χ^2_k and χ^2_{n-k} distributed respectively. Introducing the "hat" matrix $V = \hat{J}(\hat{J}'\hat{J})^{-1}\hat{J}'$ it follows that

$$\frac{(n-k)\hat{e}'V\hat{e}}{k\hat{e}'[I - V]\hat{e}} \sim F_{k,n-k,1-\alpha}$$

This method is computationally even more expensive than the likelihood method since it requires the calculation of f_i and J at a sufficient number of points to produce a contour.

Donaldson and Schnabel (1987) examined these three approaches in a Monte Carlo simulation study. Three variants of the linearization method were used. The first is the usual one:

(a) $$\text{Var}(\hat{\boldsymbol{\theta}}) = s^2(\hat{J}'\hat{J})^{-1}$$

The other two use second-order derivative information:

(b) $$\text{Var}(\hat{\boldsymbol{\theta}}) = s^2[\nabla^2 S_2(\hat{\boldsymbol{\theta}})]^{-1}$$

(c) $$\text{Var}(\hat{\boldsymbol{\theta}}) = s^2[\nabla^2 S_2(\hat{\boldsymbol{\theta}})]^{-1}\hat{J}'\hat{J}[\nabla^2 S_2(\hat{\boldsymbol{\theta}})]^{-1}$$

Formula (b) is based on log-likelihood considerations and (c) on sensitivity considerations. Donaldson and Schnabel conclude that the linearization approach (a) is not only computationally the least expensive but also superior to the other two variants. However, even the best linearization method can be very poor and often grossly underestimates the confidence regions and sometimes significantly underestimates the confidence intervals. The likelihood and lack-of-fit approaches were very reliable and were consistent with the Bates and Watts (1980) curvature measures. These measures or regression diagnostics were quite successful in predicting whether the linearization confidence regions are likely to be poor.

A further problem arises when the sample size n is small. In this case the asymptotic theory is potentially unreliable. Duncan (1978) examined the jackknife procedure as a robust alternative to the conventional procedure (5.4), since it is both independent of the linearizing approximations (i.e., the use of J as a first-order linear approximation) and insensitive to specification of the error distribution. It is considered as a rough-and-ready statistical tool for two types of inferential problems, viz bias reduction and interval estimation. The procedure is as follows:

[1] Let $\hat{\boldsymbol{\theta}}_{(i)}$ be the least squares estimate of $\boldsymbol{\theta}$ when the ith case is deleted from the sample.

[2] Calculate the *pseudo-values* as the vectors

$$\hat{\boldsymbol{\Theta}}_i = n\hat{\boldsymbol{\theta}} - (n-1)\hat{\boldsymbol{\theta}}_{(i)}$$

The sample mean and variance of $\hat{\boldsymbol{\Theta}}_i$ are given by

$$\hat{\boldsymbol{\Theta}}_J = \frac{1}{n}\sum_{i=1}^{n}\hat{\boldsymbol{\Theta}}_i$$

$$\text{Var}\,(\hat{\boldsymbol{\Theta}}_i) = \frac{1}{n-1}\sum_{i=1}^{n}(\hat{\boldsymbol{\Theta}}_i - \hat{\boldsymbol{\Theta}}_J)(\hat{\boldsymbol{\Theta}}_i - \hat{\boldsymbol{\Theta}}_J)' = S$$

[3] A $100(1-\alpha)\%$ joint confidence region for $\boldsymbol{\theta}$ is

$$\{\boldsymbol{\theta} \mid (\hat{\boldsymbol{\Theta}}_i - \boldsymbol{\theta})'S^{-1}(\hat{\boldsymbol{\Theta}}_i - \boldsymbol{\theta}) \leq \frac{k}{n-k}F_{k,n-k,1-\alpha}\} \tag{5.5}$$

Using the diagonal elements of S a $100(1-\alpha)\%$ confidence intervals for θ_i can be constructed, i.e.,

$$\hat{\Theta}_{J,i} \pm \sqrt{\frac{k}{n-k}F_{k,n-k,1-\alpha}S_{ii}} \tag{5.6}$$

The variance-covariance matrix of the jackknife estimator $\hat{\Theta}_J$ is given by S/n. The general jackknife approach seems to be related to the bias reduction technique of Quenouille (1949,1956).

Duncan considered the general case where subsets of observations are deleted. His simulation results support Miller (1974) who observed that omitting a single observation at a time eliminates arbitrariness of the groups and is therefore probably the best form of jackknife to use in any given problem. Computationally the burden of single observation deletion is higher than when groups of observations are omitted at a time.

Fox et al. (1980) describe a linear jackknife procedure which only requires a single nonlinear estimation as opposed to the $n+1$ nonlinear estimations of Duncan. Instead of using $\hat{\boldsymbol{\theta}}_{(i)}$ in calculating the pseudo-value they use the first-order approximation

$$\hat{\boldsymbol{\theta}}^1_{(i)} = \hat{\boldsymbol{\theta}} - \frac{(\hat{J}'\hat{J})^{-1}\hat{J}_i'\hat{e}_i}{1-v_{ii}} \tag{5.7}$$

where $\hat{J}_i'$ denotes the ith-row of $\hat{J}$ and v_{ii} is the ith diagonal element of the "hat" matrix V evaluated at $\hat{\boldsymbol{\theta}}$. They compared the linear jackknife to the exact jackknife in a Monte Carlo simulation study and concluded that the latter may be inferior to the linear jackknife.

Let $\hat{y}_0 = f_0(\hat{\boldsymbol{\theta}})$. To construct a $100(1-\alpha)\%$ confidence interval for the true mean response of y given $\boldsymbol{x}_0$ we need an expression for $\text{Var}(\hat{y}_0)$. By the delta method (first-order Taylor series expansion) of Kendall and Stuart (1963:232):

$$\begin{aligned}\text{Var}(\hat{y}_0) &= \text{Var}[f_0(\hat{\boldsymbol{\theta}})] \\ &= \sum_{i=1}^{k}\sum_{j=1}^{k} \frac{\partial f_0}{\partial \theta_i}\frac{\partial f_0}{\partial \theta_j}\text{Cov}(\theta_i, \theta_j)\Big|_{\boldsymbol{\theta}=\hat{\boldsymbol{\theta}}} \\ &= s^2 \nabla f_0(\hat{\boldsymbol{\theta}})'(\hat{J}'\hat{J})^{-1}\nabla f_0(\hat{\boldsymbol{\theta}})\end{aligned}$$

Hence a $100(1-\alpha)\%$ confidence interval for the true mean response of y given $\boldsymbol{x}_0$ is

$$\hat{y}_0 \pm z_{\alpha/2}\sqrt{s^2 \nabla f_0(\hat{\boldsymbol{\theta}})'(\hat{J}'\hat{J})^{-1}\nabla f_0(\hat{\boldsymbol{\theta}})} \tag{5.8}$$

Similarly a $100(1-\alpha)\%$ prediction interval for the true value of a single observation y given $\boldsymbol{x}_0$ is:

$$\hat{y}_0 \pm z_{\alpha/2}\sqrt{s^2\{1 + \nabla f_0(\hat{\boldsymbol{\theta}})'(\hat{J}'\hat{J})^{-1}\nabla f_0(\hat{\boldsymbol{\theta}})\}} \tag{5.9}$$

5.1.1.2 Hypothesis testing.

Hypothesis testing [see Judge et al. (1985:Chapter 6)] in the nonlinear regression case is approached in basically the same way as in linear regression and may be carried out using one of the following

(1) the Wald test

(2) the Lagrange multiplier test

(3) the likelihood ratio test

(4) Hartley's test.

We shall only consider the **Wald test** here. Given a general hypothesis of the form

$$H_0 : r(\boldsymbol{\theta}) = 0 \qquad \text{against} \qquad H_a : r(\boldsymbol{\theta}) \neq 0$$

where $r(\boldsymbol{\theta})$ is an l-dimensional vector, the Wald test statistic is given by:

$$W = r(\hat{\boldsymbol{\theta}})'[\nabla r(\hat{\boldsymbol{\theta}})s^2(\hat{J}'\hat{J})^{-1}\nabla r(\hat{\boldsymbol{\theta}})']^{-1}r(\hat{\boldsymbol{\theta}}) \tag{5.10}$$

and has a χ^2 distribution with l degrees of freedom under H_0.

5.1.1.3 Bias.

Box (1971) derived the following asymptotic expression for calculating the bias in the estimator $\hat{\boldsymbol{\theta}}$:

$$Bias(\hat{\boldsymbol{\theta}}) = -\frac{1}{2}\sigma^2(\hat{J}'\hat{J})^{-1}\sum_{j=1}^{n}\nabla f_j \operatorname{tr}\{(\hat{J}'\hat{J})^{-1}\nabla^2 f_j\} \tag{5.11}$$

where $\nabla f_j \equiv \nabla f_j(\boldsymbol{x}_j, \hat{\boldsymbol{\theta}})$ and $\nabla^2 f_j \equiv \nabla^2 f_j(\boldsymbol{x}_j, \hat{\boldsymbol{\theta}})$. The % bias of $\hat{\theta}_i$ is defined as:

$$\frac{Bias(\hat{\theta}_i)}{\hat{\theta}_i} \times 100 \tag{5.12}$$

Quenouille (1949,1956) suggested the following calculation to reduce the bias in $\hat{\boldsymbol{\theta}}$ to order of $1/n^2$:

$$\hat{\boldsymbol{\Theta}} = n\hat{\boldsymbol{\theta}} - \frac{n-1}{n}\sum_{i=1}^{n}\hat{\boldsymbol{\theta}}_{(i)} \tag{5.13}$$

Summary

Clarke (1982) summarized the properties of $\hat{\boldsymbol{\theta}}$ as follows:

- The least squares estimate $\hat{\boldsymbol{\theta}}$ that minimizes $S_2(\boldsymbol{\theta})$ also maximizes the likelihood function.
- Contours of constant likelihood for given σ^2 may be defined by $S_2(\boldsymbol{\theta}) = c$ where c is some constant.
- The estimates $\hat{\boldsymbol{\theta}}$ are not sufficient statistics for $\boldsymbol{\theta}$ (an exception occurs when all of the parameters enter linearly).
- The maximum likelihood estimator $\hat{\boldsymbol{\theta}}$ is consistent and asymptotically normally distributed with mean $\boldsymbol{\theta}$ and covariance matrix I where I is Fisher's information matrix.
- $s^2 = S_2(\hat{\boldsymbol{\theta}})/n$ is a consistent estimator of σ^2.

- The likelihood ratio statistic is defined as

$$T = \frac{(n-k)\{S_2(\boldsymbol{\theta}) - S_2(\hat{\boldsymbol{\theta}})\}}{kS_2(\hat{\boldsymbol{\theta}})}$$

where $\boldsymbol{\theta}$ is the value of the parameter vector hypothesized under the test. For large samples, T can be assumed to follow an F distribution with k and $n-k$ degrees of freedom.

5.2. Nonlinear L_1-norm estimation

5.2.1. Statistical inference

For the linear model (Chapter 1, §1.4), Bassett and Koenker (1978) and Nyquist (1980) derived the asymptotic distribution of the L_1-norm estimator. Dielman and Pfaffenberger (1982) used these results to construct confidence intervals for a conditional mean response as well as a prediction interval for a single observation.

The results of Jennrich and Nyquist together with the fact that the error terms are additive, led Gonin and Money (1985a) to conjecture that the L_1-norm estimator $\sqrt{n}(\hat{\boldsymbol{\theta}} - \boldsymbol{\theta})$ for the nonlinear model (2.1) is asymptotically normally distributed with mean zero and variance $\omega_1^2 Q^{-1}$ where the moment ratio parameter

$$\omega_1^2 = \frac{1}{[2\mathcal{F}'(m)]^2}$$

and $\mathcal{F}'(m)$ is the ordinate of the error distribution e at the median, m. ω_1^2 can be estimated by the Cox and Hinkley (1974) estimator (see §1.5, Chapter 1) or by the α-percent nonparametric confidence interval approach [see McKean and Schrader (1987)].

5.2.1.1 Confidence intervals.

In the nonlinear case these intervals hold only asymptotically. Specifically, a $100(1-\alpha)\%$ confidence interval for θ_j is given by

$$\hat{\theta}_j \pm z_{\alpha/2}\sqrt{\omega_1^2(\hat{J}'\hat{J})^{jj}} \tag{5.14}$$

The jackknife procedure can be used when the sample size is small.

A $100(1-\alpha)\%$ confidence interval for the true mean response of y given $\boldsymbol{x_0}$ is

$$\hat{y}_0 \pm z_{\alpha/2}\sqrt{\omega_1^2 \nabla f_0(\hat{\boldsymbol{\theta}})'(\hat{J}'\hat{J})^{-1}\nabla f_0(\hat{\boldsymbol{\theta}})} \tag{5.15}$$

Similarly a $100(1-\alpha)\%$ prediction interval for the true value of a single observation y given $\boldsymbol{x_0}$ is:

$$\hat{y}_0 \pm z_{\alpha/2}\sqrt{\omega_1^2\{1 + \nabla f_0(\hat{\boldsymbol{\theta}})'(\hat{J}'\hat{J})^{-1}\nabla f_0(\hat{\boldsymbol{\theta}})\}} \tag{5.16}$$

5.2.1.2 Hypothesis testing.

Koenker and Bassett (1982) investigated three test statistics viz: the Wald, pseudo-likelihood ratio and Lagrange multiplier tests. They observe that if $\mathcal{F}'(m) > 0$, all three test statistics have limiting central χ^2 behaviour under the null hypothesis and noncentral χ^2 behaviour under local alternatives to the null hypothesis. When σ^2 is bounded similar results are known for these three tests based on least squares methods. This work was extended by Koenker (1987) to the general linear hypothesis and the test statistics were again shown to have asymptotically the same χ^2-distribution.

Judge et al. (1985) considered these tests for the nonlinear least squares case and in view of this and the above we assume analogous results hold in the case of nonlinear L_1-norm estimation. Furthermore, in view of the conclusions reached by Dielman and Pfaffenberger (1984) for the linear L_1 case, we shall only consider the **Wald test** here.

Given a general hypothesis of the form

$$H_0 : r(\boldsymbol{\theta}) = 0 \qquad \text{against} \qquad H_a : r(\boldsymbol{\theta}) \neq 0$$

where $r(\boldsymbol{\theta})$ is an l-dimensional vector, the Wald test statistic is given by:

$$W = r(\hat{\boldsymbol{\theta}})'[\nabla r(\hat{\boldsymbol{\theta}})\omega_1^2(\hat{J}'\hat{J})^{-1}\nabla r(\hat{\boldsymbol{\theta}})']^{-1}r(\hat{\boldsymbol{\theta}}) \tag{5.17}$$

and has a χ^2 distribution with l degrees of freedom under H_0.

5.2.1.3 Bias and consistency.

It is conjectured that for nonlinear L_1-norm estimation the bias in $\hat{\boldsymbol{\theta}}$ can be calculated by using ω_1^2 in place of σ^2 in the formula of Box (1971), hence

$$Bias(\hat{\boldsymbol{\theta}}) = -\frac{1}{8[\mathcal{F}'(m)]^2}(\hat{J}'\hat{J})^{-1}\sum_{j=1}^{n}\nabla f_j \operatorname{tr}\{(\hat{J}'\hat{J})^{-1}\nabla^2 f_j\} \tag{5.18}$$

Definition 5.1 A random variable X_n is $o_p(\alpha)$ if for any $\epsilon > 0$

$$\lim_{n\to\infty} Pr(|X_n|/\alpha > \epsilon) = 0$$

Definition 5.2 A random variable $\hat{\boldsymbol{\theta}}$ is weakly consistent for $\boldsymbol{\theta}$ if $\hat{\boldsymbol{\theta}} = \boldsymbol{\theta} + o_p(1)$.

Oberhofer (1982) investigated the consistency of the L_1-norm estimator of the parameters $\boldsymbol{\theta}$ in the nonlinear model (2.1) and proved the following **Consistency Theorem:**

Theorem 5.1 The L_1-norm estimator $\hat{\boldsymbol{\theta}}$ for model (2.1) is *weakly consistent* under the following six assumptions:

(1) $\boldsymbol{\theta}$ is an inner point of a compact set $\Theta \subset \Re^k$.

(2) The model functions $f_i(\boldsymbol{x_i}, \boldsymbol{\theta})$ are well defined and continuous in $\boldsymbol{\theta} \in \Theta$.

(3) The errors $\{e_i\}$ are independent random variables.

(4) The distribution function of e_i is denoted by $F(u) = \int_{-\infty}^{u} f(e_i)de_i$ and $F(0) = \frac{1}{2}$.

(5) The function

$$\frac{1}{n}\sum_{i=1}^{n}\{f_i(\boldsymbol{x_i}, \boldsymbol{\theta}) - f_i(\boldsymbol{x_i}, \tilde{\boldsymbol{\theta}})\}^2$$

is continuous in $\boldsymbol{\theta} \in \Theta$ for every $\tilde{\boldsymbol{\theta}} \in \Theta$ and uniformly continuous in n.

(6) Let $g_i(\boldsymbol{\theta}) = f_i(\boldsymbol{x_i}, \boldsymbol{\theta}) - f_i(\boldsymbol{x_i}, \tilde{\boldsymbol{\theta}}_0)$. If $\Theta_0 \subset \Theta$ is a closed set and $\boldsymbol{\theta}_0 \notin \Theta_0$ then there exist numbers $\epsilon > 0$ and n_0 such that for all $n \geq n_0$

$$\inf_{\Theta_0} \frac{1}{n}\sum_{i=1}^{n} T|g_i(\boldsymbol{\theta})| \geq \epsilon$$

where $\quad T = \min\{F(\frac{|g_i(\boldsymbol{\theta})|}{2}) - \frac{1}{2}, \frac{1}{2} - F(-\frac{|g_i(\boldsymbol{\theta})|}{2})\}$

Remarks:

(1) The first three assumptions were first used by Jennrich (1969) and Malinvaud (1970) to prove consistency of nonlinear least squares estimators.

(2) Dupačová (1987) discusses some techniques for obtaining asymptotic properties of the L_1-norm estimators in nonlinear regression where the unknown parameters $\boldsymbol{\theta}$ are restricted to a convex set. Consistency as well as asymptotic normality of the L_1-norm estimator were also derived [see also Withers (1987)].

5.3. Nonlinear L_p-norm estimation

5.3.1. Asymptotic distribution of L_p-norm estimators (additive errors)

Nyquist (1980,1983) shows that, under the assumptions A1 to A4b (Chapter 1) for the linear L_p-norm estimator $\hat{\boldsymbol{\theta}}$, asymptotically

$$\sqrt{n}(\hat{\boldsymbol{\theta}} - \boldsymbol{\theta}) \sim N(O, \omega_p^2 Q^{-1}) \tag{5.19}$$

where the *moment ratio parameter of residuals*

$$\omega_p^2 = \begin{cases} \frac{1}{[2f'(0)]^2} & \text{if } p = 1; \\ \frac{E[|e_i|^{2p-2}]}{[(p-1)E(|e_i|^{p-2})]^2} & \text{if } 1 < p < \infty. \end{cases}$$

and $Q = \lim_{n\to\infty} X'X/n$ is positive definite with rank$(Q) = k$. (Assumption A3).

The results of Jennrich and Nyquist together with the fact that the error terms are additive, led Gonin and Money (1985a) to conjecture that the L_p-norm estimator $\sqrt{n}(\hat{\boldsymbol{\theta}} - \boldsymbol{\theta})$ for the nonlinear model (2.1) is asymptotically normally distributed with mean zero and variance $\omega_p^2 Q^{-1}$. In the subsequent sections we consider applications of this proposal.

5.3.2. Statistical inference

5.3.2.1 Confidence intervals.

Specifically, a $100(1-\alpha)\%$ confidence interval for θ_j is given by

$$\hat{\theta}_j \pm z_{\alpha/2}\sqrt{\omega_p^2(\hat{J}'\hat{J})^{jj}} \tag{5.20}$$

where ω_p^2 is estimated by

$$\hat{\omega}_p^2 = \frac{m_{2p-2}}{[(p-1)m_{p-2}]^2}$$

with $m_r = \frac{1}{n}\sum_{i=1}^{n} |\hat{e}_i|^r$ and $\hat{e}_i$ is the ith residual. (see §1.5.1).

A $100(1-\alpha)\%$ confidence interval for the true mean response of y given $\boldsymbol{x}_0$ is

$$\hat{y}_0 \pm z_{\alpha/2}\sqrt{\omega_p^2 \nabla f_0(\hat{\boldsymbol{\theta}})'(\hat{J}'\hat{J})^{-1}\nabla f_0(\hat{\boldsymbol{\theta}})} \tag{5.21}$$

Similarly a $100(1-\alpha)\%$ prediction interval for the true value of a single observation y given $\boldsymbol{x}_0$ is:

$$\hat{y}_{new} \pm z_{\alpha/2}\sqrt{\omega_p^2\{1 + \nabla f_0(\hat{\boldsymbol{\theta}})'(\hat{J}'\hat{J})^{-1}\nabla f_0(\hat{\boldsymbol{\theta}})\}} \tag{5.22}$$

5.3.2.2 Hypothesis testing.

Given a general hypothesis of the form

$$H_0 : r(\boldsymbol{\theta}) = 0 \qquad \textbf{against} \qquad H_a : r(\boldsymbol{\theta}) \neq 0$$

where $r(\boldsymbol{\theta})$ is an l-dimensional vector. The Wald test statistic is then given by:

$$W = r(\hat{\boldsymbol{\theta}})'[\nabla r(\hat{\boldsymbol{\theta}})\omega_p^2(\hat{J}'\hat{J})^{-1}\nabla r(\hat{\boldsymbol{\theta}})']^{-1}r(\hat{\boldsymbol{\theta}}) \tag{5.23}$$

and has a χ^2 distribution with l degrees of freedom under H_0.

5.3.2.3 Bias.

For nonlinear L_p-norm estimation the bias in $\hat{\boldsymbol{\theta}}$ can be calculated by:

$$Bias(\hat{\boldsymbol{\theta}}) = -\frac{1}{2}\omega_p^2(\hat{J}'\hat{J})^{-1}\sum_{j=1}^{n}\nabla f_j \,\mathrm{tr}\{(\hat{J}'\hat{J})^{-1}\nabla^2 f_j\} \tag{5.24}$$

5.4. The choice of the exponent p

In this section we consider guidelines for choosing the exponent p. Three formulae for choosing p will be presented. One formula is motivated from theoretical considerations while the other two are established through Monte Carlo simulation.

5.4.1. The simulation study

In biological problems exponential models often occur, it was therefore decided to use such a model in this simulation study. Other models may prove equally constructive.

The following two-parameter exponential model was used in the study

$$y = 5 + 4\exp(\theta_1 x_1) + 3\exp(\theta_2 x_2) + e \tag{5.25}$$

The coefficients 5, 4 and 3 were arbitrarily chosen. It was assumed that $E(e) = 0$ and $\text{Var}(e) = \sigma^2 = 25$. The generation of random deviates from these distributions is discussed in Appendix 5A. The formulae for generating random deviates from the continuous symmetric distributions used in this study are given in Table 5.1 (r, s are uniform $[0, 1]$ random numbers).

Table 5.1:

Formulae for generating random deviates from the uniform, parabolic, triangular, normal, contaminated normal, Laplace and Cauchy distributions.

Distribution	β_2	Random Deviate
Uniform	1.8	$x = 8.6603(2r - 1)$
Parabolic	2.14	$x = 22.3607\cos(\arccos(1 - 2r)/3 + 240°)$
Triangular	2.4	$x = 12.2474(r + s - 1)$
Normal	3.0	$x = 5\Phi^{-1}(r)$
Contaminated Normal	4.0	$x_1 = 6.2796\Phi^{-1}(r)$; $x_2 = 3.2506\Phi^{-1}(s)$
Contaminated Normal	5.0	$x_1 = 6.7389\Phi^{-1}(r)$; $x_2 = 2.1419\Phi^{-1}(s)$
Laplace	6.0	$x = 3.5355\ln 2r$ if $0 < r \leq 0.5$
		$x = -3.5355\ln 2(1 - r)$ if $0.5 < r < 1$
Cauchy	—	$x = -1.5522\cot \pi r$

In the simulation study three algorithms were used. For the nonlinear L_1-norm estimation problem the Anderson-Osborne-Watson algorithm (Chapter 2) was used, while for $1 < p < \infty$ the first-order gradient algorithm (Chapter 4) was used and for L_∞-norm estimation the Anderson-Osborne-Watson (Chapter 3) algorithm was used. The values θ_1 and θ_2 in model (5.25) were arbitrarily fixed at 1.0 and 1.5 respectively. Fifty values for x_{1i}, x_{2i}, $i = 1, \ldots, 50$ were selected from a $[\frac{1}{2}, \frac{3}{2}]$ uniform distribution and held fixed in all 500 samples for all error distributions and L_p-norm estimations. The values for (x_{1i}, x_{2i}) are given in Table 1 (Appendix 5B).

5.4.2. Simulation results and selection of p

The information present in the sample covariance matrix can be summarized by its generalized variance, the determinant $|\mathrm{Cov}(\hat{\theta}_1, \hat{\theta}_2)|$. For this reason the selection of p is based on the minimum sample generalized variance. To find this minimum value would entail an exhaustive search over all values of p. We therefore performed simulations only for the following values of p: 1.0, 1.25, 1.5, 1.75, 2.0, 2.25, 2.5, 2.75, 3.0 and ∞. *In the case of the uniform, parabolic and triangular distributions other values of p had to be considered (see Table 2, Appendix 5B).* The generalized variances of the 500 sample estimates are given in Table 2 (Appendix 5B) for various values of p and the 8 symmetric error distributions. A plot of the optimum value of p versus the kurtosis is given in Figure 5.1.

These numerical results also support the theoretical results outlined in §5.3. Recall that the asymptotic variance of $\sqrt{n}(\hat{\boldsymbol{\theta}} - \boldsymbol{\theta})$ was postulated to be $\omega_p^2 Q^{-1}$. Should Q^{-1} be independent of p, the expression ω_p^2 (Table 2, Appendix 5C) will yield the optimal value of p for the given error distribution. Figure 5.2 displays the ω_p^2 curve for some of the distributions.

The minimum value of ω_p^2 with respect to p for each distribution can be obtained from the graph. A numerical procedure may also be used to calculate this minimum value. For the normal distribution we see that the minimum lies at $p = 2$, for the Laplace distibution we find that $p = 1$ and for the uniform, $p \to \infty$. Except for the Laplace distribution the above is in agreement with our simulation results, and agrees with the results found by Money et al. (1982) and Sposito et al. (1983) in the linear case. The following was observed by Johnson and Kotz (1970:Vol 2 p 26) in connection with the location model: "Although the median is a maximum likelihood estimator of θ and unbiased, it is not a minimum variance unbiased estimator of θ. Indeed, for small values of n (the sample size) it is possible to construct unbiased estimators with smaller variance than the median". This could explain the reason

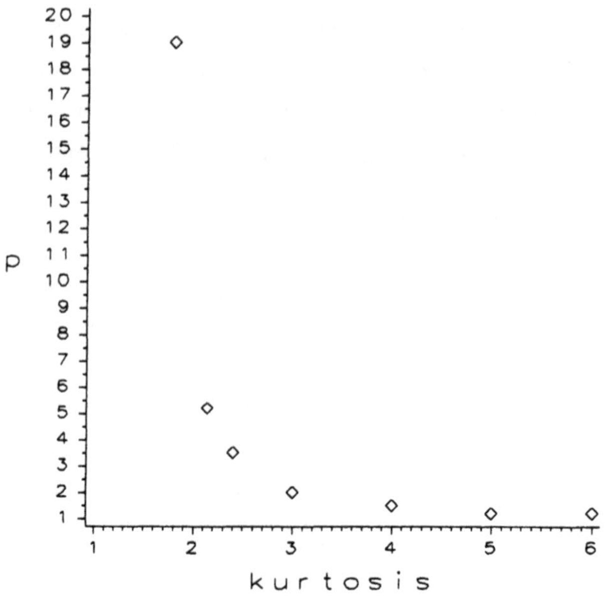

Figure 5.1 Value of p vs kurtosis.

why $p = 1.25$ is chosen by the simulation study (see Table 2). This value of p was also suggested by Ekblom (1974).

5.4.3. The efficiency of L_p-norm estimators for varying values of p

Definition 5.3 The relative efficiency of L_p-norm estimators is defined by the ratio:

$$R = \frac{|\mathrm{Cov}(\hat{\theta}_1, \hat{\theta}_2)| \text{ using the "optimal" exponent } p}{|\mathrm{Cov}(\hat{\theta}_1, \hat{\theta}_2)| \text{ using another exponent } \tilde{p}}$$

The generalized variances in Table 2 are used in the formula for R. The analytical results of §5.3.1 suggest an analogous ratio: $\tilde{R} = \omega_p^2/\omega_{\tilde{p}}^2$.

The ratios R and $\tilde{R}$ are displayed in Table 3 (Appendix 5B). We observe, using the generalized variance as a basis, that these ratios indicate the efficiency of L_p-norm estimation. It is important to note that R is based on values of $n = 50$, whilst $\tilde{R}$ is based on the asymptotic generalized variance. This probably accounts for the

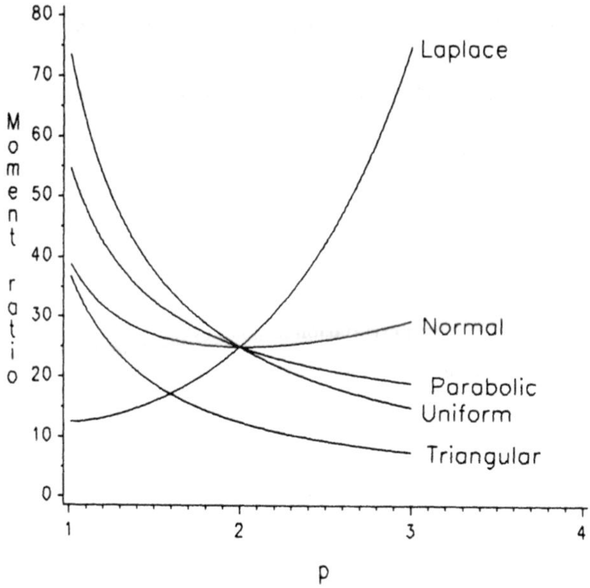

Figure 5.2 ω_p^2 for five distributions vs p.

differences between R and $\tilde{R}$ in Table 3. In the construction of the table we used the "optimal" value of p (indicated by the simulation study) as a basis for comparison of R and $\tilde{R}$. *To be specific in the case of the uniform distribution we used the optimal value of* $p = 19$ *(Table 2). Similarly, for the parabolic distribution we used* $p = 5.25$ *as a basis value for comparison. For the triangular distribution the basis value of* $p = 3.5$ *was used.*

For the uniform distribution there is fair agreement between R and $\tilde{R}$ for all values of p. For parabolic errors there is good agreement between R and $\tilde{R}$ for values of p between 2.5 and 3. For normally distributed errors the agreement between R and $\tilde{R}$ for values of p between 1.75 and 2.5 (differences were all less than 5%) is also good. For the Laplace distribution the agreement between R and $\tilde{R}$ is reasonable and $\tilde{R}$ indicates that $p = 1.25$ and $p = 1$ in L_p-norm estimation will be equally efficient.

5.4.4. The relationship between p and the sample kurtosis

The simulation results of §5.4.3 indicate that the following empirical formulae

$$p = \frac{9}{\beta_2^2} + 1 \qquad \text{for } 1 \leq p < \infty \tag{5.26}$$

and

$$p = \frac{6}{\beta_2} \qquad \text{for } 1 \leq p < 2 \tag{5.27}$$

can be used for nonlinear L_p-norm estimation [see Gonin and Money (1985a)]. Both these formulae are, however, ad hoc. Another theoretically motivated formula due to Gonin and Money (1985b) is based on the pth order exponential distribution which has the density function:

$$f_X(x) = \frac{1}{\phi\Gamma(1+1/p)2^{1+1/p}} \exp\left\{-\frac{1}{2}\left|\frac{x-\theta}{\phi}\right|^p\right\} \qquad -\infty < x < \infty \tag{5.28}$$

with $0 < \theta < \infty$ a location and $\phi > 0$ a scale parameter [see Box and Tiao (1973:157) and Turner (1960)]. If the residual distribution belongs to this class of exponential distributions then the maximum likelihood estimate of our parameter vector can be obtained by simply minimizing the sum of the pth power of the absolute residuals. We can also obtain p by maximizing the likelihood function over p. This approach though theoretically sound, is not very practicable.

It can be shown that the rth central moment of the distribution in (5.28) is:

$$\mu_r = 2^{\frac{r}{p}}\phi^r\Gamma(\frac{r+1}{p})/\Gamma(\frac{1}{p}) \qquad \text{for } r \text{ even}$$

with $\Gamma(\cdot)$ the usual Gamma function. The kurtosis β_2 is then given by

$$\beta_2 = \frac{\Gamma(5/p)\Gamma(1/p)}{\{\Gamma(3/p)\}^2} \tag{5.29}$$

Box and Tiao (1973) use an alternative measure of kurtosis, β. In particular Box and Tiao set $p = 2/(1+\beta)$ and observe that $-1 < \beta \leq 1$ whenever $1 \leq p < \infty$. For the normal distribution, $p = 2$ implies $\beta = 0$ and for the Laplace, $p = 1$ implies $\beta = 1$ while for the uniform distribution, $p \to \infty$ implies $\beta \to -1$.

This approach therefore overcomes the inadequacy of the Money et al. (1982) and Sposito et al. (1983) formulae which do not select a large enough p when the errors are uniformly distributed ($\beta_2 = 1.8$). Using the alternative measure β, we find as $\beta \to -1$ that $p = 2/(1+\beta) \to \infty$.

In §5.5 it will be shown that although formulae (5.26), (5.27) and (5.29) predict different values of p, these estimators are consistent in that asymptotically the mean square error tends to zero. We shall also demonstrate the performance of these three formulae when we discuss the adaptive algorithm in §5.6.

5.5. The asymptotic distribution of p

All three formulae are functions of the kurtosis β_2. In the following theorem and corollary we show that $\hat{p} = f(\hat{\beta}_2)$ is asymptotically normally distributed. A rigorous proof may be found in Cramér (1946:§27.7, 28.4) or Theil (1971:§8.3). We shall only sketch the proof. The rth central sample and population moments are denoted by m_r and μ_r respectively.

Theorem 5.2 Let $H(m_r, m_s)$ be a function of two sample central moments m_r and m_s. If $H(m_r, m_s)$ is continuous and has continuous first- and second-order partial derivatives with respect to m_r and m_s in a neighbourhood of the point (μ_r, μ_s) and $H(m_r, m_s)$ is assumed independent of n, then asymptotically

$$\sqrt{n}[H(m_r, m_s) - H(\mu_r, \mu_s)] \sim \boldsymbol{N}(0, \sigma_H^2)$$

where

$$\sigma_H^2 = n\left\{ \mathrm{Var}(m_r)\{\frac{\partial H}{\partial m_r}\}^2 + \mathrm{Var}(m_s)\{\frac{\partial H}{\partial m_s}\}^2 + 2\mathrm{Cov}(m_r, m_s)\frac{\partial H}{\partial m_r}\frac{\partial H}{\partial m_s} \right\}$$

where it is assumed that all partial derivatives are evaluated at (μ_r, μ_s).

Proof A Taylor series expansion of $H(m_r, m_s)$ about (μ_r, μ_s) yields

$$H(m_r, m_s) - H(\mu_r, \mu_s) = \frac{\partial H}{\partial m_r}(m_r - \mu_r) + \frac{\partial H}{\partial m_s}(m_s - \mu_s) + R \qquad (5.30)$$

with remainder term

$$R = \frac{1}{2}\left[\frac{\partial^2 H}{\partial m_r^2}(m_r - \mu_r)^2 + \frac{\partial^2 H}{\partial m_s^2}(m_s - \mu_s)^2 + \frac{\partial^2 H}{\partial m_r \partial m_s}(m_r - \mu_r)(m_s - \mu_s) \right]$$

All second derivatives are evaluated at a point (ξ, ς) where

$$\begin{aligned} \xi &= \lambda m_r + (1-\lambda)\mu_r \\ \varsigma &= \lambda m_s + (1-\lambda)\mu_s \end{aligned} \qquad 0 < \lambda < 1$$

and all first derivatives are evaluated at μ_r and μ_s. Ignoring R and taking the expectation of (5.30) yields $\mathcal{E}[H(m_r, m_s)] = H(\mu_r, \mu_s)$ and hence it follows that $\sqrt{n}[H(m_r, m_s) - H(\mu_r, \mu_s)]$ has mean 0. Squaring both sides of (5.30) and taking expectations yields

$$\begin{aligned} \mathcal{E}[H(m_r, m_s) - H(\mu_r, \mu_s)]^2 &= (\frac{\partial H}{\partial m_r})^2 \mathcal{E}(m_r - \mu_r)^2 + (\frac{\partial H}{\partial m_s})^2 \mathcal{E}(m_s - \mu_s)^2 \\ &+ 2\frac{\partial H}{\partial m_r}\frac{\partial H}{\partial m_s}\mathcal{E}(m_r - \mu_r)(m_s - \mu_s) \end{aligned}$$

Hence

$$\mathrm{Var}[H(m_r, m_s)] = (\frac{\partial H}{\partial m_r})^2 \mathrm{Var}(m_r) + (\frac{\partial H}{\partial m_s})^2 \mathrm{Var}(m_s) + 2\frac{\partial H}{\partial m_r}\frac{\partial H}{\partial m_s}\mathrm{Cov}(m_r, m_s)$$

From the Lindeberg-Lévy Central Limit theorem [Cramér (1946:215)] it follows that the asymptotic distribution of $\sqrt{n}[H(m_r, m_s) - H(\mu_r, \mu_s)]$ is normal with mean and variance as derived above. ∎

Remark: In the case of formulae (5.26) and (5.27) H is simply

$$H(\hat{p}) = H(m_4/m_2^2)$$

The analytical form for H in the case of formula (5.29) cannot be derived. The following Corollary follows (see Appendix 5D):

Corollary 5.1 The formulae $\hat{p} = 9/\hat{\beta}_2^2 + 1$ and $\hat{p} = 6/\hat{\beta}_2$ are asymptotically normally distributed with means and variances as given in the following table:

Formula	$\mathcal{E}(\hat{p})$	$\mathrm{Var}(\hat{p})$
Money et al.	$\frac{9}{\beta_2^2} + 1$	$324\mathrm{Var}(\hat{\beta}_2)/\beta_2^6$
Sposito et al.	$\frac{6}{\beta_2}$	$36\mathrm{Var}(\hat{\beta}_2)/\beta_2^4$

where $\text{Var}(\hat{\beta}_2)$ is given by

$$\frac{4\mu_4^3 - \mu_2^2\mu_4^2 + \mu_2^2\mu_8 - 4\mu_2\mu_4\mu_6 - 8\mu_2^2\mu_3\mu_5 + 16\mu_2\mu_3^2\mu_4 + 16\mu_2^3\mu_3^2}{n\mu_2^6}$$

Proof The expected values follow immediately. For (5.26):

$$\text{Var}(\hat{p}) = \text{Var}(\frac{9}{\hat{\beta}_2^2}) = 81\text{Var}(\frac{1}{\hat{\beta}_2^2}) = 81(-2/\beta_2^3)^2\text{Var}(\hat{\beta}_2) = \frac{324\text{Var}(\hat{\beta}_2)}{\beta_2^6}$$

$$\text{Similarly for (5.27),} \quad \text{Var}(\hat{p}) = \text{Var}(\frac{6}{\hat{\beta}_2}) = \frac{36\text{Var}(\hat{\beta}_2)}{\beta_2^4}$$

∎

Remark: Analytical expressions for the expected value and variance of $\hat{p}$ for formula (5.29) cannot be obtained. We observe that for $\beta_2 > 3$ the Money et al. formula has a smaller asymptotic variance than the Sposito et al. formula. For $\beta_2 < 3$ the reverse occurs. For $\beta_2 = 3$ asymptotic variances are the same.

Explicit expressions for $\text{Var}(\hat{p})$ for several symmetric error distributions are derived in Appendix 5D and summarized in Table 5.2. We do observe, however, that the Sposito et al. formula was only suggested for $p \in [1, 2]$.

Table 5.2:

$\mathcal{E}(\hat{p})$ and $\text{Var}(\hat{p})$ for various error distributions for formulae (5.4) and (5.5)

Error Distribution	Formula (5.4)		Formula (5.5)	
	$\mathcal{E}(\hat{p})$	$\text{Var}(\hat{p})$	$\mathcal{E}(\hat{p})$	$\text{Var}(\hat{p})$
Uniform	3.78	$12.5/n$	3.33	$4.5/n$
Parabolic	2.97	$9/n$	2.80	$4.6/n$
Triangular	2.56	$7.25/n$	2.50	$4.64/n$
Normal	2.00	$10.7/n$	2.00	$10.7/n$
C. Normal ($\beta_2 = 4$)	1.56	$6.9/n$	1.50	$12.2/n$
C. Normal ($\beta_2 = 5$)	1.36	$3.0/n$	1.20	$8.4/n$
Laplace	1.25	$8.25/n$	1.00	$33/n$

5.6. The adaptive algorithm for L_p-norm estimation

Suppose least squares estimation is used to fit a model to a set of data. The kurtosis of the residuals is computed and then a prediction of the optimal exponent p is made. Will this necessarily be the best value of p? The choice of p has been considered by e.g., Harter (1977), Money et al. (1982) and Sposito et al. (1983) in linear L_p-norm and by Gonin and Money (1985a,1985b,1987) and Militký and Čáp (1987) in nonlinear L_p-norm problems. Some of these authors proposed certain selection rules for p based on the kurtosis of the error distribution. Instead of using a selection rule the following adaptive procedure was suggested by Gonin and Money (1985b):

> Fit a curve using least squares (any finite value of $p \geq 1$ may be used initially). Compute the sample kurtosis $\hat{\beta}_2$ of the resulting residuals and make a prediction of the optimal exponent p using any of the 3 formulae. Use this estimated value of p and fit a new curve to the data. Subsequently compute the sample kurtosis of the resulting residuals and make a new prediction of the true exponent p. Repeat the process until no further change in the values of p is detected.

The Algorithm

Step 0: Set $i = 0$ with $p^{(i)} = 2$.

Step 1: Fit the model to the data using $p^{(i)}$.

Step 2: Compute the residuals and $\hat{\beta}_2$ Use either formula (5.26) or (5.27) or (5.29) to calculate $p^{(i+1)}$.

Step 3: Repeat Steps 1 and 2 until $|p^{(i+1)} - p^{(i)}| < 10^{-4}$.

The algorithm consists of an inner iteration (minimization over the parameters) and an outer iteration (calculation of p). The adaptive algorithm therefore calculates the limit of the sequence $\{p^{(i)}\}$. Theoretical convergence, however, has not been proved. It need not necessarily converge to a global solution (the true value of p). Using examples where the true p is known, it has been found that the successive $p^{(i)}$'s converged in about 4 iterations. We have found a derivative-free numerical procedure such as Steffenson's algorithm [Dahlquist and Björck (1974:230)] efficient as a rootfinder of equation (5.29). The unbiased estimates of the second and fourth order sample moments were used:

$$\hat{\mu}_2 = \frac{1}{n-1}\sum_{i=1}^{n}(\hat{e}_i - \bar{e})^2$$

$$\hat{\mu}_4 = \frac{(n^2 - 2n + 3)}{(n-1)(n-2)(n-3)}\sum_{i=1}^{n}(\hat{e}_i - \bar{e})^4 - \frac{3(n-1)(2n-3)}{n(n-2)(n-3)}\hat{\mu}_2^2$$

In order to gain some insight into the behaviour of the exponent p when the sample size is finite a restricted simulation study was undertaken using the model (5.25). Since we know that values of p are asymptotically normally distributed only two error distributions (the normal and the Laplace) with mean zero and variance 9 were considered.

In the simulation study we fix the values of θ_1 and θ_2 in model (5.25) at 1.0 and 1.5 respectively. A sample of n values for x_{1i} and x_{2i} from a $\left[\frac{1}{2}, \frac{3}{2}\right]$ uniform distribution was selected and held fixed in all simulations. In each case 500 experiments were simulated such that the error term was a random variable from a specified distribution. Various sample sizes n were used in order to study the asymptotic properties of the exponent p. The adaptive procedure described previously was used to determine the best p for each simulation experiment. The Cramér-von Mises goodness-of-fit test (Appendix 5E) can be used to determine whether the "optimal" p is normally distributed. The simulation results based on additive normally distributed errors and using the 3 prediction formulae for p are given in Table 4 (appendix 5B).

In the Box and Tiao case when $n = 30$, we observe a fairly large variance when compared with the other formulae. In small samples it so happens that the kurtosis is quite variable and sometimes quite small so that formula (5.4) will yield quite a large estimate of p. (In these simulation runs, experiments with an estimated value of $p \geq 10$ were, however, excluded). This would account for the large variance observed. For larger sample sizes this phenomenon no longer occurred. The values in parenthesis are obtained if we use $n = 400$ in the formulae given in Table 3. In Table 5 (Appendix 5B) analogous results for the Laplace distribution are given.

5.7. Critique of the model and regression diagnostics

The logical step after parameter estimation and inference is critique of the model. This involves the use of regression diagnostics to detect outliers and identify influential observations.

5.7.1. Regression diagnostics (nonlinear least squares)

It seems reasonable to assume that most of the diagnostics and residual analyses for linear least squares will apply approximately in the nonlinear least squares case. We shall follow the approach as set out in Cook and Weisberg (1982).

Reconsider the nonlinear model

$$y_i = f_i(\boldsymbol{x}_i, \boldsymbol{\theta}) + e_i, \qquad i = 1, \ldots, n \quad (n \geq k) \tag{5.31}$$

According to Hoaglin and Welsch (1978) criterion the diagonal elements v_{ii} of $\hat{J}(\hat{J}'\hat{J})^{-1}\hat{J}'$ may serve as diagnostics to identify "high-leverage" points.

The approach is to fit the model to the data deleting one case, for example $(y_i, x_{i1}, \ldots, x_{im})$, at a time. The resulting parameter estimate is denoted by $\hat{\boldsymbol{\theta}}_{(i)}$. The case i would be influential should the difference between $\hat{\boldsymbol{\theta}}$ and $\hat{\boldsymbol{\theta}}_{(i)}$ be "large". Recall that Fox et al. (1980) gave a first-order approximation of $\hat{\boldsymbol{\theta}}_{(i)}$ viz:

$$\hat{\boldsymbol{\theta}}^1_{(i)} = \hat{\boldsymbol{\theta}} - \frac{(\hat{J}'\hat{J})^{-1}\hat{J}_i'\hat{e}_i}{1 - v_{ii}} \tag{5.32}$$

where $\hat{J}_i'$ denotes the ith-row of $\hat{J}$. Cook and Weisberg (1982) define the generalized distance measure (elliptical norm):

$$D_i = \frac{(\hat{\boldsymbol{\theta}} - \hat{\boldsymbol{\theta}}_{(i)})'\hat{J}'\hat{J}(\hat{\boldsymbol{\theta}} - \hat{\boldsymbol{\theta}}_{(i)})}{ks^2} \tag{5.33}$$

Assuming that D_i behaves like an F-distribution with k and $n-k$ degrees of freedom, then one would be interested in the lower (5% and 10%) percentiles of the F-distribution. Any D_i larger than these lower percentiles should be cause for concern. Should $\hat{\boldsymbol{\theta}}_{(i)}$ be approximated by $\hat{\boldsymbol{\theta}}^1_{(i)}$ then this distance measures becomes

$$\tilde{D}_i = \frac{\hat{r}_i^2 v_{ii}}{k(1 - v_{ii})} \tag{5.34}$$

where

$$\hat{r}_i = \frac{\hat{e}_i}{s\sqrt{1 - v_{ii}}} \tag{5.35}$$

is the approximate Studentized residual.

This type of norm is applicable only if $S_2(\hat{\boldsymbol{\theta}})$ has approximate elliptical contours. If not, these elliptical confidence regions can be badly biased [Beale (1960)]. For this reason Cook and Weisberg give three alternative norms for $\hat{\boldsymbol{\theta}} - \hat{\boldsymbol{\theta}}_{(i)}$ or $\hat{\boldsymbol{\theta}} - \hat{\boldsymbol{\theta}}^1_{(i)}$. We shall only consider one alternative norm (the other two are based on maximum likelihood considerations).

$$FD_i^1 = \frac{1}{ks^2}\sum_{j=1}^{n}[f_j(\boldsymbol{x}_j, \hat{\boldsymbol{\theta}}) - (f_j(\boldsymbol{x}_j, \hat{\boldsymbol{\theta}}^1_{(i)})]^2 \tag{5.36}$$

$$FD_i = \frac{1}{ks^2}\sum_{j=1}^{n}[f_j(\boldsymbol{x}_j, \hat{\boldsymbol{\theta}}) - (f_j(\boldsymbol{x}_j, \hat{\boldsymbol{\theta}}_{(i)})]^2 \tag{5.37}$$

These are norms of the change in the vector of fitted values. Outlier detection is usually based on the Studentized residual. If both $\hat{r}_i$ and v_{ii} are "large" then the ith case will be an outlier and influential. If $v_{ii} \approx 1$ then from the fact that V is idempotent and symmetric it follows that

$$\sum_{j=1}^{n} v_{ij}^2 = v_{ii} \qquad i = 1, \ldots, n$$

If $v_{ii} \rightarrow 1$ then $v_{ij} \rightarrow 0$ for all $i \neq j$. In this instance $e_i \approx (1 - v_{ii})y_i$ so that

$$\hat{r}_i \approx \frac{(1 - v_{ii})y_i}{s\sqrt{1 - v_{ii}}} = \frac{\sqrt{1 - v_{ii}}\,y_i}{s}$$

Outliers are therefore difficult to detect if v_{ii} is close to 1 and y_i is moderately large.

A "large" $\hat{r}_i$ and a "small" v_{ii} would probably indicate a bona fide outlier. On the other hand a "large" v_{ii} or $v_{ii}/tr(V)$ and a "small" $\hat{r}_i$ would probably indicate that the case is influential.

5.7.2. Regression diagnostics (nonlinear L_p-norm case)

In this case we suggest that the generalised distance measure of Cook and Weisberg be adapted by substituting ω_p^2 for s^2, the p-Jacobian matrix J_p for J in (5.33) and denoting it by $D_{p,i}$. Again a "large" value of $D_{p,i}$ would indicate that the ith case is influential. Denote the "hat" matrix by $V_p = \hat{J}_p(\hat{J}_p'\hat{J}_p)^{-1}\hat{J}_p'$. In (5.32) we substitute the p-residual vector $(\hat{\boldsymbol{y}} - \boldsymbol{f})_p$ for $\hat{\boldsymbol{e}}$, $v_{p,ii}$ for v_{ii} and J_p for J. The measures in (5.36) and (5.37) can be adapted in a similar fashion. Further work in this regard is being undertaken by Schall and Gonin (1988).

CONCLUSION: *In this Chapter we made an attempt to summarise the statistical aspects of L_p-norm estimators available in the literature. In a number of instances conjectures were made, this apparent lack of analytical results needs to be addressed in future research. In the next chapter we consider a number of practical problems.*

Appendix 5A: Random deviate generation

In all the symmetric error distributions to be considered assume $E(e) = 0$ and $\text{Var}(e) = \sigma^2 = 25$. Properties of these distributions may be found in Johnson and Kotz (1970).

The uniform distribution ($\beta_2 = 1.8$)

It has the density function,

$$f_X(x) = \begin{cases} \frac{1}{2b}, & -b \le x \le b \\ 0, & \text{otherwise} \end{cases}$$

the distribution function

$$F(x) = \frac{x+b}{2b}$$

and moments

$$\mu_r = \mu'_r = \begin{cases} \frac{b^r}{r+1}, & r \text{ even} \\ 0, & r \text{ odd} \end{cases}$$

$$\mu_2 = \sigma^2 = \frac{b^2}{3}, \qquad \mu_4 = \frac{b^4}{5} \qquad \text{hence} \quad \beta_2 = \frac{\mu_4}{\mu_2^2} = 1.8$$

To generate a random deviate from this distribution we use the inverse transform method which requires $x = F^{-1}(r)$ where r is a uniform $[0, 1]$ random number. Thus setting

$$x = b(2r - 1) \Rightarrow x \sim \text{ uniform } [-b, b]$$

For variance $\sigma^2 = 25$ we have $b = 8.6603$ and

$$x = 8.6603(2r - 1) \tag{1}$$

The parabolic distribution ($\beta_2 = \frac{15}{7}$)

It has the density function

$$f_X(x) = \begin{cases} a(b - x^2), & -\sqrt{b} \le x \le \sqrt{b} \\ 0, & \text{otherwise} \end{cases}$$

and distribution function

$$F(x) = -\frac{ax^3}{3} + abx + \frac{2ab^{\frac{3}{2}}}{3}$$

The moments are

$$\mu_r = \mu_r' = \begin{cases} \frac{4ab^{(r+3)/2}}{(r+1)(r+3)}, & r \text{ even} \\ 0, & r \text{ odd} \end{cases}$$

Thus

$$\mu_2 = \sigma^2 = \frac{4ab^{\frac{5}{2}}}{15}, \qquad \mu_4 = \frac{4ab^{\frac{7}{2}}}{35}$$

Hence

$$\beta_2 = \frac{15}{7} \quad \text{since} \quad \mu_0 = \frac{4ab^{\frac{3}{2}}}{3} = 1$$

We shall use the inverse transform method to calculate a random deviate from this distribution. Set

$$r = -\frac{ax^3}{3} + abx + \frac{2ab^{\frac{3}{2}}}{3}$$

The analytical roots (real) may be calculated by the Ferreo-Tartaglia-Cardano formula (see Spiegel (1968) formula 9.4) as:

$$\begin{aligned} \alpha &= \arccos(+R/\sqrt{-Q^3}) \\ &= \arccos(1-2r) \end{aligned} \tag{2}$$

where $R = b^{\frac{3}{2}}(1-2r)$ and $Q = -b$. The three *real* roots of this cubic are

$$x_i = 2\sqrt{b}\cos(\frac{\alpha}{3} + 120(i-1)^\circ), \quad i = 1, 2, 3$$

By inspection we see that $-\sqrt{b} \le x_3 \le \sqrt{b}$ for $0 \le r \le 1$:

r	α	x_1	x_2	x_3
0.0	0°	$2\sqrt{b}$	$-\sqrt{b}$	$-\sqrt{b}$
0.5	90°	$\sqrt{3b}$	$-\sqrt{3b}$	0
1.0	180°	$\sqrt{b}$	$-2\sqrt{b}$	$\sqrt{b}$

It follows that $\sigma^2 = \frac{b}{5} = 25$ hence $b = 125$ and

$$x = 22.3607\cos(\frac{\alpha}{3} + 240^\circ) \tag{3}$$

where α is given by (2).

Remarks: There are a number of errors in Spiegel (1968:32):

(1) The formulae for S and T contain third roots i.e.,

$$S = \sqrt[3]{R + \sqrt{Q^3 + R^2}} \qquad T = \sqrt[3]{R - \sqrt{Q^3 + R^2}}$$

(2) Formula (9.4) should read:

$$x_i = 2\sqrt{-Q}\cos(\frac{\alpha}{3} + 120(i-1)^\circ) - \frac{a_1}{3}, \quad i = 1, 2, 3 \quad \text{where } \cos\theta = +R/\sqrt{-Q^3}$$

The triangular distribution ($\beta_2 = 2.4$)

It has the density function

$$f_X(x) = \begin{cases} a(b - |x|), & -b \le x \le b,\ a > 0 \\ 0, & \text{otherwise} \end{cases}$$

and distribution function

$$F(x) = \begin{cases} \frac{1}{2} + \frac{x}{b} + \frac{1}{2}(\frac{x}{b})^2, & -b \le x \le 0 \\ \frac{1}{2} + \frac{x}{b} - \frac{1}{2}(\frac{x}{b})^2, & 0 \le x \le b \end{cases}$$

The moments are

$$\mu_r = \mu_r' = \begin{cases} \frac{2ab^{r+2}}{(r+1)(r+2)}, & r \text{ even} \\ 0 & r \text{ odd} \end{cases}$$

Thus

$$\mu_2 = \frac{ab^4}{6} = \sigma^2, \qquad \mu_4 = \frac{ab^6}{15}$$

Hence

$$\beta_2 = 2.4 \quad \text{since} \quad \mu_0 = ab^2 = 1$$

(a) If we use the inverse transform method to generate a random deviate from this distribution, we set

$$x = \begin{cases} -b \pm b\sqrt{2r}, & 0 \le r \le .5 \\ b \pm b\sqrt{2(1-r)}, & .5 < r \le 1 \end{cases}$$

Again, by inspection we can show that

$$x = \begin{cases} -b + b\sqrt{2r}, & 0 \le r \le .5 \\ b - b\sqrt{2(1-r)}, & .5 < r \le 1 \end{cases}$$

so that $-b \le x \le b$. Since it follows that $b^2 = 6\sigma^2$, or $b = 12.2474$, we set:

$$x = \begin{cases} 12.2474(\sqrt{2r} - 1, & 0 \le r \le .5 \\ 12.2474[1 - \sqrt{2(1-r)}], & .5 < r \le 1 \end{cases} \tag{4}$$

(b) Using the convolution theorem it can be shown that the sum of two uniform random numbers: $x_1 \sim [-2b, -b]$ and $x_2 \sim [b, 2b]$ or $x = x_1 + x_2$ will have the required triangular distribution. So we choose:

$$x = 12.2474(r + s - 1) \tag{5}$$

where r, s are uniform over $[0, 1]$.

From a computational point of view (5) is preferred to (4), since it does not involve the taking of square roots nor is selection based on the value of r as in (4).

The normal distribution ($\beta_2 = 3$)

It has the density function

$$f_X(x) = \frac{1}{\sqrt{2\pi}\sigma} \exp\left\{ \frac{(x - \mu)^2}{2\sigma^2} \right\} \qquad -\infty < x < \infty$$

For the normal distribution the central moments are

$$\mu_r = \begin{cases} \frac{(r)!\sigma^r}{2^{r/2}(r/2)!} & r \text{ even} \\ 0, & r \text{ odd} \end{cases}$$

To generate random normal deviates, we used the Odeh and Evans (1974) approximation to the inverse error function $x_r = \Phi^{-1}(r)$ which is accurate to 7 decimals. It is a rational fraction approximation of $|x_r|$ for $10^{-20} < r < 0.5$. The procedure is as follows:

The Algorithm

Step 1: Set $\epsilon = 10^{-20}$.

$p_0 = -0.322232431088$

$p_1 = -1$

$p_2 = -0.342242088547$

$p_3 = -0.0204231210245$

$p_4 = -0.453642210148 \times 10^{-4}$

$q_0 = 0.099348462606$

$q_1 = 0.588581570495$

$q_2 = 0.531103462366$

$q_3 = 0.103537752850$

$q_4 = 0.0038560700634$

Step 2: Set $x_r = 0$ and $r_s = r$.

Step 3: If $r_s > 0.5$ set $r_s := 1 - r_s$.

Step 4: If $r_s < \epsilon$ return with $x_r = 0$.

Step 5: If $r_s = 0.5$ return with $x_r = 0$.

Step 6: Set

$$x_i = \frac{1}{\sqrt{\ln r_s^2}}$$

$$\begin{aligned} x_r = & x_i + ((((x_i p_4 + p_3)x_i + p_2)x_i + p_1)x_i + p_0) \\ & /((((x_i q_4 + q_3)x_i + q_2)x_i + q_1)x_i + q_0) \end{aligned}$$

Step 7: If $r < 0.5$ then set $x_r = -x_r$, return with x_r.

For variance $\sigma^2 = 25$ we set

$$x = 5x_r \tag{6}$$

The contaminated normal distribution ($\beta_2 = 4$ and 5)

It has the density function

$$f_X(x) = \frac{w_1}{\sqrt{2\pi}\sigma_1} \exp\left\{\frac{(x-\mu_1)^2}{2\sigma_1^2}\right\} + \frac{w_2}{\sqrt{2\pi}\sigma_2} \exp\left\{\frac{(x-\mu_2)^2}{2\sigma_2^2}\right\}$$

with $w_1 + w_2 = 1$ and $-\infty < x < \infty$. We shall choose $w_1 = w_2 = \frac{1}{2}$ and $\mu_1 = \mu_2 = 0$. The central moments are

$$\mu_r = \begin{cases} \frac{(r)!}{2^{r/2}(r/2)!}[w_1\sigma_1^r + w_2\sigma_2^r] & r \text{ even} \\ 0, & r \text{ odd} \end{cases}$$

Thus

$$\mu_2 = \mu_2' = \frac{\sigma_1^2 + \sigma_2^2}{2}$$

$$\mu_4 = \frac{3}{2}(\sigma_1^4 + \sigma_2^4)$$

$$\beta_2 = 6\,\frac{\sigma_1^4 + \sigma_2^4}{(\sigma_1^2 + \sigma_2^2)^2}$$

A contaminated normal deviate with mean $\mu = 0$, variance $\sigma^2 = 25$ and kurtosis β_2 can be obtained by solving the following two equations for σ_1 and σ_2:

$$\sigma_1^2 + \sigma_2^2 = 2\sigma^2$$

$$\sigma_1^4 + \sigma_2^4 = \frac{2}{3}\sigma^4\beta_2$$

so that

$$\sigma_1^2 = \sigma^2(1 + \sqrt{\beta_2/3 - 1}) \quad \text{and} \quad \sigma_2^2 = \sigma^2(1 - \sqrt{\beta_2/3 - 1})$$

Then generate

$$x_1 = \sigma_1 x_r, \qquad x_2 = \sigma_1 x_s \tag{7}$$

where $r, s \sim$ uniform $[0, 1]$ whilst x_r, x_s are calculated as in the normal distribution.

For variance $\sigma^2 = 25$ we set

$$x_1 = 6.2796x_r, \qquad x_2 = 3.2506x_s \quad \text{if } \beta_2 = 4 \tag{8}$$

$$x_1 = 6.7389x_r, \qquad x_2 = 2.1419x_s \quad \text{if } \beta_2 = 5 \tag{9}$$

We then merge the random deviates from the two differing normal distributions in accordance with the weights (fractions) w_1 and w_2. This combination will then yield random deviates from the required contaminated normal distribution.

The Laplace distribution ($\beta_2 = 6$)

It has the density function

$$f_X(x) = \frac{1}{2b}\exp\{-\frac{|x-a|}{b}\} \qquad -\infty < x < \infty$$

and distribution function

$$F(x) = \begin{cases} \frac{1}{2}\exp\{(x-a)/b\}, & x \le a \\ 1 - \frac{1}{2}\exp\{-(x-a)/b\}, & x > a \end{cases}$$

Let mean $\mu = a = 0$ then the moments are:

$$\mu_r = \begin{cases} r!b^r, & r \text{ even} \\ 0, & r \text{ odd} \end{cases}$$

Thus

$$\mu_2 = \sigma^2 = 2b^2 \qquad \mu_4 = 4!b^4$$

and $\beta_2 = 6$ whilst $b = 3.5355$.

By the inverse transform method we choose

$$x = \begin{cases} 3.5355\ln 2r, & 0 \le r \le .5 \\ -3.5355\ln 2(1-r), & .5 < r \le 1 \end{cases} \tag{10}$$

The Cauchy distribution (β_2 undefined)

It has the density function

$$f_X(x) = \frac{b}{\pi[b^2 + (x-a)^2]} \qquad -\infty < x < \infty$$

It is symmetric about a and hence for mean $\mu = 0$ we choose $a = 0$. It has the distribution function

$$F(x) = \frac{1}{2} + \frac{1}{\pi}\arctan(\frac{x}{b})$$

By the inverse transform method we choose

$$x = -b\cot\pi r \tag{11}$$

and b so that the 95% percentile of the Cauchy distribution coincides with the 95% percentile of the Normal (0,5) distribution:

$$\frac{1}{2} + \frac{1}{\pi}\arctan(\frac{x}{b}) = 0.95 \Rightarrow x = 6.3138b = 5 \times 1.96 \Rightarrow b = 1.5522$$

Appendix 5B: Tables

Table 1:

Table of uniform $[\frac{1}{2}, \frac{3}{2}]$ random numbers (x_{1i}, x_{2i}) $i = 1, \ldots, 50$.

x_{1i}	x_{2i}	x_{1i}	x_{2i}
0.5973	0.8510	0.5533	0.7646
1.2652	1.2740	0.9891	1.2452
1.3635	0.6443	1.2548	0.8730
0.9877	0.5210	1.3287	0.9596
1.4091	0.6642	0.9913	1.0826
0.7927	0.7860	1.0202	1.0573
0.9227	1.3932	0.6491	1.1790
0.6965	0.6423	0.5722	1.3755
0.5323	0.6446	0.5559	0.9432
0.7560	1.1789	0.5934	0.5426
0.8348	1.2788	1.0007	1.4275
0.5803	0.8821	0.9530	1.0238
1.3080	0.8369	0.5629	1.2321
1.3016	0.7694	1.3314	0.8769
0.9764	1.0870	0.5078	0.6416
1.1654	0.9136	1.0406	0.7297
0.6169	0.5333	1.2118	1.0228
0.6718	0.5508	1.0947	1.0500
0.5601	1.0507	1.3752	1.1405
0.9557	0.6110	1.1992	1.0996
0.5635	0.9013	1.3929	0.8897
0.7615	0.9698	1.0614	0.7824
1.4074	0.9894	0.9783	0.7873
0.9031	1.2073	0.6244	1.3081
1.0733	0.6969	0.6773	0.9543

Table 2:
$|\text{Cov}(\hat{\theta}_1, \hat{\theta}_2)|$ $(n = 50, \sigma^2 = 25$, estimates = values below $\times 10^{-4})$

p

Distribution	1.0	1.25	1.50	1.75	2.0	2.25	2.50	2.75	3.0	∞
Uniform †	1.127	0.475	0.270	0.178	0.129	0.099	0.079	0.065	0.055	0.016
Parabolic ‡	0.682	0.327	0.211	0.158	0.129	0.110	0.097	0.088	0.082	0.333
Triangular §	0.444	0.235	0.171	0.144	0.129	0.120	0.115	0.111	0.109	0.777
Normal	0.345	0.200	0.150	0.132	0.127	0.128	0.135	0.145	0.159	8.239
C.Normal(4)	0.178	0.110	0.095	0.099	0.113	0.136	0.169	0.215	0.274	24.84
C.Normal(5)	0.055	0.044	0.051	0.071	0.108	0.164	0.245	0.355	0.498	44.14
Laplace	0.072	0.059	0.067	0.088	0.124	0.183	0.276	0.420	0.639	large
Cauchy	0.011	0.025	0.833	31.84	162.9	537.8	1039	819.3	1287	∞

Table 3:
Efficiencies based on $|\text{Cov}(\hat{\theta}_1, \hat{\theta}_2)|$ ($\tilde{R}$ in parenthesis)

p

Distribution	1.0	1.25	1.50	1.75	2.0	2.25	2.50	2.75	3.0	∞
Uniform †	0.01	0.02	0.03	0.05	0.07	0.09	0.11	0.14	0.16	-
	(0.03)	(0.04)	(0.05)	(0.07)	(0.08)	(0.09)	(0.11)	(0.12)	(0.14)	
Parabolic ‡	0.10	0.20	0.31	0.41	0.50	0.59	0.67	0.74	0.79	0.20
	(0.27)	(0.37)	(0.46)	(0.53)	(0.60)	(0.65)	(0.70)	(0.74)	(0.78)	
Triangular §	0.24	0.46	0.63	0.74	0.83	0.89	0.93	0.96	0.98	0.14
	(0.17)	(0.25)	(0.33)	(0.42)	(0.50)	(0.58)	(0.67)	(0.75)	(0.83)	
Normal	0.37	0.64	0.85	0.96	1.00	0.99	0.94	0.88	0.80	0.02
	(0.64)	(0.82)	(0.93)	(0.98)	(1.0)	(0.99)	(0.95)	(0.91)	(0.85)	
C.Normal	0.53	0.86	1.00	0.96	0.84	0.70	0.56	0.44	0.35	0.00
4.0	(0.78)	(0.95)	(1.0)	(0.98)	(0.90)	(0.81)	(0.71)	(0.61)	(0.52)	
C.Normal	0.80	1.00	0.86	0.61	0.41	0.27	0.18	0.12	0.09	0.00
5.0	(0.93)	(1.0)	(0.92)	(0.77)	(0.62)	(0.48)	(0.38)	(0.31)	(0.25)	
Laplace	0.82	1.00	0.88	0.67	0.48	0.32	0.21	0.14	0.09	0.00
	(1.08)	(1.0)	(0.85)	(0.69)	(0.54)	(0.42)	(0.32)	(0.24)	(0.18)	
Cauchy	1.00	0.44	0.01	0.00	0.00	0.00	0.00	0.00	0.00	0.00

†$p = 19$ yields a generalized variance of 0.009×10^{-4} ‡$p = 5.25$ yields 0.065×10^{-4}
§$p = 3.5$ yields 0.107×10^{-4}.

Table 4:

Means and variances of p for varying sample sizes based on 500 experiments (normal errors)

Sample size	formula (5.26)		formula (5.27)		formula (5.29)	
n	Mean	Variance	Mean	Variance	Mean	Variance
30	2.373	0.345	2.242	0.260	2.565	1.647
50	2.261	0.203	2.219	0.164	2.479	1.619
100	2.117	0.101	2.097	0.093	2.228	0.351
200	2.070	0.048	2.056	0.047	2.105	0.122
400	2.040	0.028	2.019	0.028	2.055	0.058
(400)		(0.027)		(0.027)		

Table 5:

Means and variances of p for varying sample sizes based on 500 experiments (Laplace errors)

Sample size	formula (5.26)		formula (5.27)		formula (5.29)	
n	Mean	Variance	Mean	Variance	Mean	Variance
30	1.685	0.218	1.569	0.279	1.506	0.498
50	1.536	0.106	1.419	0.192	1.325	0.195
100	1.406	0.042	1.281	0.115	1.205	0.082
200	1.334	0.021	1.134	0.065	1.116	0.037
400	1.291	0.012	1.059	0.041	1.068	0.021
(400)		(0.021)		(0.083)		

Appendix 5C: ω_p^2 for symmetric error distributions

In this Appendix we show how ω_p^2 can be calculated for various distributions.

The uniform distribution

$$E[|e|^{2p-2}] = 2\int_0^b \frac{|e|^{2p-2}}{2b}de = \frac{b^{2p-2}}{2p-1}$$

$$E[|e|^{p-2}] = 2\int_0^b \frac{|e|^{p-2}}{2b}de = \frac{b^{p-2}}{p-1}$$

Thus

$$\omega_p^2 = \frac{b^2}{2p-1} = \frac{3\sigma^2}{2p-1}$$

The parabolic distribution

$$E[|e|^{2p-2}] = 2\int_0^{\sqrt{b}} |e|^{2p-2}a(b-e^2)de = \frac{3b^{p-1}}{4p^2-1}$$

$$E[|e|^{p-2}] = 2\int_0^{\sqrt{b}} |e|^{p-2}a(b-e^2)de = \frac{3b^{\frac{p-2}{2}}}{p^2-1}$$

$$\omega_p^2 = \frac{b(p+1)^2}{12p^2-3} = \frac{5\sigma^2(p+1)^2}{12p^2-3}$$

The triangular distribution

$$E[|e|^{2p-2}] = 2\int_0^b |e|^{2p-2}a(b-e)de = \frac{ab^{2p}}{p(2p-1)}$$

$$E[|e|^{p-2}] = 2\int_0^b |e|^{p-2}a(b-e)de = \frac{2ab^p}{p(p-1)}$$

$$\omega_p^2 = \frac{pb^2}{4(2p-1)} = \frac{3p\sigma^2}{4p-2}$$

The normal distribution

$$E[|e|^{2p-2}] = 2\int_0^\infty |e|^{2p-2}\frac{1}{\sqrt{2\pi}\sigma}\exp\left\{-\frac{e^2}{2\sigma^2}\right\}de = \frac{(2\sigma^2)^{p-1}\Gamma(p-\frac{1}{2})}{\sqrt{\pi}} \quad (12)$$

$$E[|e|^{p-2}] = 2\int_0^\infty |e|^{p-2}\frac{1}{\sqrt{2\pi}\sigma}\exp\left\{-\frac{e^2}{2\sigma^2}\right\}de = \frac{(2\sigma^2)^{\frac{p-2}{2}}\Gamma(\frac{p-1}{2})}{\sqrt{\pi}} \quad (13)$$

$$\omega_p^2 = \frac{2\sqrt{\pi}\sigma^2\Gamma(p-\frac{1}{2})}{\{(p-1)\Gamma(\frac{p-1}{2})\}^2}$$

The contaminated normal distribution

Denote the standard normal density function by $\phi(\cdot)$. Then it follows from (12) and (13) :

$$E[|e|^{2p-2}] = \frac{2^{p-1}\Gamma(p-\frac{1}{2})}{\sqrt{\pi}}[w_1\sigma_1^{2p-2} + w_2\sigma_2^{2p-2}]$$

$$E[|e|^{p-2}] = \frac{2^{\frac{p-2}{2}}\Gamma(\frac{p-1}{2})}{\sqrt{\pi}}[w_1\sigma_1^{p-2} + w_2\sigma_2^{p-2}]$$

$$\omega_p^2 = \frac{4\sqrt{\pi}\Gamma(p-\frac{1}{2})}{\{(p-1)\Gamma(\frac{p-1}{2})\}^2}\frac{\sigma_1^{2p-2}+\sigma_2^{2p-2}}{(\sigma_1^{p-2}+\sigma_2^{p-2})^2} \qquad \text{since } w_1 = w_2 = \frac{1}{2}$$

The Laplace distribution

$$E[|e|^{2p-2}] = \frac{1}{b}\int_0^\infty |e|^{2p-2}\exp\left(-\frac{e}{b}\right)de = b^{2p-2}\Gamma(2p-1)$$

$$E[|e|^{p-2}] = \frac{1}{b}\int_0^\infty |e|^{p-2}\exp\left(-\frac{e}{b}\right)de = b^{p-2}\Gamma(p-1)$$

$$\omega_p^2 = \frac{b^2\Gamma(2p-1)}{\{(p-1)\Gamma(p-1)\}^2} = \frac{\sigma^2\Gamma(2p-1)}{2\{(p-1)\Gamma(p-1)\}^2}$$

Note that $\omega_p^2 = \sigma^2$ when $p = 2$ for all the distributions. When $p = 1$ the following Table results:

Table 1:

ω_1^2 for the uniform, triangular, parabolic, normal, contaminated normals and Laplace distribution.

Distribution	ω_1^2
Uniform	$3\sigma^2$
Parabolic	$2.22\sigma^2$
Triangular	$1.5\sigma^2$
Normal	$1.57\sigma^2$
C. Normal	$[\pi\sigma^2(2-\beta_2/3)]/(1+\sqrt{2-\beta_2/3})$
Laplace	$0.5\sigma^2$

Table 2:

Expressions for the moment ratio parameter ω_p^2 for the uniform, triangular, parabolic, normal, contaminated normal and Laplace distributions.

Distribution	$\omega_p^2 \quad (1 < p < \infty)$
Uniform	$3\sigma^2/(2p-1)$
Parabolic	$5\sigma^2(p+1)^2/(12p^2-3)$
Triangular	$3p\sigma^2/(4p-2)$
Normal	$2\sqrt{\pi}\sigma^2\Gamma(p-\frac{1}{2})/\{(p-1)\Gamma[\frac{1}{2}(p-1)]\}^2$
C. Normal	$4\sqrt{\pi}\Gamma(p-\frac{1}{2})(\sigma_1^{2p-2}+\sigma_2^{2p-2})/\{(p-1)\Gamma[\frac{1}{2}(p-1)](\sigma_1^{p-2}+\sigma_2^{p-2})\}^2$
Laplace	$\sigma^2\Gamma(2p-1)/(2\{(p-1)\Gamma(p-1)\}^2)$

In the above formulae $(p-1)\Gamma(p-1) \to 1$ and $(p-1)\Gamma[(p-1)/2]) \to 2$ as $p \to 1$.

Appendix 5D: Var($\hat{p}$) for symmetric error distributions

In §5.5 we showed that Var($\hat{p}$) for formulae (5.26) and (5.27) are $324\text{Var}(\hat{\beta}_2)/\beta_2^6$ and $36\text{Var}(\hat{\beta}_2)/\beta_2^4$ respectively where Var($\hat{\beta}_2$) is given by

$$\frac{4\mu_4^3 - \mu_2^2\mu_4^2 + \mu_2^2\mu_8 - 4\mu_2\mu_4\mu_6 - 8\mu_2^2\mu_3\mu_5 + 16\mu_2\mu_3^2\mu_4 + 16\mu_2^3\mu_3^2}{n\mu_2^6}$$

for symmetric distributions this expression simplifies to

$$\text{Var}(\hat{\beta}_2) = \frac{4\mu_4^3 - \mu_2^2\mu_4^2 + \mu_2^2\mu_8 - 4\mu_2\mu_4\mu_6}{n\mu_2^6}$$

The uniform distribution

For the uniform distribution the moments of even order are

$$\mu_{2r} = \frac{b^{2r}}{2r+1}$$

Hence

$$\text{Var}(\hat{\beta}_2) = \frac{3^6}{n}\left[\frac{4}{125} - \frac{1}{225} + \frac{1}{81} - \frac{4}{105}\right] = \frac{1.31657}{n}$$

Thus for the Money et al. (1982) formula:

$$\text{Var}(\hat{p}) = \frac{324 \times 1.31657}{(1.8)^6 n} = \frac{12.5416}{n}$$

For the Sposito et al. (1983) formula we have

$$\text{Var}(\hat{p}) = \frac{36 \times 1.31657}{(1.8)^4 n} = \frac{4.5150}{n}$$

The parabolic distribution

For the parabolic distribution the moments of even order are

$$\mu_{2r} = \frac{3b^r}{(2r+1)(2r+3)}$$

Hence

$$\text{Var}(\hat{\beta}_2) = \frac{5^6}{n}\left[\frac{108}{42875} - \frac{9}{30625} + \frac{1}{825} - \frac{12}{3675}\right] = \frac{2.68575}{n}$$

$$\text{Money et al. (1982):} \qquad \text{Var}(\hat{p}) = \frac{8.9878}{n}$$

$$\text{Sposito et al. (1983):} \qquad \text{Var}(\hat{p}) = \frac{4.5856}{n}$$

The triangular distribution

For the triangular distribution the moments of even order are

$$\mu_{2r} = \frac{b^{2r}}{(r+1)(2r+1)}$$

Hence

$$\text{Var}(\hat{\beta}_2) = \frac{6^6}{n}\left[\frac{4}{3375} - \frac{1}{8100} + \frac{1}{1620} - \frac{4}{2520}\right] = \frac{4.27886}{n}$$

$$\text{Money et al. (1982):} \qquad \text{Var}(\hat{p}) = \frac{7.2545}{n}$$

$$\text{Sposito et al. (1983):} \qquad \text{Var}(\hat{p}) = \frac{4.6429}{n}$$

The normal distribution

For the normal distribution the moments of even order are

$$\mu_{2r} = \frac{(2r)!\sigma^{2r}}{2^r r!}$$

$$\text{Hence} \qquad \text{Var}(\hat{\beta}_2) = \frac{24}{n}$$

Thus for both the Money et al. (1982) and Sposito et al. (1983) formulae:

$$\text{Var}(\hat{p}) = \frac{10.667}{n}$$

The contaminated normal distributions

For the contaminated normal distribution the moments of even order are

$$\mu_{2r} = \frac{(2r)!w_1\sigma_1^{2r}}{2^r r!} + \frac{(2r)!w_2\sigma_2^{2r}}{2^r r!}$$

Since $w_1 = w_2 = .5$ we have

$$\mu_2 = \frac{\sigma_1^2 + \sigma_2^2}{2}$$

$$\mu_4 = \frac{3}{2}(\sigma_1^4 + \sigma_2^4)$$

$$\mu_6 = \frac{15}{2}(\sigma_1^6 + \sigma_2^6)$$

$$\mu_8 = \frac{105}{2}(\sigma_1^8 + \sigma_2^8)$$

We can solve for σ_1^2 and σ_2^2 in terms of the overall variance σ^2. For $\beta_2 = 4$

$$\sigma_1^2 + \sigma_2^2 = \sigma^2$$

$$\sigma_1^4 + \sigma_2^4 = \frac{8}{3}\sigma^4$$

$$\sigma_1^6 + \sigma_2^6 = 4\sigma^6$$

$$\sigma_1^8 + \sigma_2^8 = \frac{56}{9}\sigma^8$$

and $\quad \mu_2 = \sigma^2 \quad \mu_4 = 4\sigma^4 \quad \mu_6 = 30\sigma^6 \quad \mu_8 = 326.67\sigma^8$

Hence

$$\text{Var}(\hat{\beta}_2) = \frac{86.6667}{n}$$

Money et al. (1982): $\quad \text{Var}(\hat{p}) = \dfrac{6.8555}{n}$

Sposito et al. (1983): $\quad \text{Var}(\hat{p}) = \dfrac{12.1875}{n}$

Similarly for $\beta_2 = 5$

Money et al. (1982): $\quad \text{Var}(\hat{p}) = \dfrac{3.0413}{n}$

Sposito et al. (1983): $\quad \text{Var}(\hat{p}) = \dfrac{8.4480}{n}$

The Laplace distribution

For the Laplace distribution the moments of even order are

$$\mu_{2r} = (2r)!b^{2r}$$

Hence

$$\text{Var}(\hat{\beta}_2) = \frac{55296 - 2304 + 161280 - 138240}{64n} = \frac{1188}{n}$$

Money et al. (1982): $\quad \text{Var}(\hat{p}) = \dfrac{8.25}{n}$

Sposito et al. (1983): $\quad \text{Var}(\hat{p}) = \dfrac{33}{n}$

Appendix 5E: The Cramér-von Mises goodness-of-fit test

Goodness-of-fit tests use statistics based on the observed empirical distribution function (EDF) of the data. EDF statistics are discussed by Stephens (1974). The Cramér-von Mises test statistic $W^*(W^2)$ is appropriate when testing for normality where the two unknown parameters (μ and σ^2) are estimated by the sample mean $\bar{r}$ and variance s^2.

Suppose the observations have been arranged in ascending order,

$$r_1 < r_2 < \cdots < r_n$$

Let the standardised normal distribution function be given by $F(w)$. The test proceeds as follows:

Step 1: Calculate the quantities $w_i = (r_i - \bar{r})/s$.

Step 2: Calculate the standardized normal values $z_i = F(w_i)$.

Step 3: Calculate the modified Cramér-von Mises statistic

$$W^* = \left(\sum_{i=1}^{n}\left[\frac{z_i - (2i-1)}{2n}\right]^2 + \frac{1}{12n}\right)(1 + \frac{1}{2n})$$

Step4: Check for significance of W^* in the following table:

Critical values of W^* for the α-level of significance

α	Critical value
0.150	0.091
0.100	0.104
0.050	0.126
0.025	0.148
0.001	0.178

If W^* is greater than the critical value at level α then we conclude, at the α-level of significance, that the data do not follow a normal distribution.

Chapter 5: Additional notes

§5.1

The asymptotic normality of $\hat{\boldsymbol{\theta}}$ was noted by Box and Lucas (1959) [see also Malinvaud (1970), Goldfeldt and Quandt (1972) and Gallant (1975)].

§5.1.1

For further detail on the jackknife procedure see Miller (1974).

§5.1.1.1

1 Gallant (1975) compared these tests in a simulation study and suggested that the likelihood ratio test be used in nonlinear regression.

2 Milliken and DeBruin (1978) also derived a procedure for testing general hypotheses about $\boldsymbol{\theta}$.

3 Khorasani and Milliken (1982) used the confidence regions of $\boldsymbol{\theta}$ in conjunction with optimization procedures to determine a conservative simultaneous confidence band around the nonlinear model.

4 Johnson and Milliken (1983) described a procedure which tests linear hypotheses about the parameters $\boldsymbol{\theta}$. Their procedure for hypothesis testing was carried out using an approximate weighted least squares method. For example, if we wish to know whether a particular nonlinear relationship holds for different populations, then the parameters of the model for each population group may be compared by testing a linear hypothesis about the parameter vector $\boldsymbol{\theta}$.

6 The bias formula according to Ratkowsky (1983) is a good rule of thumb for indicating nonlinear behaviour. The interested reader is referred to Bates and Watts (1980) who derived curvature measures of nonlinearity.

7 Curtis (1986) describes a process by which ill-determined parameters may be rejected. This problem occurs when there is high correlation between parameters.

8 Cook and Tsai (1985) investigate the behaviour of ordinary residuals in nonlinear regression. These residuals may produce misleading results when used in diagnostic methods analogous to those in linear regression. A new type of residual, the projected residual is introduced which overcomes many of the shortcomings of the usual residual.

§5.2 and §5.3

The expressions for the confidence intervals for the true mean response and true value of a single observation appear to be new.

§5.4

1 The RAND Corporation's 1 million random digits table (1979) are available on magnetic tape and since these numbers are considered to be "truly" random and free of autocorrelation and periodicity they were used instead of a pseudo-random number generator.

2 The reader is also referred to Chapter 6 of Kennedy and Gentle (1980) for a thorough discussion of random number generation.

§5.7

Regression diagnostics for the linear L_1-norm case are considered by Narula and Wellington (1985).

Bibliography Chapter 5

Bates, D.M. and Watts, D.G. (1980). Relative curvature measures of nonlinearity. *J.R. Statist. Soc., Ser. B* **42**, pp 1–25.

Bassett, G. and Koenker, R. (1978). Asymptotic theory of least absolute error regression. *J. Amer. Statist.* **73**, pp 618–622.

Beale, E.M.L. (1960). Confidence regions in non-linear estimation. *J.R. Statist. Soc.* **B 22**, pp 41–88.

Box, M.J. (1971). Bias in nonlinear estimation. *J.R. Statist. Soc., Ser. B* **33**, pp 171–201.

Box, G.E.P. and Lucas, H.L. (1959). Design of experiments in non-linear situations. *Biometrika* **46**, pp 77–90.

Box, G.E.P. and Tiao, G.C. (1973). Bayesian inference in statistical analysis. *Addison-Wesley, Reading, Massachusetts.*

Clarke, G.P.Y. (1980). Moments of the least squares estimators in a non-linear regression model. *J.R. Statist. Soc. B,* **42**, pp 227–237.

Clarke, G.P.Y. (1982). The analysis of nonlinear regression models. PhD Thesis. University of London.

Clarke, G.P.Y. (1987). Approximate confidence limits for a parameter function in nonlinear regression. *J. Amer. Statist.* **82**, pp 221–230.

Cook, R.D. and Tsai, C-H. (1985). Residuals in nonlinear regression. *Biometrika,* **72**, pp 23–29.

Cook, R.D. and Weisberg, S. (1982). Residuals and influence in regression. *Chapman and Hall, New York.*

Cox, D.R. and Hinkley, D.V. (1974). Theoretical statistics, *Chapman and Hall, London.*

Cramér, H. (1946). Mathematical methods of statistics. *Princeton University Press, Princeton.*

Curtis, A.R. (1986). Analysis of covariance after nonlinear least-squares fitting. *IMA J. Numer. Anal.* **6**, pp 453–461.

Dahlquist, G. and Björck, A. (1974). Numerical methods. *Translated by N. Anderson, Prentice Hall, New Jersey.*

Dielman, T.E. and Pfaffenberger, R.C. (1982). LAV (Least Absolute Value) estimation in linear regression: A review. In: S. Zanakis and J. Rustagi, eds. *TIMS Studies in Management Science, North Holland, Amsterdam,* pp 31–52.

Dielman, T.E. and Pfaffenberger, R.C. (1984). Tests of linear Hypotheses and L_1 estimation: A Monte Carlo comparison. *Amer. Statist. Assoc., Business and Economic Statistics Proc.*, pp 644–647.

Donaldson, J.R. and Schnabel, R.B. (1987). Computational experience with confidence regions and confidence intervals for nonlinear least squares. *Technometrics* **29**, pp 67–82.

Duncan, G.T. (1978). An empirical study of jackknife - constructed confidence regions in nonlinear regression. *Technometrics* **20**, pp 123–129.

Dupačova, J. (1987). Asymptotic properties of restricted L_1-estimates of regression In: Y. Dodge, ed. *Statistical data analysis – Based on the L_1-norm and related methods, North Holland, Amsterdam*, pp 263–274.

Ekblom, H. (1974). L_p-methods for robust regression. *BIT* **14**, pp 22–32.

Fox, T., Hinkley, D.V. and Larntz, K. (1980). Jackknifing in nonlinear regression. *Technometrics* **22**, pp 29–33.

Gallant, A.R. (1975). Nonlinear regression. *The American Statistician*, **29**, pp 73–81.

Goldfeldt, S.M. and Quandt, R.E. (1972). Nonlinear methods in econometrics. *North-Holland, Amsterdam.*

Gonin, R. (1986). Numerical algorithms for solving nonlinear L_p-norm estimation problems: Part I - A first-order gradient algorithm for well-conditioned small residual problems. *Commun. Statist.-Simula. Computa.* **15**, pp 801–813.

Gonin, R. and Money, A.H. (1985a). Nonlinear L_p-norm estimation: Part I – On the choice of the exponent, p, where the errors are additive. *Commun. Statist.– Theor. Meth.* **14**, pp 827–840.

Gonin, R. and Money, A.H. (1985b). Nonlinear L_p-norm estimation: Part II – The asymptotic distribution of the exponent, p, as a function of the sample kurtosis. *Commun. Statist.– Theor. Meth.* **14**, pp 841–849.

Gonin, R. and Money, A.H. (1987). Outliers in physical processes: L_1- or adaptive L_p-norm estimation? In: Y. Dodge, ed. *Statistical data analysis – Based on the L_1-norm and related methods, North Holland, Amsterdam*, pp 447–454.

Harter, H.L. (1977). The nonuniqueness of absolute values regression. *Commun. Statist.-Simula. Computa.* **6**, pp 829–838.

Jennrich, R.I. (1969). Asymptotic properties of non–linear least squares estimators. *Ann. Math. Statist.* **40**, pp 633–643.

Hoaglin, D.C. and Welsch, R.E. (1978). The hat matrix in regression and Anova. *Amer. Statistician* **32**, pp 17–22.

Johnson, N.L. and Kotz, S. (1970). Distributions in statistics: Continuous univariate distributions Vols. 1 and 2. *Houghton Mifflin, New York.*

Johnson, P. and Milliken, G. (1983). A simple procedure for testing linear hypotheses about the parameters of a nonlinear model using weighted least squares. *Commun. Statist.-Simula. Computa.*, **12**, pp 135–145.

Judge, G.G., Griffiths, W.E., Hill R.C., Lütkepohl, H. and Lee, T.C. (1985). The theory and practice of econometrics. *2nd edition, John Wiley, New York.*

Kendall, M.G. and Stuart, A. (1963). The advanced theory of statistics, Volume 1, Distribution theory. *Charles Griffin, London.*

Kennedy, W.J. and Gentle, J.E. (1980). Statistical Computing. *Marcel Dekker, New York.*

Khorasani, F. and Milliken, G.A. (1982). Simultaneous confidence bands for nonlinear models. *Commun. Statist.-Theor. Meth.*, **11**, pp 1241–1253.

Koenker, R. (1987). A comparison of asymptotic testing methods for L_1 regression. In: Y. Dodge, ed. *Statistical data analysis – Based on the L_1-norm and related methods, North Holland, Amsterdam*, pp 287–295.

Koenker, R. and Bassett, G. (1982). Tests of linear hypotheses and ℓ_1 estimation. *Econometrica* **50**, pp 43–61.

Malinvaud, E. (1970). The consistency of nonlinear regressions. *Ann. Math. Statist.* **41**, pp 956–969.

McKean, J.W. and Schrader, R.M. (1987). Least absolute errors analysis of variance. In: Y. Dodge, ed. *Statistical data analysis – Based on the L_1-norm and related methods, North Holland, Amsterdam*, pp 297–305.

Miller, R. (1974). The jackknife - a review. *Biometrika* **61**, pp 1–16.

Militký, J. and Čáp, J. (1987). Application of the Bayes approach to adaptive L_p nonlinear regression. *Computational Statistics & Data Analysis*, **5**, pp 381–389.

Milliken, G.A. and DeBruin, R.L. (1978). A procedure to test hypotheses for nonlinear models. *Commun. Statist.-Theor. Meth.*, **7**, pp 65–79.

Money, A.H., Affleck-Graves, J.F., Hart, M.L. and Barr, G.D.I. (1982). The linear regression model: L_p norm estimation and the choice of p. *Commun. Statist.-Simula. Computa.* **11**, pp 89–109.

Narula, S.C. and Wellington, J.F. (1985). Interior analysis for the minimum sum of absolute errors regression. *Technometrics* **27**, pp 181–188.

Nyquist, H. (1980). Recent studies on L_p-norm estimation. *PhD Thesis. University of Umeå, Sweden.*

Nyquist, H. (1983). The optimal L_p norm estimator in linear regression models, *Commun. Statist.- Theor. Meth.* **12**, pp 2511–2524.

Oberhofer, W. (1982). The consistency of nonlinear regression minimizing the L_1-norm. *Ann. Stat.* **10**, pp 316–319.

Odeh, R.E. and Evans, J.O. (1974). Algorithm AS 70: Percentage points of the normal distribution. *Appl. Stat.*, **23**, pp 96–97.

Osborne, M.R. and Watson, G.A. (1969). An algorithm for minimax approximation in the nonlinear case. *Comput. J.* **12**, pp 63–68.

Osborne, M.R. and Watson, G.A. (1971). On an algorithm for discrete nonlinear L_1 approximation. *Comput. J.* **14**, pp 184–188.

Quenouille, M.H. (1949). Approximate tests of correlation in time series. *J.R. Statist. Soc., Ser. B* **11**, pp 18–84.

Quenouille, M.H. (1956). Notes on bias in estimation. *Biometrika* **43**, pp 353–360.

Random Digits Table. (1979). *RAND Computation Centre. Santa Monica, California* 90406.

Ratkowsky, D.A. (1983). Nonlinear regression modeling. *Marcel Dekker, New York.*

Schall, R. and Gonin (1988). Diagnostics for nonlinear L_p-norm estimation. *Unpublished manuscript.*

Spiegel, M.R. (1968). Mathematical handbook of formulas and tables. *Schaum's Outline Series, McGraw-Hill, New York.*

Sposito, V.A., Hand, M.L. and Skarpness, B. (1983). On the efficiency of using the sample kurtosis in selecting optimal L_p estimators. *Commun. Statist.-Simula. Computa.* **12**, pp 265–272.

Stephens, M.A. (1974). EDF Statistics for goodness of fit and some comparisons. *JASA* **69**, pp 730–737.

Theil, H. (1971). The principles of econometrics. *John Wiley, New York.*

Turner, M.E. (1960). An heuristic estimation method. *Biometrics* **16**, pp 299–301.

Withers, C.S. (1987). The bias and skewness of L_1-estimates in regression. *Computational Statistics & Data Analysis,* **5**, pp 301–303.

6

Application of L_p-Norm Estimation

In this chapter the application of nonlinear L_p-norm estimation to practical problems will be considered. Applications to be considered include the modelling of certain physiological processes of interest in medical research, for example respiratory physiology and pharmacology, as well as a two-equation econometric model of production data. *It is interesting that all these nonlinear models were generated as the analytical solutions to ordinary differential equations.*

Not only is the estimation of the parameters of the model important but also the construction of confidence intervals of these parameters as well as the testing of simple hypotheses on these parameters. We shall also show how useful the adaptive procedure (§5.6) of Chapter 5 is in identifying outlying observations. We also demonstrate that the adaptive L_p-norm estimation procedure can select a value of p close to 1. The advantage of this procedure is that we do not have to decide à priori to use L_1-norm estimation but instead allow the data to prescribe which value of p to use. The regression diagnostic procedures of §5.7 will be applied to identify possible outlying observations or the inherent inadequacy of the model.

6.1. Compartmental models, bioavailability studies in Pharmacology

Weiner (1974) states that *"the object of a bioavailability study is to quantify the relative amount and rate of absorption of the administered drug which reaches the general circulation intact."* and defines comparative bioavailability as the comparison of the " *bioavailability of the new drug to a standard or reference drug.*" In the analysis of data obtained from a bioavailability study there are two distinct phases:

(i) Nonlinear estimation of the bioavailability parameters.

(ii) The use of these estimates in an experimental design, such as a cross-over design, to test whether the new drug differs significantly from the standard drug with respect to its bioavailability.

We shall only concern ourselves with the first phase. The interested reader is referred to Buncher and Tsay (1974) or Winer (1971) for more detail regarding experimental design.

6.1.1. Mathematical models of drug bioavailability

In bioavailability studies compartmental models are used. One such compartment, the vascular compartment contains the drug and carries it to another compartment the tissue. These compartments form a system characterised by the transfer rate of the drug between compartments.

Consider a two compartment system where the one is open to the environment (for example oral administration or intravenous injection). Schematically:

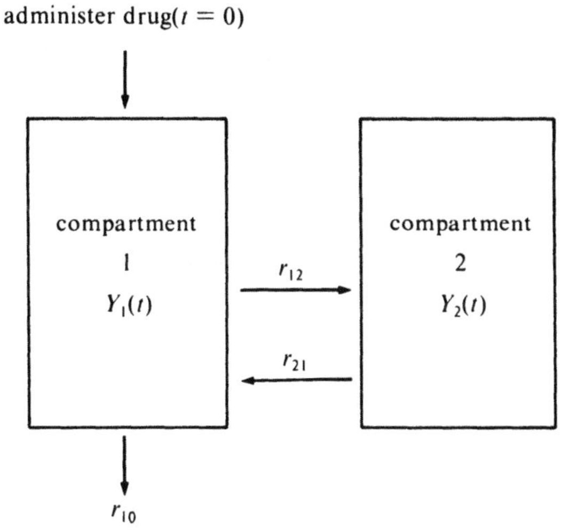

Figure 6.1 Compartmental models.

r_{ij} = constant transfer rate from compartment $i \to j$

y_i = drug concentration in compartment i at time t

Assuming that the rate of change in concentration is proportional to the amount present, we can derive the following system of ordinary differential equations:

$$Y_1'(t) = -(r_{10} + r_{12})Y_1(t) + r_{21}Y_2(t)$$
$$Y_2'(t) = r_{12}Y_1(t) - r_{21}Y_2(t)$$

This can be solved by the standard analytical techniques of linear differential equations to yield:

$$Y_1(t) = P_1 \exp(-\lambda_1 t) + P_2 \exp(-\lambda_2 t) \tag{6.1}$$

where P_1 and P_2 are arbitrary constants and λ_i are given by:

$$\frac{1}{2}\{-(r_{10} + r_{12} + r_{21}) \pm \sqrt{(r_{10} + r_{12} + r_{21})^2 - 4r_{10}r_{21}}\}$$

Since the concentration of the drug is measured in the vascular compartment we are only interested in $Y_1(t)$ (the transfer rates cannot be obtained directly from the differential equations). We can therefore fit a model which resembles the analytical solution (6.1). Denote the observed concentrations $Y_1(t)$ by $y(t)$ and use the model

$$y(t) = \theta_1 \exp(-\theta_3 t) + \theta_2 \exp(-\theta_4 t) \tag{6.2}$$

The pharmacologist is usually interested in three quantities or bioavailability parameters which can be calculated once the model describing the drug concentrations has been fitted. These are

(1) Area Under the concentration Curve:

$$AUC = \int_0^\infty y(t)\,dt = \frac{\theta_1}{\theta_3} + \frac{\theta_2}{\theta_4}$$

which will have units of mass×time e.g., μg-hours.

(2) Time till maximum concentration:

$$t_{\max} = \frac{\ln(-\theta_1\theta_3/\theta_2\theta_4)}{\theta_3 - \theta_4}$$

(3) Maximum concentration $C_{\max} = y(t_{\max})$.

6.1.2. Outlying observations in pharmacokinetic models

Frome and Yakatan (1980) [see also Gonin and Money (1987)] considered the one compartment model which can be formulated as:

$$y(t) = \frac{\theta_1\theta_3}{\theta_1 - \theta_2}[\exp(-\theta_2 t) - \exp(-\theta_1 t)] \tag{6.3}$$

The parameter θ_1 is the absorption rate constant, θ_2 the elimination rate constant whilst $\theta_3 = fD/V$ where D is the initial amount of drug administered, V the volume of distribution and f the fraction of the drug absorbed from the gastro-intestinal tract. $y(t)$ is the drug concentration in the serum where t is measured in hours. Note that θ_3 can therefore be interpreted as physical density (mass/volume).

We are only interested in one of the bioavailability parameters

$$AUC = \int_0^\infty y(t)\,dt = \frac{\theta_3}{\theta_2}$$

Rodda et al. (1975) considered the effect of outliers on least squares estimation in the one compartment model. Frome and Yakatan adjusted their data to contain certain patterns of outliers and suggested that L_1-norm estimation should be used. The data are given in Table 6.1.

Table 6.1: Frome and Yakatan's data

t	Pattern 0	Pattern 1	Pattern 2	Pattern 6
0.083	10.9	10.9	10.9	10.9
0.167	19.1	19.1	19.1	19.1
0.250	25.3	25.3	25.3	25.3
0.500	35.4	15.0	15.0	55.0
0.750	38.5	38.5	20.0	60.0
1.000	38.4	38.4	38.4	60.0
1.500	34.8	34.8	34.8	34.8
2.250	28.2	28.2	28.2	28.2
3.000	22.6	22.6	22.6	22.6
4.000	16.7	16.7	16.7	16.7
6.000	9.2	9.2	9.2	9.2
8.000	5.0	5.0	5.0	5.0
10.000	2.8	2.8	2.8	2.8
12.000	1.5	1.5	1.5	1.5

We shall apply the adaptive procedure (see §5.6, Chapter 5) to a few of these data sets (patterns 0,1,2 and 6). The mixture algorithm of Chapter 4 (§4.3.3) was used to fit model (6.3) to the data. Similarly for the L_1-norm and L_∞-norm estimations, the Anderson-Osborne-Watson algorithms (§2.3.2.1 and §3.5.1.1) were used. The same initial value $\boldsymbol{\theta}^0 = (25, 1, 10)$ used by Frome and Yakatan was used. (True values are $\boldsymbol{\theta}^* = (3, 0.3, 50)$ and $AUC = 166.7$).

Pattern 0: The optimal value of p as predicted by formulae (5.26), (5.27) or (5.29) yields estimates of $\boldsymbol{\theta}$ which do not differ from those yielded by least squares. This is probably due to the fact that it is a small residual problem. The results are given in Table 6.2.

Table 6.2: Pattern 0 – Optimal solutions for differing optimal values of p

	p		$\boldsymbol{\theta}^*$		AUC
least squares	2.00	2.995	0.300	50.014	166.7
Money et al.	3.24	2.993	0.300	50.012	166.7
Sposito et al.	2.89	2.993	0.300	50.013	166.7
Box and Tiao	∞	2.995	0.300	49.978	166.6
LAV	1	2.990	0.301	50.000	166.1

For these different values of p the parameter $\boldsymbol{\theta}^*$ and AUC remain unaffected.

Pattern 1: This example contains one outlier and its effect on the parameter estimates is given in Table 6.3.

Table 6.3: Pattern 1 – Optimal solutions for differing optimal values of p

	p		$\boldsymbol{\theta}^*$		AUC
least squares	2.00	2.144	0.306	49.490	161.7
Money et al.	1.03	2.994	0.301	50.032	166.2
Sposito et al.	0.4	2.998	0.301	50.026	166.2
Box and Tiao	0.6	3.000	0.300	50.001	166.7
LAV	1	3.003	0.300	49.965	166.6

Observe the marked deviation of $\boldsymbol{\theta}^*$ and AUC from their true values in the case of least squares. The adaptive L_p-norm procedure estimates $\boldsymbol{\theta}^*$ and AUC close to the true values. A plot of the data, observed values and fitted values for the various values of p are displayed in Fig. 6.2 and Table 6.4 which also demonstrate the ability of our adaptive procedure in identifying the outlier.

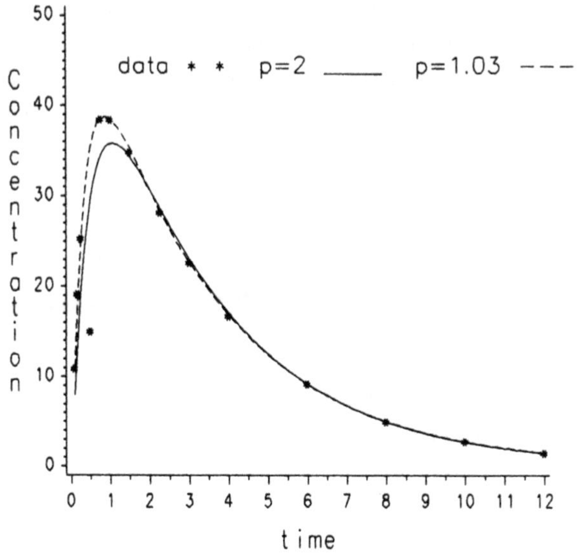

Figure 6.2 Pattern 1 data with fitted curves.

Table 6.4: Pattern 1: Data, fitted values and residuals ($p = 2, 1.03, 0.4$)

p		2		1.03		0.4	
t_i	y_i	$\hat{y}_i$	$\hat{y}_i - y_i$	$\hat{y}_i$	$\hat{y}_i - y_i$	$\hat{y}_i$	$\hat{y}_i - y_i$
0.083	10.900	7.964	-2.936	10.868	-0.032	10.879	-0.021
0.167	19.100	14.500	-4.600	19.162	0.062	19.179	0.079
0.250	25.300	19.703	-5.597	25.281	-0.019	25.300	0.000
0.500	15.000†	29.778	14.778	35.408	20.408	35.423	20.423
0.750	38.500	34.326	-4.174	38.500	0.000	38.508	0.008
1.000	38.400	35.741	-2.659	38.387	-0.013	38.389	-0.011
1.500	34.800	34.154	-0.646	34.800	0.000	34.797	-0.003
2.250	28.200	28.520	0.320	28.204	0.004	28.200	0.000
3.000	22.600	22.942	0.342	22.554	-0.046	22.552	-0.048
4.000	16.700	16.947	0.247	16.700	0.000	16.700	0.000
6.000	9.200	9.190	-0.010	9.151	-0.049	9.153	-0.047
8.000	5.000	4.981	-0.019	5.015	0.015	5.016	0.016
10.000	2.800	2.699	-0.101	2.748	-0.052	2.749	-0.051
12.000	1.500	1.463	-0.037	1.506	0.006	1.507	0.007

†outlier (true value was 35.4)

With $p = 1.03$ or less, the outlier at $t = 0.5$ with a residual of 20.4 is clearly identified (also true when $p = 2$) while all the remaining residuals are small. When we examine the residuals in the least squares case we find that the first 6 residuals are all larger than 2.5 in magnitude. For $p = 1.03$ all the residuals (except the fourth) are less than 0.07 in magnitude. L_p-norm estimation is therefore useful in not only identifying an outlier but is also sufficiently robust to cope with such an outlier.

If we use L_1-norm estimation then $\hat{\theta}_1 = 3.003, \hat{\theta}_2 = 0.300, \hat{\theta}_3 = 49.965$.

$$\hat{J}'\hat{J} = \begin{pmatrix} 102.218 & 2.03571 & 10.7617 \\ 2.03571 & 18098.1 & -166.335 \\ 10.7617 & -166.335 & 3.30641 \end{pmatrix}$$

$$(\hat{J}'\hat{J})^{-1} = \begin{pmatrix} 0.0271433 & -0.00151592 & -0.164607 \\ -0.00151592 & 0.000187434 & 0.0143632 \\ -0.164607 & 0.0143632 & 1.56078 \end{pmatrix}$$

We use the asymptotic formula for the variance of the parameter estimates to gain an indication of the variability of these estimates. In order to determine $\mathrm{Var}(\hat{\boldsymbol{\theta}})$ we need to estimate the quantity

$$\lambda = \frac{1}{2f(m)} \tag{6.4}$$

where $f(m)$ is the ordinate of the error distribution e at the median, m. We shall use the Cox and Hinkley (1974) procedure for estimating λ (§1.5, Chapter 1). Let the residuals be $\hat{e}_i$ then $f(m)^{-1}$ can be estimated by:

$$\hat{f}(m)^{-1} = \frac{\hat{e}_{(i)} - \hat{e}_{(j)}}{(i-j)/\tilde{n}} \tag{6.5}$$

where $\hat{e}_{(i)}$ denotes the ordered (nonzero) residuals $\hat{e}_i$ and $\tilde{n} = n - k = 11$ is the adjusted sample size.

Gonin and Money (1985) conjectured that asymptotically, $\mathrm{Var}(\hat{\boldsymbol{\theta}})$ is given by $\lambda^2[J(\hat{\boldsymbol{\theta}})'J(\hat{\boldsymbol{\theta}})]^{-1}$ where $J(\hat{\boldsymbol{\theta}})$ is the usual Jacobian matrix. We find that for $v = 1, 2$ and 3 that the $\hat{\lambda}$'s are about the same, hence we choose $v = 1$ for which $\hat{\lambda} = 0.063$ so that

$$\hat{\sigma}_{\theta_1} = 0.013 \qquad \hat{\sigma}_{\theta_2} = 0.001 \qquad \hat{\sigma}_{\theta_3} = 0.095$$

We also computed the jackknife confidence intervals and these are compared with the asymptotic confidence intervals. Alternatively using the α-percent nonparametric confidence interval formula (McKean and Schrader) we would find an estimate $\hat{\lambda} = 0.044$.

	95 % confidence intervals		
	Jackknife	Asymptotic	α-percent
θ_1	2.88 – 3.11	2.98 – 3.02	2.99 – 3.02
θ_2	0.29 – 0.31	0.30 – 0.30	0.30 – 0.30
θ_3	49.54 – 50.52	49.81 – 50.12	49.86 – 50.07

Pattern 2: In this example there are two outliers present, both lying below their true values. The results are given in Tables 6.5 and 6.6.

Table 6.5: Pattern 2 – Optimal solutions for differing optimal values of p

p		θ^*			AUC
least squares	2.00	1.845	0.288	45.697	158.7
Money et al.	1.21	2.721	0.300	49.399	164.7
Sposito et al.	0.9	2.982	0.301	50.034	166.2
Box and Tiao	0.9	2.982	0.301	50.034	166.2
LAV	1	2.990	0.301	50.000	166.1

Table 6.6: Pattern 2: Data, fitted values and residuals ($p = 2, 1.21, 0.9$)

p		2		1.21		0.9	
t_i	y_i	$\hat{y}_i$	$\hat{y}_i - y_i$	$\hat{y}_i$	$\hat{y}_i - y_i$	$\hat{y}_i$	$\hat{y}_i - y_i$
0.083	10.900	6.410	-4.490	9.861	-1.039	10.828	-0.072
0.167	19.100	11.817	-7.283	17.566	-1.534	19.100	0.000
0.250	25.300	16.249	-9.051	23.394	-1.906	25.210	-0.090
0.500	15.000†	25.366	10.366	33.553	18.553	35.346	20.346
0.750	20.000†	30.066	10.066	37.131	17.131	38.461	18.461
1.000	38.400	32.053	-6.347	37.490	-0.910	38.367	-0.033
1.500	34.800	31.769	-3.031	34.482	-0.318	34.800	0.000
2.250	28.200	27.495	-0.705	28.170	-0.030	28.208	0.008
3.000	22.600	22.635	0.035	22.583	-0.017	22.556	-0.044
4.000	16.700	17.106	0.406	16.748	0.048	16.700	0.000
6.000	9.200	9.644	0.444	9.201	0.001	9.149	-0.051
8.000	5.000	5.427	0.427	5.054	0.054	5.012	0.012
10.000	2.800	3.054	0.254	2.776	-0.024	2.745	-0.055
12.000	1.500	1.718	0.218	1.525	0.025	1.504	0.004

†outliers (true values were 35.4 and 38.5)

A plot of the data, observed values and fitted values for the various values of p is given in Fig. 6.3.

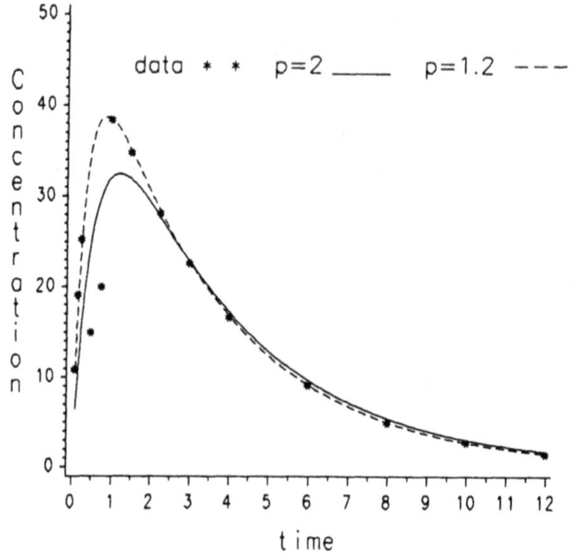

Figure 6.3 Pattern 2 data with fitted curves.

Pattern 6: In this example there are three outliers present all three lying above their true values. The results are given in Tables 6.7 and 6.8. The fitted and observed values are plotted in Figure 6.4.

Table 6.7: Pattern 6 – Optimal solutions for differing optimal values of p

	p		θ^*		AUC
least squares	2.0	2.265	0.525	83.401	158.7
Money et al.	2.8	2.222	0.562	84.976	151.2
Sposito et al.	2.7	2.227	0.556	84.763	152.5
Box and Tiao	4.4	2.160	0.608	87.241	143.5
LAV	1	2.980	0.302	50.200	166.2

Observe that the curve underestimates the maximum and hence the Box and Tiao formula tries to minimize this maximum deviation (Chebychev). Assuming that the observations around the peak concentration are realistic then one would intuitively expect that L_p-norm estimation would be appropriate. In this case L_1-norm estimation would give erroneous estimates.

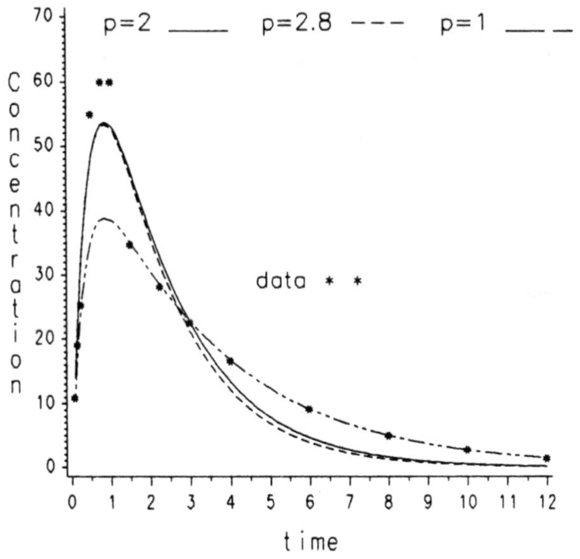

Figure 6.4 Pattern 6 data with fitted curves.

Table 6.8: Pattern 6: Data, fitted values and residuals ($p = 2, 2.8$)

p		2		2.8	
t_i	y_i	$\hat{y}_i$	$\hat{y}_i - y_i$	$\hat{y}_i$	$\hat{y}_i - y_i$
0.083	10.900	13.980	3.080	13.972	3.072
0.167	19.100	25.083	5.983	25.071	5.971
0.250	25.300	33.590	8.290	33.570	8.270
0.500	55.000 †	48.525	-6.475	48.436	-6.564
0.750	60.000 †	53.379	-6.621	53.146	-6.854
1.000	60.000 †	52.957	-7.043	52.528	-7.472
1.500	34.800	45.770	10.970	44.921	10.121
2.250	28.200	32.661	4.461	31.378	3.178
3.000	22.600	22.360	-0.240	20.952	-1.648
4.000	16.700	13.288	-3.412	12.017	-4.683
6.000	9.200	4.656	-4.544	3.914	-5.286
8.000	5.000	1.630	-3.370	1.273	-3.727
10.000	2.800	0.571	-2.229	0.414	-2.386
12.000	1.500	0.200	-1.300	0.135	-1.365

†outliers (true values were 35.4, 38.5 and 38.4)

CONCLUSION: *In practice observations that appear to be aberrant may actually be valid. In such cases L_1-norm estimation is inappropriate. The following scheme is therefore suggested:*

(1) Determine both the L_1 and adaptive L_p-norm estimates.

(2) Should both be in agreement then either value of p is appropriate. (see for example Pattern 0, 1, and 2).

(3) Should they disagree (see e.g. Pattern 6) as to those observations which are possible outliers, then it should be determined whether the outliers are in fact "wild points" by re-examining the data.

Should these points be valid data points then the L_p-norm estimates should be used since this procedure weights the observations appropriately. Alternatively modify the model appropriately.

If they are invalid points then L_1-norm parameter estimates could be used or the data should be corrected accordingly.

6.2. Oxygen saturation in respiratory physiology

6.2.1. Problem background

Pulmonary disorders such as asthma, chronic bronchitis and emphysema produce specific patterns when lung function tests are performed on the patient. One such lung function test involves the relationship between oxygen saturation in arterial blood (SO_2) and the partial pressure of oxygen in the pulmonary capillary (PO_2). SO_2 is expressed as a percentage (%) whilst PO_2 is measured in torr (mm Hg). PO_2 can only be measured directly in arterial blood. This requires catheterization, an invasive procedure which carries the risk of infection as well as causing discomfort to the severely ill patient.

Over the last decade a new device, the fibre optic ear oximeter has been developed. The oximeter which is attached to the lobe of the patient's ear is able to measure the absorption spectrum of oxyhaemoglobin in the arterial blood of the ear. From this spectrum the oxygen saturation can be calculated. Once the saturation is known, the PO_2 can be calculated by means of the oxygen dissociation curve. This obviates the use of invasive techniques to measure PO_2.

6.2.2. The data

The measurement of SO_2 (y) and PO_2 (x) values in blood samples is standard practice in respiratory units all over the world. One such source of data was reported by Severinghaus (1979) and is presently being used as a standard by the respiratory unit at Tygerberg Hospital in Cape Town. The data are as follows:

x	y	x	y	x	y	x	y	x	y
4	2.56	24	43.14	44	79.55	70	94.06	140	98.62
6	4.37	26	48.27	46	81.71	75	95.10	150	98.77
8	6.68	28	53.16	48	83.52	80	95.84	175	99.03
10	9.58	30	57.54	50	85.08	85	96.42	200	99.20
12	12.96	32	61.69	52	86.59	90	96.88	225	99.32
14	16.89	34	65.16	54	87.70	95	97.25	250	99.41
16	21.40	36	68.63	56	88.93	100	97.49		
18	26.50	38	71.94	58	89.95	110	97.91		
20	32.12	40	74.69	60	90.85	120	98.21		
22	37.60	42	77.29	65	92.73	130	98.44		

6.2.3. Modelling justification

The statistician was asked to derive a mathematical model which would describe the data adequately such that the absolute deviations between the observations and the model would not exceed 1%. Additional requirements of the model were stipulated as follows:

(1) that it be invertible with respect to the variables SO_2 and PO_2.

(2) that it be implementable on a portable calculator for clinical use; and

(3) that the model parameters be physiologically meaningful.

The purpose of the study is to find a mathematical relationship between oxygen saturation SO_2 (response variable) and oxygen tension PO_2 (explanatory variable). In practice once the oxygen saturation (explanatory variable) is known, the PO_2 (response variable) can be calculated by means of the oxygen dissociation curve. This obviates the use of invasive techniques to measure PO_2 and hence the importance of the invertibility of the model. A plot of the data revealed a nonlinear relationship between oxygen saturation and oxygen tension (see Fig. 6.5).

A family of growth curves which has proven to be useful in practical situations is the Richards (1959) family of growth curves. Du Toit (1979) has written this family as a four-parameter function:

$$f(x) = \alpha(1 + sign(\beta\rho^x)\beta\rho^x)^\lambda, \quad x > 0 \tag{6.6}$$

where λ determines the point of inflection of the curve; x is the value of the explanatory variable (e.g., time, pressure); $f(x)$ is the characteristic being measured; α is the asymptote or limiting value of f, i.e., $f \to \alpha$ as $x \to \infty$; β is the potential increase or decrease (change) in f as x varies and finally ρ is a slope parameter which characterises the rate at which f varies with a change in x. In general a decrease in ρ coincides with an increase in growth rate and vice versa. If $f(t)$ increases monotonically then $sign(\beta\rho^x) = -1$ for $\lambda \geq 0$ and $sign(\beta\rho^x) = 1$ for $\lambda < 0$. The following bounds are placed on the parameters:

$$\begin{aligned} &\alpha \geq 0 \\ &\infty > \beta \geq 0 \qquad \text{if } sign(\beta\rho^x) = 1 \\ &0 \leq \beta \leq 1 \qquad \text{if } sign(\beta\rho^x) = -1 \\ &0 \leq \rho \leq 1 \end{aligned}$$

Under these bounds $f(x)$ is a monotonic function in x and possesses a uniquely determined inverse.

There are 3 special cases of interest:

$$f(x) = \begin{cases} \alpha(1-\beta\rho^x) & \text{if } \lambda = 1 \ \text{(modified exponential);} \\ \alpha/(1+\beta\rho^x) & \text{if } \lambda = -1 \ \text{(logistic);} \\ \alpha\exp(-\beta^*\rho^x) & \text{if } |\lambda| \to \infty,\ \beta^* = |\beta\lambda| \ \text{(Gompertz);} \end{cases} \tag{6.7}$$

Initially [see Du Toit and Gonin (1984)] a Richards curve was fitted to the data. The estimated value of the shape parameter λ was very large. This suggested that a Gompertz curve [see (6.6)] should be refitted to the data. This equation, however, yielded residuals in excess of 1%. The researchers felt that the inadequate fit was probably an artefact of the possible dependency of the parameters β and ρ on PO_2. This was subsequently borne out by the following investigation.

The data set was subdivided into 41 overlapping subsets of 6 observations:

$$\begin{pmatrix} (x_1, y_1), & \cdots & (x_6, y_6) \\ (x_2, y_2), & \cdots & (x_7, y_7) \\ \vdots & \vdots & \vdots \\ (x_{41}, y_{41}), & \cdots & (x_{46}, y_{46}) \end{pmatrix}$$

A Gompertz curve was then fitted to each subset. From this it was concluded that a Gompertz curve with varying values of β and ρ would yield a good fit over the region of PO_2-values. Equivalently this can be achieved by fitting the following extended Richards curve:

$$f(x) = f(g(x)) = \alpha(1-\beta\rho^{g(x)})^\lambda \tag{6.8}$$

with $g(x)$ a continuous function in x.

It was shown by Du Toit and Gonin (1984) that the relationship between $\bar{x}_i$ (mean of x_j's in the ith subset) and $g(\bar{x}_i)$ can be described by a Richards curve. Hence the resulting Richards curve can be substituted for $g(x)$ in (6.8). This resulted in the following model

$$y = C_5 C_1^{C_2^{(1-C_3^x)^{C_4}}} \tag{6.9}$$

where

$$C_i = \frac{1}{1+e^{\theta_i}} \qquad i = 1,2,3 \qquad C_4 = 1 + e^{\theta_4} \qquad C_5 = \theta_5$$

One other aspect that still has to be addressed is that of inversion of the model. The inverse of (6.9) is

$$x = \frac{\ln(1 - K(y)^{1/K_3})}{K_2}$$

where

$$K(y) = \ln[-\ln(y/\theta_5)/B]K_1$$

$$B = \ln[1 + \exp(\theta_1)] = \ln(1/C_1)$$
$$K_1 = \ln[1 + \exp(\theta_2)] = \ln(1/C_2)$$
$$K_2 = \ln[1 + \exp(\theta_3)] = \ln(1/C_3)$$
$$K_3 = 1 + \exp(\theta_4) = C_4$$

6.2.4. Problem solution

A plot of the observed data is given in Fig. 6.5 below.

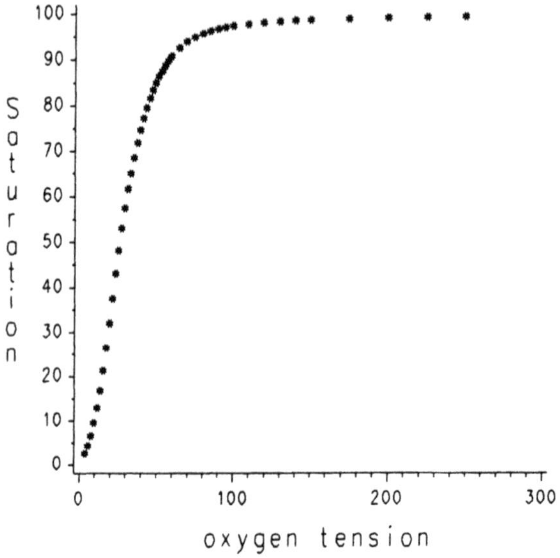

Figure 6.5 Relationship between SO_2 (%) and PO_2 (mm Hg).

From this scatterplot the nonlinear relationship between SO_2 and PO_2 is obvious. Model (6.9) was fitted to the data using both nonlinear least squares and adaptive L_p-norm estimation. The starting value $\theta^0 = (4.5, 6.0, -4.2, -1.0, 100)$ was used in least squares. The least squares solution was used to initiate the adaptive L_p-norm estimation. The results of the various fits are given in Tables 6.9 and 6.10.

- In practice a high number of significant digits (9 decimal places) in the values of the parameter estimates have to be retained to ensure the desired degree of accuracy required for clinical application.

Table 6.9: Optimal solutions for different values of p

	p	θ^*				
least squares	2.00	4.3702	7.2070	-4.1385	-1.2702	99.6080
Money et al.	2.42	4.3632	7.1272	-4.1224	-1.2489	99.6467
Sposito et al.	2.38	4.3642	7.1356	-4.1241	-1.2513	99.6426
Box and Tiao	3.02	4.3501	7.0281	-4.1020	-1.2199	99.6960

For the different values of p the parameter estimates are essentially the same. This is to be expected since we are dealing with a small residual problem. The observed values, fitted values and residuals for $p = 2$ and 2.38 are displayed in Table 6.10. Observe that the fitted values for $p = 2$ and 2.38 only differ in the second decimal. The residuals are all less than $\frac{1}{2}\%$ in magnitude.

We are now in a position to construct confidence intervals for the individual parameters. Should the nonlinear least squares estimates be used then the asymptotic $100(1-\alpha)\%$ confidence interval for θ_j is given by

$$\hat{\theta}_j \pm z_{\alpha/2}\sqrt{\sigma^2(\hat{J}'\hat{J})^{jj}}$$

Where $(\hat{J}'\hat{J})^{jj}$ denotes the jth diagonal element of $[J(\hat{\theta})'J(\hat{\theta})]^{-1}$, $z_{\alpha/2}$ is the appropriate percentile of the standard normal distribution and $\hat{\theta}$ is the estimate of θ. If σ^2 is unknown we use s^2. The matrix $\hat{J}'\hat{J}$ and its inverse $(\hat{J}'\hat{J})^{-1}$ from the least squares fit are given in lower triangular form below:

$$\hat{J}'\hat{J}$$

$$\begin{pmatrix} 896.004611 & & & & \\ -987.936567 & 1307.038790 & & & \\ -6993.356431 & 8862.032145 & 60973.807960 & & \\ 2017.566480 & -2344.490482 & -16498.687178 & 4660.156620 & \\ -80.812411 & 132.612218 & 811.951174 & -190.736122 & 27.39596 \end{pmatrix}$$

$$(\hat{J}'\hat{J})^{-1}$$

$$\begin{pmatrix} 0.265883 & & & & \\ 0.789663 & 3.625890 & & & \\ -0.172853 & -0.756897 & 0.159306 & & \\ -0.341897 & -1.261869 & 0.271159 & 0.493955 & \\ -0.295528 & -1.574820 & 0.320355 & 0.502146 & 0.789262 \end{pmatrix}$$

Table 6.10: Oxysaturation: Fitted values and residuals ($p = 2, 2.38$)

p		2		2.38	
x_i	y_i	$\hat{y}_i$	$\hat{y}_i - y_i$	$\hat{y}_i$	$\hat{y}_i - y_i$
4	2.56	2.772	0.212	2.772	0.212
6	4.37	4.299	-0.071	4.295	-0.075
8	6.68	6.459	-0.221	6.450	-0.230
10	9.58	9.313	-0.267	9.301	-0.279
12	12.96	12.866	-0.094	12.851	-0.109
14	16.89	17.056	0.166	17.041	0.151
16	21.40	21.772	0.372	21.758	0.358
18	26.50	26.865	0.365	26.853	0.353
20	32.12	32.175	0.055	32.167	0.047
22	37.60	37.548	-0.052	37.543	-0.057
24	43.14	42.849	-0.291	42.847	-0.293
26	48.27	47.969	-0.301	47.971	-0.299
28	53.16	52.829	-0.331	52.832	-0.328
30	57.54	57.374	-0.166	57.379	-0.161
32	61.69	61.575	-0.115	61.580	-0.110
34	65.16	65.420	0.260	65.425	0.265
36	68.63	68.913	0.283	68.917	0.287
38	71.94	72.065	0.125	72.068	0.128
40	74.69	74.896	0.206	74.898	0.208
42	77.29	77.430	0.140	77.430	0.140
44	79.55	79.691	0.141	79.689	0.139
46	81.71	81.704	-0.006	81.700	-0.010
48	83.52	83.494	-0.026	83.489	-0.031
50	85.08	85.085	0.005	85.079	-0.001
52	86.59	86.499	-0.091	86.491	-0.099
54	87.70	87.754	0.054	87.746	0.046
56	88.93	88.870	-0.060	88.861	-0.069
58	89.95	89.863	-0.087	89.853	-0.097
60	90.85	90.746	-0.104	90.735	-0.115
65	92.73	92.558	-0.172	92.546	-0.184
70	94.06	93.926	-0.134	93.915	-0.145
75	95.10	94.969	-0.131	94.959	-0.141
80	95.84	95.773	-0.067	95.764	-0.076
85	96.42	96.398	-0.022	96.390	-0.030
90	96.88	96.890	0.010	96.883	0.003
95	97.25	97.280	0.030	97.275	0.025
100	97.49	97.593	0.103	97.589	0.099
110	97.91	98.053	0.143	98.052	0.142
120	98.21	98.365	0.155	98.366	0.156
130	98.44	98.583	0.143	98.586	0.146
140	98.62	98.740	0.120	98.745	0.125
150	98.77	98.855	0.085	98.861	0.091
175	99.03	99.035	0.005	99.044	0.014
200	99.20	99.133	-0.067	99.142	-0.058
225	99.32	99.189	-0.131	99.199	-0.121
250	99.41	99.223	-0.187	99.234	-0.176

The residuals for the respective fits are plotted against their predicted values in Figs. 6.6 and 6.7.

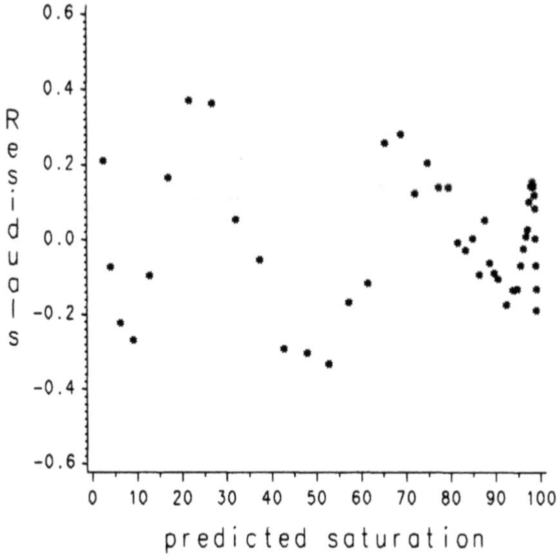

Figure 6.6 $\hat{y}_i - y_i$ vs $\hat{y}_i$ for the least squares fit.

Matrix $\hat{J}'\hat{J}$ and its inverse $(\hat{J}'\hat{J})^{-1}$ from $L_{2.38}$ estimation are given below:

$$\hat{J}'\hat{J}$$

$$\begin{pmatrix} 899.426611 & & & & \\ -999.948015 & 1334.401538 & & & \\ -7008.190701 & 8950.466854 & 60956.475385 & & \\ 2036.387996 & -2383.831742 & -16613.036602 & 4727.984276 & \\ -81.103486 & 134.283179 & 811.602091 & -191.874558 & 27.375632 \end{pmatrix}$$

$$(\hat{J}'\hat{J})^{-1}$$

$$\begin{pmatrix} 0.264982 & & & & \\ 0.778035 & 3.526261 & & & \\ -0.171808 & -0.743033 & 0.157849 & & \\ -0.337992 & -1.233251 & 0.267412 & 0.484569 & \\ -0.306775 & -1.607203 & 0.330294 & 0.516391 & 0.838507 \end{pmatrix}$$

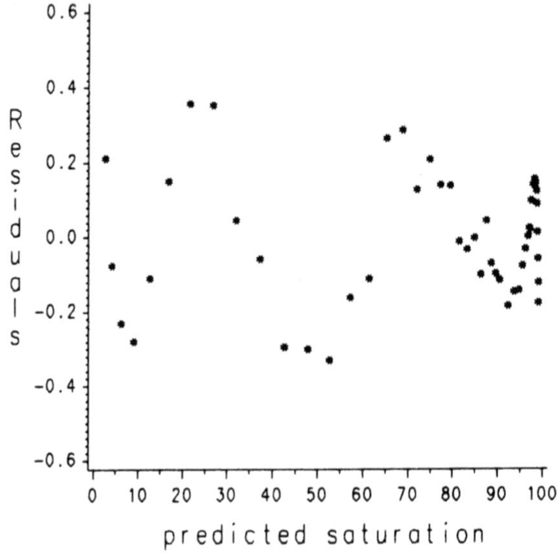

Figure 6.7 $\hat{y}_i - y_i$ vs $\hat{y}_i$ for the $L_{2.38}$ fit.

Should the adaptive scheme be used then an asymptotic $100(1-\alpha)\%$ confidence interval for θ_j is given by

$$\hat{\theta}_j \pm z_{\alpha/2}\sqrt{\hat{\omega}_p^2(\hat{J}'\hat{J})^{jj}}$$

where $\hat{\omega}_p^2$ is calculated as follows:

$$\hat{\omega}_p^2 = \frac{m_{2p-2}}{[(p-1)m_{p-2}]^2}$$

with $m_r = \frac{1}{n}\sum_{i=1}^{n}|\hat{e}_i|^r$ where $\hat{e}_i$ is the ith residual. 95% confidence intervals for least squares ($s^2 = 0.03206$) and $p = 2.38$ ($\omega_{2.38}^2 = 0.02672$) are given in Table 6.11.

Table 6.11: 95% confidence intervals for $\theta_j, j = 1, \ldots, 5$

	least squares	$L_{2.38}$
θ_1	$[4.18, 4.56]$	$[4.19, 4.53]$
θ_2	$[6.52, 7.90]$	$[6.52, 7.76]$
θ_3	$[-4.28, -3.99]$	$[-4.26, -3.99]$
θ_4	$[-1.52, -1.02]$	$[-1.48, -1.02]$
θ_5	$[99.29, 99.93]$	$[99.34, 99.94]$

Table 6.12: Jackknife estimates of $\hat{\boldsymbol{\theta}}$

$\hat{\theta}_1^{(i)}$	$\hat{\theta}_2^{(i)}$	$\hat{\theta}_3^{(i)}$	$\hat{\theta}_4^{(i)}$	$\hat{\theta}_5^{(i)}$	D_i
4.28994	7.00523	-4.09336	-1.17712	99.6858	0.4448
4.39210	7.25396	-4.14897	-1.29395	99.5922	0.0142
4.41992	7.28581	-4.15646	-1.31700	99.5845	0.1007
4.40593	7.22452	-4.14312	-1.29288	99.6078	0.1044
4.37485	7.19063	-4.13507	-1.26780	99.6160	0.0115
4.37371	7.26390	-4.15084	-1.28850	99.5849	0.0415
4.39717	7.36483	-4.17336	-1.33129	99.5505	0.3727
4.40714	7.35794	-4.17261	-1.33642	99.5561	0.3933
4.37600	7.22479	-4.14266	-1.27907	99.6020	0.0041
4.36545	7.19692	-4.13594	-1.26385	99.6108	0.0028
4.35076	7.19197	-4.13324	-1.24954	99.6074	0.0731
4.35868	7.23132	-4.14184	-1.26415	99.5925	0.0770
4.36697	7.27059	-4.15043	-1.27902	99.5783	0.0815
4.37294	7.25255	-4.14766	-1.28114	99.5886	0.0197
4.37452	7.24414	-4.14624	-1.28105	99.5931	0.0092
4.35651	7.11950	-4.11956	-1.24128	99.6429	0.0856
4.35303	7.11598	-4.11833	-1.23728	99.6426	0.0988
4.36230	7.17113	-4.13035	-1.25601	99.6206	0.0132
4.35762	7.15876	-4.12715	-1.24903	99.6237	0.0345
4.36247	7.18295	-4.13251	-1.25797	99.6146	0.0126
4.36359	7.19233	-4.13440	-1.26054	99.6106	0.0112
4.37045	7.20724	-4.13857	-1.27048	99.6080	0.0000
4.37090	7.20620	-4.13850	-1.27088	99.6089	0.0003
4.37014	7.20753	-4.13854	-1.27015	99.6076	0.0000
4.37058	7.19280	-4.13617	-1.26889	99.6156	0.0041
4.37058	7.21838	-4.14047	-1.27197	99.6024	0.0016
4.36920	7.19170	-4.13568	-1.26718	99.6150	0.0018
4.36793	7.18114	-4.13361	-1.26445	99.6194	0.0039
4.36663	7.17278	-4.13192	-1.26197	99.6227	0.0056
4.36153	7.14131	-4.12548	-1.25243	99.6345	0.0154
4.36223	7.15377	-4.12775	-1.25476	99.6279	0.0096
4.36196	7.15675	-4.12819	-1.25484	99.6253	0.0089
4.36601	7.18324	-4.13355	-1.26259	99.6153	0.0023
4.36892	7.20019	-4.13704	-1.26790	99.6097	0.0002
4.37073	7.20949	-4.13901	-1.27105	99.6075	0.0000
4.37160	7.21313	-4.13982	-1.27249	99.6073	0.0004
4.37421	7.22248	-4.14200	-1.27660	99.6086	0.0045
4.37351	7.21323	-4.14035	-1.27477	99.6166	0.0089
4.37140	7.19865	-4.13749	-1.27074	99.6249	0.0117
4.36932	7.18703	-4.13511	-1.26704	99.6298	0.0116
4.36800	7.18142	-4.13387	-1.26489	99.6307	0.0095
4.36774	7.18344	-4.13412	-1.26480	99.6268	0.0055
4.36995	7.20486	-4.13807	-1.26965	99.6095	0.0000
4.37438	7.23899	-4.14459	-1.27844	99.5867	0.0051
4.37935	7.27571	-4.15163	-1.28816	99.5637	0.0219
4.38411	7.31069	-4.15832	-1.29747	99.5426	0.0509

Another question that may well be asked is whether the parameter estimate $\hat{\theta}_5$ is significantly less than 100%. From the confidence intervals this is the clearly the case. This result is in agreement with the physiological conjecture that the upper limit of oxygen saturation cannot reach 100%.

What about critique of the model? One way of looking at this question is to see whether there are any influential observations. To achieve this we delete one observation at a time and refit the model to the remaining observations. The basic idea is due to Quenouille (1949,1956). The results are summarised in Table 6.12. Recall that $\hat{\theta}_l^{(i)}$ is the estimate of θ_l with the ith observation deleted.

D_i is the Cook distance measure (5.33) discussed in Chapter 5. Assuming that D_i behaves like an F-distribution with k and $n-k$ degrees of freedom, then one would be interested in the lower (5% and 10%) percentiles of the F-distribution. Any D_i larger than these lower percentiles should be cause for concern. We performed a least squares fit throughout. The value $F_{5,41,0.1} = 0.317$, hence observations 1, 7 and 8 could be influential.

CONCLUSION: *We can conclude that least squares provides a good fit and the model is adequate since all the residuals are less than 1% in magnitude (no outliers). It would appear as if a few of the observations are influential. Adaptive L_p-norm estimation selects larger values of p although this does not markedly affect the parameter estimates .*

6.3. Production models in Economics

Arrow et al. (1961) derived the following general formulation of the production function:

$$Q = \gamma[\delta K^{-\rho} + (1-\delta)L^{-\rho}]^{-\frac{1}{\rho}} \tag{6.10}$$

where Q is real output, K is capital input, L is labour input, and γ, ρ and δ are the unknown parameters of the model. The model (6.10) is the solution to a differential equation of the form:

$$Q = \Phi\left(Q - L\frac{dQ}{dL}\right)$$

The model has the properties of homogeneity, constant elasticity of substitution (CES), $\sigma = \frac{1}{1+\rho}$ between Capital and Labour and the possibility of different elasticities for different industries. Special cases are the Cobb-Douglas (CES=1) and the Leontief (CES $\to$ 0) production functions.

Bodkin and Klein (1967) consider a more general additive model which is given by:

$$Q_t = A10^{\lambda t}[\delta K_t^{-\rho} + (1-\delta)L_t^{-\rho}]^{-\frac{\mu}{\rho}} + \epsilon_t \tag{6.11}$$

where the parameters of the model are: A, λ, ρ, δ, and μ, t is time and ϵ_t the additive error term. Model (6.11) is referred to as the CES production function. The case $\mu = 1$ gives the constant returns to scale model which coincides with (6.10).

Another relationship, the marginal productivity relationship, is assumed to hold simultaneously with (6.11). This relationship is written as:

$$PR_t = \frac{\delta}{1-\delta}\left(\frac{K_t}{L_t}\right)^{-\rho-1} + \epsilon_t \tag{6.12}$$

where PR is the ratio of the price of capital services to the wage rate. The United States production data from 1909 to 1949 are given in Bard (1974:134) where $K_t \equiv z_1$, $L_t \equiv z_2$, $Q_t \equiv z_3$, $t \equiv z_4$ and $PR_t \equiv z_5$. A plot of Q_t vs K_t and L_t is given in Fig. 6.8. while in Fig. 6.9 we plot PR_t vs K_t and L_t.

6.3.1. The two-step approach

A number of approaches are possible for estimating the parameters, one such procedure consists of two steps. In the first step δ and ρ are estimated from (6.12), and in the second step these estimates are substituted into (6.11), and then the resulting relationship involving the three unknown parameters A, λ, μ is fitted.

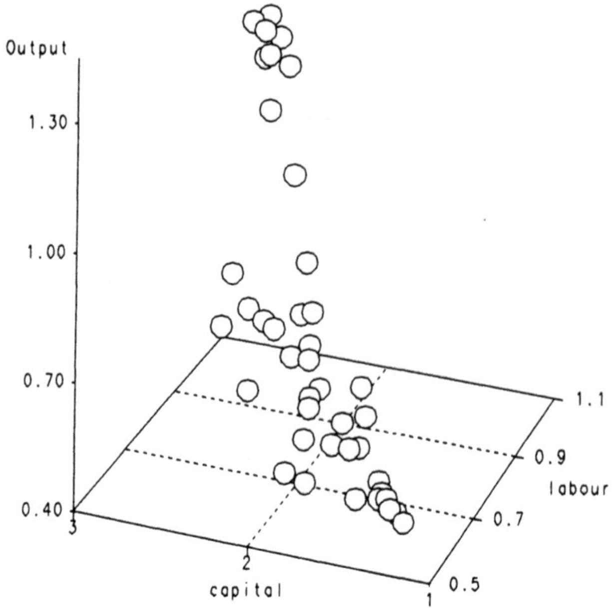

Figure 6.8 Q_t (real output) vs K_t (capital input) and L_t (labour input).

6.3.1.1 Two-step nonlinear least squares.

We now estimate the parameters of the CES production function using nonlinear least squares. If we fit (6.12) to the data we obtain the following solution:

$$\delta = 0.322454 \qquad \rho = -0.090533$$

With these values substituted into (6.11) the following solution,

$$A = 0.632596 \quad \mu = 1.274209 \qquad \lambda = 0.006401$$

was obtained. We are now in a position to construct confidence intervals for the individual parameters. An asymptotic $100(1-\alpha)\%$ confidence interval for θ_j is given by

$$\hat{\theta}_j \pm z_{\alpha/2}\sqrt{\sigma^2(\hat{J}'\hat{J})^{jj}}$$

The matrix $\hat{J}'\hat{J}$ and its inverse $(\hat{J}'\hat{J})^{-1}$ from the least squares fit are given in lower triangular form below. A value of $s^2 = 2.39046 \times 10^{-4}$ was used.

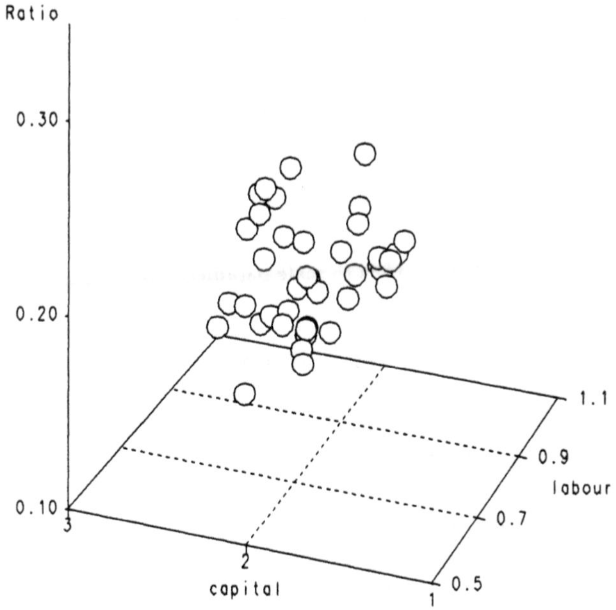

Figure 6.9 PR_t ratio vs K_t (capital input) and L_t (labour input).

Step 1: (δ, ρ):

$$\hat{J}'\hat{J} = \begin{pmatrix} 34.809305 & \\ -7.090014 & 1.469052 \end{pmatrix}$$

$$(\hat{J}'\hat{J})^{-1} = \begin{pmatrix} 1.691729 & \\ 8.164709 & 40.085652 \end{pmatrix}$$

95% confidence intervals for δ and ρ are

$$\delta \in [0.2818, 0.3631] \qquad \rho \in [-0.2886, 0.1075]$$

It would appear that ρ is not significantly different from 0 for this set of data.

Step 2: (A, λ, μ).

$$\hat{J}'\hat{J} = \begin{pmatrix} 60.146984 & & \\ 698.455492 & 22581.629206 & \\ 6.232362 & 153.048506 & 1.304006 \end{pmatrix}$$

$$(\hat{J}'\hat{J})^{-1} = \begin{pmatrix} 0.033196 & & \\ 0.000237 & 0.000218 & \\ -0.186516 & -0.026745 & 4.797352 \end{pmatrix}$$

A value of $s^2 = 1.38824 \times 10^{-3}$ was used. 95% confidence intervals for A, λ and μ are:

$$A \in [0.6189, 0.6463] \qquad \lambda \in [0.0053, 0.0075] \qquad \mu \in [1.1090, 1.4395]$$

It would seem as if the constant returns to scale parameter, μ is significantly larger than 1.

6.3.1.2 Two-step adaptive L_p-norm estimation.

In the first step we estimate δ and ρ using the adaptive scheme. The value of $p = 1.76$ was obtained using formula (5.26) whereby yielding:

$$\delta = 0.321543 \qquad \rho = -0.091014$$

With these values substituted into (6.11) the following solution with $p = 1.42$ was obtained:

$$A = 0.636887 \quad \mu = 1.290232 \quad \lambda = 0.005927$$

was obtained. The matrix $\hat{J}'\hat{J}$ and its inverse $(\hat{J}'\hat{J})^{-1}$ from the $L_{1.76}$-norm fit are given in lower triangular form below.

Step 1: δ and ρ:

$$\hat{J}'\hat{J} = \begin{pmatrix} 34.653601 & \\ -7.047921 & 1.458182 \end{pmatrix}$$

$$(\hat{J}'\hat{J})^{-1} = \begin{pmatrix} 1.699352 & \\ 8.213579 & 40.384973 \end{pmatrix}$$

with $\omega_p^2 = 2.3328 \times 10^{-4}$. 95% confidence intervals for δ and ρ are

$$\delta \in [0.2812, 0.3618] \qquad \rho \in [-0.2873, 0.1053]$$

Step 2: A, λ and μ.

$$\hat{J}'\hat{J} = \begin{pmatrix} 59.300957 & & \\ 675.669766 & 22399.817584 & \\ 6.066671 & 150.747584 & 1.282482 \end{pmatrix}$$

$$(\hat{J}'\hat{J})^{-1} = \begin{pmatrix} 0.032999 & & \\ 0.000264 & 0.000216 & \\ -0.187118 & -0.026610 & 4.792766 \end{pmatrix}$$

with $\omega_p^2 = 1.4042 \times 10^{-3}$.

95% confidence intervals for A, λ and μ are

$$A \in [0.6231, 0.6507] \qquad \lambda \in [0.0048, 0.0070] \qquad \mu \in [1.1241, 1.4564]$$

The fitted values and residuals for the least squares and adaptive L_p-norm estimation ($p = 1.42$) are given in Table 6.13. The residuals vs the fitted values in the least squares case are displayed in Fig. 6.10.

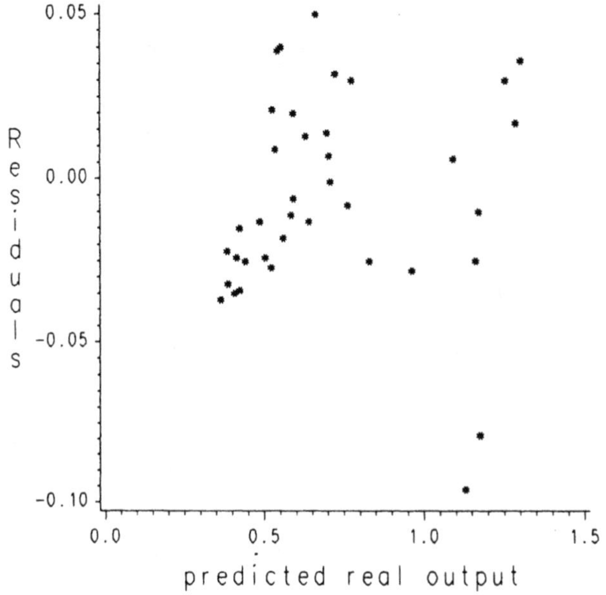

Figure 6.10 $\hat{y}_i - y_i$ vs $\hat{y}_i$ for the least squares fit.

We are also interested in the influential observations and the possibility of any outliers. To investigate this we used $\delta = 0.321543$ and $\rho = -0.091014$ in (6.11). We then deleted one observation at a time and used the adaptive scheme on the resulting reduced data sets, Table 6.14 resulted.

In the calculation of the Cook's distances, $D_{p,i}$ we used $\hat{J}_p'\hat{J}_p$ obtained in Step 2 with $\omega_p^2 = 1.4042 \times 10^{-3}$. The value $F_{3,38,0.05} = 0.116$, hence observations 1–14, 18–27, 29–41 appear to be influential.

Table 6.13: US Production data: Fitted values and residuals ($p = 2, 1.42$)

p			2		1.42
t	Q_t	$\hat{Q}_t$	$\hat{Q}_t - Q_t$	$\hat{Q}_t$	$\hat{Q}_t - Q_t$
-20	0.40	0.366	-0.037	0.375	-0.028
-19	0.41	0.387	-0.022	0.396	-0.012
-18	0.42	0.390	-0.032	0.399	-0.023
-17	0.44	0.415	-0.024	0.425	-0.014
-16	0.46	0.426	-0.034	0.435	-0.025
-15	0.44	0.410	-0.035	0.418	-0.027
-14	0.44	0.424	-0.015	0.432	-0.007
-13	0.50	0.487	-0.013	0.496	-0.004
-12	0.53	0.535	0.009	0.545	0.019
-11	0.58	0.561	-0.018	0.572	-0.007
-10	0.55	0.523	-0.027	0.531	-0.018
-9	0.51	0.526	0.021	0.534	0.028
-8	0.47	0.443	-0.025	0.448	-0.020
-7	0.53	0.504	-0.024	0.510	-0.018
-6	0.60	0.593	-0.006	0.600	0.001
-5	0.60	0.586	-0.011	0.592	-0.004
-4	0.66	0.642	-0.013	0.649	-0.006
-3	0.69	0.699	0.014	0.707	0.022
-2	0.69	0.725	0.032	0.732	0.039
-1	0.71	0.760	0.054	0.767	0.061
0	0.76	0.822	0.065	0.829	0.072
1	0.68	0.768	0.089	0.773	0.094
2	0.61	0.664	0.050	0.666	0.053
3	0.51	0.554	0.040	0.554	0.040
4	0.50	0.543	0.039	0.542	0.038
5	0.57	0.591	0.020	0.590	0.019
6	0.62	0.631	0.013	0.630	0.012
7	0.71	0.710	-0.001	0.709	-0.002
8	0.75	0.776	0.030	0.775	0.028
9	0.70	0.705	0.007	0.702	0.003
10	0.77	0.765	-0.008	0.761	-0.011
11	0.86	0.831	-0.025	0.826	-0.029
12	0.99	0.964	-0.028	0.960	-0.032
13	1.09	1.093	0.006	1.089	0.001
14	1.18	1.173	-0.010	1.168	-0.016
15	1.26	1.178	-0.079	1.171	-0.085
16	1.23	1.133	-0.096	1.124	-0.105
17	1.19	1.164	-0.025	1.155	-0.034
18	1.22	1.255	0.030	1.244	0.019
19	1.27	1.303	0.036	1.291	0.024
20	1.27	1.288	0.017	1.274	0.003

Table 6.14: Jackknife estimates of $\hat{\boldsymbol{\theta}}$

$\hat{A}^{(i)}$	$\hat{\lambda}^{(i)}$	$\hat{\mu}^{(i)}$	$\hat{p}^{(i)}$	$D_{p,i}$
0.635870	0.005988	1.291043	1.435471	0.2177
0.635979	0.006004	1.288986	1.461500	0.2424
0.635862	0.005993	1.290566	1.444392	0.2324
0.635949	0.006007	1.288505	1.458744	0.2479
0.635887	0.006000	1.289027	1.439829	0.2275
0.635811	0.005965	1.293769	1.433594	0.1834
0.636067	0.005977	1.291335	1.461132	0.1565
0.636259	0.005995	1.287521	1.462285	0.1425
0.637788	0.005806	1.300277	1.417209	0.3186
0.636235	0.006035	1.282229	1.462942	0.2431
0.635972	0.006004	1.287352	1.448933	0.2032
0.638252	0.005839	1.292593	1.404162	0.3386
0.635565	0.005917	1.301068	1.440960	0.2192
0.635787	0.005944	1.295924	1.443704	0.1554
0.636892	0.005937	1.290347	1.449898	0.0092
0.636193	0.005976	1.289501	1.457888	0.1054
0.636268	0.005992	1.286574	1.456476	0.1151
0.637678	0.005813	1.302553	1.421180	0.2956
0.638124	0.005743	1.308264	1.384656	0.7094
0.638171	0.005692	1.314160	1.346020	1.0474
0.638017	0.005641	1.323884	1.340130	1.4763
0.638452	0.005745	1.309084	1.388594	0.8738
0.639069	0.005891	1.281395	1.365608	0.5445
0.638739	0.006011	1.267284	1.394283	0.4753
0.638774	0.006046	1.263061	1.400240	0.5701
0.637895	0.006044	1.270420	1.441637	0.2763
0.637676	0.006044	1.272926	1.453859	0.2414
0.636138	0.005916	1.296279	1.449573	0.0614
0.638001	0.006008	1.277441	1.433216	0.2226
0.637296	0.006035	1.277175	1.462320	0.1920
0.635769	0.005842	1.305633	1.431969	0.2086
0.635914	0.005788	1.308420	1.392758	0.3901
0.636387	0.005835	1.298846	1.387399	0.2400
0.636622	0.005963	1.292907	1.467597	0.1692
0.637069	0.005969	1.277720	1.417522	0.2114
0.637930	0.005907	1.277425	1.333124	0.5645
0.635631	0.005926	1.287428	1.504752	0.5021
0.637047	0.005825	1.295507	1.375541	0.3252
0.636419	0.006024	1.299491	1.502080	1.6668
0.636275	0.006033	1.302931	1.505181	2.3898
0.636438	0.006044	1.291211	1.504519	0.9470

We see that the resulting values of p are still much smaller than 2, this may mean that there are probably two or more outliers in the data or that the model does not provide an adequate fit.

6.3.2. The global least squares approach

Another approach is a global one. The following objective function is formed:

$$\sum_{i=1}^{41}[(Q_t - \hat{Q}_t)^2 + (PR_t - \hat{P}R_t)^2] \tag{6.13}$$

We then determine the values $A, \lambda, \rho, \mu, \delta$ which minimize (6.13). The optimal solution was found by means of a quasi-Newton method (Harwell subroutine VA10AD): $(A, \lambda, \delta, \rho, \mu) = (0.62968, 0.006396, 0.33326, -0.025318, 1.274604$.

CONCLUSION: *Both approaches described above assume Q_t and PR_t to be the response variables of the system of equations. However, Bodkin and Klein (1967) argue that K_t and L_t should be used as the response variables. This is crucial since the parameter estimates, in particular those for ρ and δ, depend on which variables are taken to be response variables. The model equations then become:*

$$K_t = B_t PR_t^{-1/(1+\rho)} \qquad L_t = B_t\left[\frac{1-\delta}{\delta}\right]^{1/(1+\rho)}$$

where

$$\text{where} \qquad B_t = \left(\frac{Q_t}{A10^{\lambda t}}\right)^{1/\mu}\left\{\delta\left[\frac{1-\delta}{\delta}\right]^{1/(1+\rho)} + PR_t^{\rho/(1+\rho)}\right\}^{1/\rho}$$

Estimates of the parameters are given in Bard (1974). The two-equation econometric model exhibits several minima. It would be of interest to locate all the local minima and from these determine the global minimum.

CONCLUDING REMARK

In this chapter three examples were considered in detail. These can therefore serve as case studies to be solved in the classroom.

The interest in nonlinear models is increasing rapidly judging by the number of books appearing on the subject. The field, though complex and sometimes daunting presents fruitful areas of research to both the statistician and numerical analyst.

Bibliography Chapter 6

Arrow, K.J., Chenery, H.B., Minhas, B.S. and Solow, R.M. (1961). Capital-labour substitution and economic efficiency. *Rev. Econ. Statist.* **XLIII**, pp 225–251.

Bard, Y. (1974). Nonlinear parameter estimation. *Academic Press, New York.*

Bodkin, R.G. and Klein, L.R. (1967). Nonlinear estimation of aggregate production functions. *Rev. Econ. Statist.* **XLIX**, pp 28–44.

Buncher, C.R. and Tsay, J. (1974). Statistics in the pharmaceutical industry. *Marcel Dekker, New York.*

Cox, D.R. and Hinkley, D.V. (1974). Theoretical statistics. *Chapman and Hall, London.*

Du Toit, S.H.C. (1979). The analysis of growth curves. *Phd Thesis, University of South Africa.*

Du Toit, S.H.C. and Gonin, R. (1984). The choice of an appropriate nonlinear model for the relationship between certain variables in respiratory physiology. *South African Statistical Journal* **18**, pp 161–176.

Frome, E.L. and Yakatan, G.J. (1980). Statistical estimation of the pharmokinetic parameters in the one compartment open model. *Commun. Statist.–Simula. Computa.*, **9**, pp 201–222.

Gonin, R. and Money, A.H. (1985). Nonlinear L_p-norm estimation: Part I – On the choice of the exponent, p, where the errors are additive. *Commun. Statist.–Theor. Meth.* **14**, pp 827–840.

Gonin, R. and Money, A.H. (1987). Outliers in physical processes: L_1- or adaptive L_p-norm estimation? In: Y. Dodge, ed. *Statistical data analysis – Based on the L_1-norm and related methods, North Holland, Amsterdam*, pp 447–454.

Richards, F.J. (1959). A flexible growth function for empirical use. *J. Experimental Botany* **10**, pp 290–300.

Rodda, B.E., Sampson, C.B. and Smith, D.W. (1975). The one–compartment open model: Some statistical aspects of parameter estimation. *Appl. Stat.* **24**, pp 309–318.

Quenouille, M.H. (1949). Approximate tests of correlation in time series. *J.R. Statist. Soc., Ser. B* **11**, pp 18–84.

Quenouille, M.H. (1956). Notes on bias in estimation. *Biometrika* **43**, pp 353–360.

Severinghaus, J.W. (1979). Simple, accurate equations for human blood O_2 dissociation computaions. *J. Appl. Physiol.* **46**, pp 599–602.

Weiner, D.L. (1974). Design and analysis of bioavailability studies. In: Buncher, C.R. and Tsay, J., eds. *Statistics in the pharmaceutical industry, Marcel Dekker, New York.*

Winer, B.J. (1971). Statistical principles in experimental design, *2nd Edition, McGraw-Hill, New York.*

Author Index

Subject Index

Milton Keynes UK
Ingram Content Group UK Ltd.
UKHW020021071024
449327UK00032B/2872

9 780367 451165